Current Topics in Membranes, Volume 45

The Eye's Aqueous Humor

From Secretion to Glaucoma

Current Topics in Membranes, Volume 45

Current Topics in Membranes, Volume 45

The Eye's Aqueous Humor

From Secretion to Glaucoma

Edited by

Mortimer M. Civan

Departments of Physiology and Medicine
University of Pennsylvania Medical Center
Philadelphia, Pennsylvania

ACADEMIC PRESS

San Diego London Boston New York Sydney Tokyo Toronto

Cover photo credit (paperback edition only): The "digits" of the ciliary body, a silhouette of several plicated processes. Scanning micrograph of the posterior structure of the rabbit iris. For more details, see Figure 1 in Chapter 8 by Sears and Sears.

This book is printed on acid-free paper. ♾

Academic Press
a division of Harcourt Brace & Company
525 B Street, Suite 1900, San Diego, California 92101-4495, USA
http://www.apnet.com

Academic Press Limited
24-28 Oval Road, London NW1 7DX, UK
http://www.hbuk.co.uk/ap/

International Standard Book Number: 0-12-153345-X (case)

International Standard Book Number: 0-12-174675-5 (paperback)

PRINTED IN THE UNITED STATES OF AMERICA
97 98 99 00 01 02 EB 9 8 7 6 5 4 3 2 1

To Judith, my wife and friend, without whose support this work would not have been completed.

Contents

Contributors

Numbers in parentheses indicate the pages on which the authors' contributions begin.

Peter Agre (105), Department of Biological Chemistry and Department of Medicine, Johns Hopkins University, School of Medicine, Baltimore, Maryland 21205

Richard F. Brubaker (233), Department of Ophthalmology, Mayo Clinic and Mayo Foundation, Rochester, Minnesota 55905

Mortimer M. Civan (1), Departments of Physiology and Medicine, The University of Pennsylvania, Philadelphia, Pennsylvania 19104

Miguel Coca-Prados (25), Department of Ophthalmology and Visual Science, Yale University School of Medicine, New Haven, Connecticut 06520

Tim J. C. Jacob (55), School of Molecular and Medical Biosciences, University of Wales, Cardiff CF1 3US, United Kingdom

Landon S. King (105), Division of Pulmonary and Critical Care Medicine, Department of Medicine, Johns Hopkins University, School of Medicine, Baltimore, Maryland 21205

M. Douglas Lee (105), Division of Pulmonary and Critical Care Medicine, Department of Medicine, Johns Hopkins University, School of Medicine, Baltimore, Maryland 21205

James L. Rae (69), Departments of Physiology and Biophysics and Ophthalmology, Mayo Foundation, Rochester, Minnesota 55905

Juan Sánchez-Torres (25), Department of Ophthalmology and Visual Science, Yale University School of Medicine, New Haven, Connecticut 06520

Michael Schütte (135), Departments of Ophthalmology and Physiology, Mount Sinai School of Medicine, New York, New York 10029

Jonathan Sears (203), Emory University Eye Center, Atlanta, Georgia 30322

Marvin Sears (203), Department of Ophthalmology and Visual Science, Yale University, New Haven, Connecticut 00520

Allan R. Shepard (69), Departments of Physiology and Biophysics and Ophthalmology, Mayo Foundation, Rochester, Minnesota 55905

Friederike Stumpff (163), Institut für Klinische Physiologie, Universitätsklinikum Benjamin Franklin, Freie Universität Berlin, 12200 Berlin, Germany

Michael Wiederholt (163), Institut für Klinische Physiologie, Universitätsklinikum Benjamin Franklin, Freie Universität Berlin, 12200 Berlin, Germany

J. Mario Wolosin (135), Departments of Ophthalmology, Physiology, and Biophysics, Mount Sinai School of Medicine, New York, New York 10029

Preface

This volume was conceived with several aims in mind. One aim was certainly to present a basic consensus of how the aqueous humor is formed and exits through the outflow pathways. Second, it seemed time to update what is currently known about the molecular biology of the transport components underlying aqueous humor dynamics. Third, I wished to provide current information about the clinical approaches to assessing the basic transport processes within the framework of a relatively brief, easily accessible volume. Fourth, I wanted to emphasize the phenomenon of the circadian rhythm of aqueous humor formation from a clinical and molecular point of view. This phenomenon provides the major indication that aqueous humor secretion is regulated. Finally, I hoped that the volume will stimulate fresh approaches to the regulation of aqueous humor dynamics and intraocular pressure.

I am grateful to the outstanding contributors who joined me in creating this volume. I also thank the general editors, Dr. Douglas Fambrough and Dr. Dale Benos, and Dr. Emelyn Eldredge and Charlotte Brabants of Academic Press for their efforts to expedite publication of the work. It is a particular pleasure to thank the reviewers who offered constructive criticism of the individual chapters. Drs. Miguel Coca-Prados, Mark I. Greene, Thomas R. Kleyman, Gregory S. Kopf, Cecilia W. Lo, Zhe Lu, Claire H. Mitchell, Amita Sehgal, and Richard A. Stone.

I particularly wish to take this opportunity to acknowledge Arnošt Kleinzeller, M.D., Ph.D., a distinguished physiologist, colleague, and friend at the University of Pennsylvania. He was the guiding editor of *Current Topics in Membranes* over many years and was a great source of encouragement for undertaking this work. I very much regret that he passed away before seeing the fruits of that encouragement.

Mortimer M. Civan

Previous Volumes in Series

Current Topics in Membranes and Transport

Volume 19 Structure, Mechanism, and Function of the Na/K Pump* (1983)
Edited by Joseph F. Hoffman and Bliss Forbush III

Volume 20 Molecular Approaches to Epithelial Transport* (1984)
Edited by James B. Wade and Simon A. Lewis

Volume 21 Ion Channels: Molecular and Physiological Aspects (1984)
Edited by Wilfred D. Stein

Volume 22 The Squid Axon (1984)
Edited by Peter F. Baker

Volume 23 Genes and Membranes: Transport Proteins and Receptors* (1985)
Edited by Edward A. Adelberg and Carolyn W. Slayman

Volume 24 Membrane Protein Biosynthesis and Turnover (1985)
Edited by Philip A. Knauf and John S. Cook

Volume 25 Regulation of Calcium Transport across Muscle Membranes (1985)
Edited by Adil E. Shamoo

Volume 26 Na^{+}–H^{+} Exchange, Intracellular pH, and Cell Function* (1986)
Edited by Peter S. Aronson and Walter F. Boron

Volume 27 The Role of Membranes in Cell Growth and Differentiation (1986)
Edited by Lazaro J. Mandel and Dale J. Benos

Volume 28 Potassium Transport: Physiology and Pathophysiology* (1987)
Edited by Gerhard Giebisch

** Part of the series from the Yale Department of Cellular and Molecular Physiology*

Volume 29 Membrane Structure and Function (1987)
Edited by Richard D. Klausner, Christoph Kempf, and Jos van Renswoude

Volume 30 Cell Volume Control: Fundamental and Comparative Aspects in Animal Cells (1987)
Edited by R. Gilles, Arnost Kleinzeller, and L. Bolis

Volume 31 Molecular Neurobiology: Endocrine Approaches (1987)
Edited by Jerome F. Strauss, III and Donald W. Pfaff

Volume 32 Membrane Fusion in Fertilization, Cellular Transport, and Viral Infection (1988)
Edited by Nejat Düzgünes and Felix Bronner

Volume 33 Molecular Biology of Ionic Channels* (1988)
Edited by William S. Agnew, Toni Claudio, and Frederick J. Sigworth

Volume 34 Cellular and Molecular Biology of Sodium Transport* (1989)
Edited by Stanley G. Schultz

Volume 35 Mechanisms of Leukocyte Activation (1990)
Edited by Sergio Grinstein and Ori D. Rotstein

Volume 36 Protein–Membrane Interactions* (1990)
Edited by Toni Claudio

Volume 37 Channels and Noise in Epithelial Tissues (1990)
Edited by Sandy I. Helman and Willy Van Driessche

Current Topics in Membranes

Volume 38 Ordering the Membrane Cytoskeleton Tri-Layer* (1991)
Edited by Mark S. Mooseker and Jon S. Morrow

Volume 39 Developmental Biology of Membrane Transport Systems (1991)
Edited by Dale J. Benos

Volume 40 Cell Lipids (1994)
Edited by Dick Hoekstra

Volume 41 Cell Biology and Membrane Transport Processes* (1994)
Edited by Michael Caplan

Volume 42 Chloride Channels (1994)
Edited by William B. Guggino

CHAPTER 1

Transport Components of Net Secretion of the Aqueous Humor and Their Integrated Regulation

Mortimer M. Civan

Departments of Physiology and Medicine, The University of Pennsylvania, Philadelphia, Pennsylvania 19104

I. INTRODUCTION

This book is concerned with the formation of the aqueous humor of the anterior chamber and its outflow from the eye into the venous circulation. The anterior chamber is the compartment bounded by the cornea, lens, and iris-ciliary body and contains ~0.25 mL of aqueous fluid in each eye, accounting for about 0.001% of the total body fluids of a 70-kg human. The importance of the circulation of this very small fluid compartment is at least fourfold (Krupin and Civan, 1995): (1) delivery of substrates to, and removal of metabolic products from, the avascular tissues of the anterior

1063-5823/98 $25.00

segment (cornea, lens, and trabecular meshwork); (2) delivery of ascorbate to the anterior segment tissues at a concentration roughly 25-fold higher than that in human plasma (several functional roles have been ascribed to this extraordinary gradient, especially an antioxidant function, but the precise importance of ascorbate remains uncertain); (3) participation in local immune responses; and (4) inflation of the globe to preserve its normal optical properties. For this especially important purpose, the normal range of intraocular pressures (IOP) is 15 ± 3 mm Hg. Sustained values appreciably higher than this range induce death of retinal ganglion cells and a distinctive type of optic atrophy characterized by cupping of the optic disk, hallmarks of clinical glaucoma. Glaucoma is one of the more common causes of blindness in virtually all population groups.

The IOP reflects the balance between inflow and outflow of aqueous humor. Outflow is addressed in Chapter 7. Aqueous humor secretion and IOP are not constant throughout the day. The rate of aqueous humor formation displays a striking circadian rhythm, falling two- to threefold during the period from midnight to 6 a.m. (Chapter 9). In principle, high IOP and glaucoma could result from sustained excessive secretion by the ciliary epithelium or from blockage of outflow. However, glaucoma has been found to result from a primary blockage of outflow and has never been rigorously documented to result from primary hypersecretion of aqueous humor (Chapter 9). The blockage can result either from limited access to the outflow tract at the angle formed by the cornea and iris (closed-angle glaucoma) or from blockage within the trabecular meshwork leading to the canal of Schlemm (open-angle glaucoma).

Although glaucoma is usually characterized by elevated IOP, two clinical observations have raised questions concerning the precise role of high IOP in producing glaucomatous atrophy (Chapter 9). First, some patients develop glaucomatous atrophy with IOP within the normal range, so-called "normal tension" or "low-tension" glaucoma. Second, some patients with well-documented histories of elevated IOP still have progressive glaucomatous optic atrophy, despite a satisfactory response of IOP to ocular hypotensive drugs, at least during part of the day. These observations lead to the question whether high IOP itself is the cause of glaucoma or only a strong risk factor in a multifactorial disease. Recent data suggest that programmed cell death (apoptosis) could be the mechanism of neural cell death in glaucoma (Quigley, 1995; Quigley *et al.*, 1995). One potential signal of apoptosis could be excitatory amino acids such as glutamate (Vorwerk *et al.*, 1996). Irrespective of the mechanisms involved in the neural cell death, an overwhelming volume of clinical data document that the vast majority of patients with open-angle glaucoma benefit from therapy aimed at normalizing IOP.

The foregoing considerations indicate the uncertainties surrounding the precise role of secretion in generating high IOP and in the mechanisms by which high IOP leads to blindness. Nevertheless, the mechanisms and regulation of aqueous humor secretion are of enormous importance because: (1) most of the effective ocular hypotensive drugs act by reducing secretion; (2) the efficacy of hypotensive therapy would be enormously enhanced by developing a drug capable of lowering secretion to the normal early morning rate; and (3) information concerning the molecular and cell biology and physiology of these mechanisms can provide a general perspective of transepithelial secretion and absorption. The aims of the present chapter are to: (1) present an overview of the current consensus concerning aqueous humor formation; (2) introduce the functional components underlying transport by the ciliary epithelium; and (3) indicate pathways that may regulate net secretion. Succeeding chapters focus on molecular aspects of these transport components, present what is currently known about outflow through the trabecular meshwork, examine the possible basis for the circadian rhythm of aqueous humor secretion, and provide an update of clinical measurements of aqueous dynamics.

II. STRUCTURE OF CILIARY EPITHELIUM

The ciliary epithelium is a bilayered structure covering the ciliary body, with the main components in the regions of the pars plicata and pars plana. The major component covers the pars plicata, consisting of some 70 villiform ciliary processes radiating inward toward the pupil. The stroma of each process contains loose connective tissue, a vascular core, and nerve endings. It is currently unclear whether these nerve endings preferentially innervate the vessels or the epithelium. In the pars plana, the topography is flatter. At its most posterior limit, the ora serrata, the ciliary epithelium is fused with the sensory retina and the retinal pigment epithelium. Observations obtained with both histochemical approaches (Flügel and Lütjen-Drecoll, 1988) and molecular probes (Ghosh *et al.*, 1990, 1991) have indicated regional differences in the expression of Na^+,K^+-ATPase and have raised the possibility of regional differences in net secretion by these areas (Ghosh *et al.*, 1990, 1991). The topography of the isozymes of Na^+,K^+-ATPase and its potential physiological significance are considered in Chapter 2.

The microscopic anatomy of the ciliary epithelium is unique (Fig. 1). Embryological invagination of the optic vesicle to form the optic cup leads to a bilayered epithelium whose apical membranes are in close contact. The basolateral membrane of the pigmented ciliary epithelial (PE) cell layer faces the stroma, and that of the nonpigmented ciliary epithelial

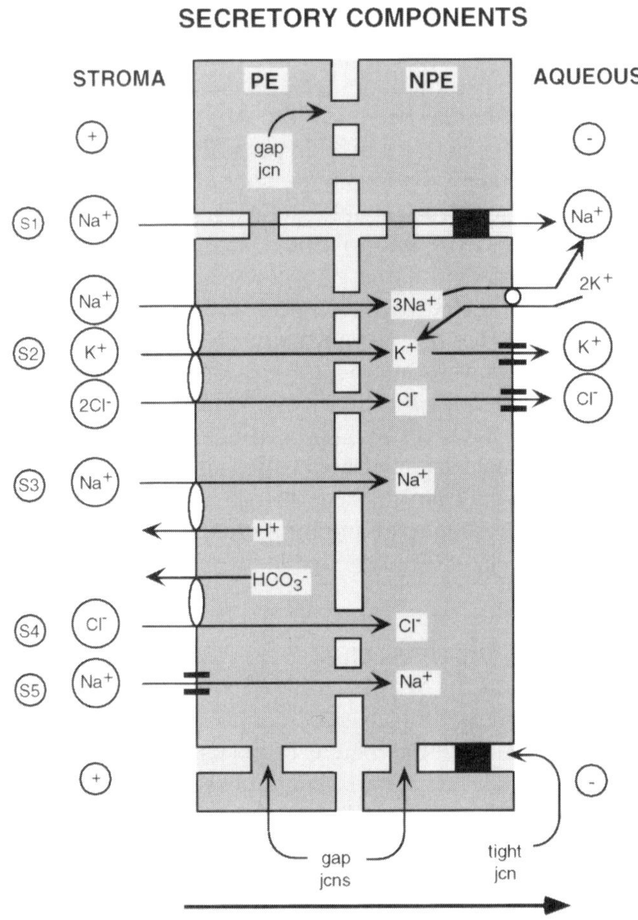

FIGURE 1 Components of unidirectional aqueous secretion of Na^+, K^+, and Cl^-. Cations, especially Na^+, are considered to cross between the cells through the tight junctions (tight jcns) through the paracellular pathway (S1) in response to the small electrical driving force across the ciliary epithelium (about −1 mV). Most of the transfer from the stromal interstitial fluid to the aqueous humor is considered to be through the transcellular route. Uptake into PE cells may proceed through a Na^+–K^+–$2Cl^-$ symport (S2), paired Na^+–H^+ (S3), and Cl^-–HCO_3^- (S4) antiports, and cation-nonselective and tetrodotoxin-sensitive Na^+ channels (S5). Ions and water pass from the PE to the NPE cells through gap junctions (gap jcns). Solutes are released from the NPE cells into the aqueous humor by the Na^+–K^+ exchange pump (3 Na^+ extruded in exchange for 2 K^+ taken up by the cell) (S6), and parallel K^+ (S7) and Cl^- channels (S8). Not included in the illustration are the NPE aquaporin-1 channels through which water is likely secreted, and the bafilomycin-inhibitable H^+-ATPase and PE T-type (Jacob, 1991a) and NPE L-type (Farahbakhsh *et al.,* 1994) Ca^{2+} channels, which may participate in regulation of aqueous humor formation.

(NPE) cell layer faces the aqueous humor. Gap junctions provide low-resistance intercellular conduits linking the cells both within each cell layer and between the PE and NPE layers (Reale and Spitznas, 1975; Raviola and Raviola, 1978; Coca-Prados *et al.*, 1992; Edelman *et al.*, 1994; Oh *et al.*, 1994). The gap junctions permit the electrical potential (Green *et al.*, 1985; Carré *et al.*, 1992) and ionic composition (Bowler *et al.*, 1996) of the two cell layers to be closely similar so that the ciliary epithelium likely functions as a syncytium under baseline conditions. Tight junctions are displayed between the cells of the NPE cell layer (Bairati and Orzalesi, 1966; Raviola and Raviola, 1978), but even when surface infoldings are taken into account, the transepithelial resistance of the ciliary epithelium is ≤ 0.6 k$\Omega \cdot$cm^2 (Krupin and Civan, 1995). Thus, the ciliary epithelium falls within the class of "leaky" epithelia (Rose and Schultz, 1971; Frömter and Diamond, 1972).

III. OVERVIEW OF NET SECRETION BY CILIARY EPITHELIUM

As for all epithelia, transmural transport can proceed either through the cells (transcellular pathway) or between the cells (paracellular pathway) (Fig. 1). In general, the primary event is the transcellular transfer of solute, which may establish electrical driving forces favoring accompanying paracellular transport (Frizzel *et al.*, 1979). The resulting osmotic gradient then favors water flow through membrane pores, diffusively through the lipid bilayers of the plasma membranes, and between the cells through the paracellular pathway. One possible general exception to this observation is the suggestion that water may be stoichiometrically coupled to the movement of two or more symports (Zeuthen *et al.*, 1996; Loo *et al.*, 1996). The formation specifically of the aqueous humor is clearly dependent on transcellular movement because inhibitors of transport or metabolism reduce net secretion by about 75% (Cole, 1960, 1977). Furthermore, Bill (1973) claimed that the passive Starling forces probably favor reabsorption (rather than secretion) of water. For these reasons, movement through the paracellular pathway has been largely neglected in recent years. Nevertheless, the small transepithelial potential (about -1 mV, aqueous humor negative to stroma) does provide a driving force for Na^+ secretion. Whether this paracellular cation movement is significant is as yet unclear.

With rare exception (Sears, 1984; Civan *et al.*, 1996, 1997), models of aqueous humor formation generally equate net secretion with unidirectional secretion. The tacit assumption is that unidirectional reabsorption from the aqueous humor back into the stroma is very much slower than unidirectional secretion. This assumption has not yet been rigorously tested. Figure 1 presents many of the components likely to be involved in unidirec-

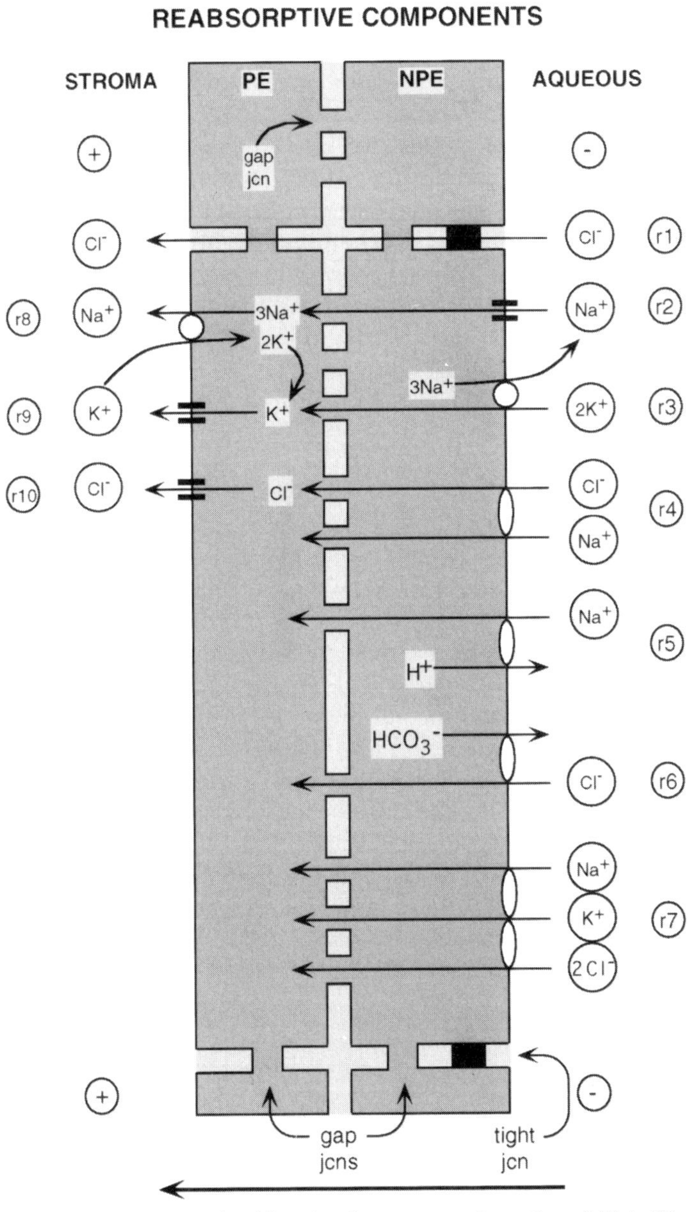

FIGURE 2 Components of unidirectional aqueous reabsorption of Na^+, K^+, and Cl^-. The small transepithelial potential favors anion reabsorption, principally Cl^- but also HCO_3^-, through the paracellular pathway (r1). The bulk of the reabsorption is considered to proceed through the transcellular route. The initial uptake step may proceed through an amiloride-sensitive Na^+ channel and cation-nonselective channel (r2), the Na^+–K^+ exchange pump (2 K^+ taken up in exchange for 3 Na^+ extruded) (r3), a thiazide-sensitive Na^+–Cl^- symport (r4),

tional secretion. Figure 2 presents the components that may underlie unidirectional reabsorption. These secretory and absorptive components are considered separately in Sections IV and V, respectively.

IV. UNIDIRECTIONAL SECRETION

The transcellular transfer of solutes and water from the stromal interstitium to the aqueous humor involves three steps: (1) uptake of solute and water at the stromal surface by PE cells, (2) transfer from PE to NPE cells through gap junctions, and (3) transfer of solute and water by NPE cells into aqueous humor. Each of these steps is addressed in turn in this section.

A. Uptake of Solute and Water at the Stromal Surface by PE Cells

1. Na^+–K^+–$2Cl^-$ Symport

The Na^+–K^+–$2Cl^-$ symport has long been recognized as a major mechanism for the uptake of Na^+, K^+, and Cl^- by many absorptive and secretory epithelial (Geck *et al.*, 1980). Two isoforms of the symport have been cloned and sequenced, displaying limited homology with Na^+–Cl^- and K^+–Cl^- symports within the family of electroneutral cation–chloride symports (Payne and Forbush, 1995; Hebert *et al.*, 1996). Under certain conditions, the PE cells are also likely to use a Na^+–K^+–$2Cl^-$ symport for solute uptake, because: (1) electrometric measurements of intact shark ciliary epithelium have demonstrated that furosemide decreases intracellular Cl^- activity (Wiederholt and Zadunaisky, 1986); (2) measurements of cell volume by electronic cell sorting have identified Na^+-, K^+-, and Cl^--dependent and bumetanide-inhibitable uptake activated by shrinking freshly dissociated bovine PE cells (Edelman *et al.*, 1994); (3) measurements of $^{86}Rb^+$ uptake by cultured human PE cells have demonstrated a bumetanide-sensitive uptake of tracer (Von Brauchitsch and Crook, 1993); and (4) electron probe X-ray microanalysis (Civan, 1983) of the intact rabbit ciliary epithelium has documented that bumetanide can reduce the intracellular Cl content

---

paired Na^+–H^+ (r5) and Cl^-–HCO_3^- (r6), and a Na^+–K^+–$2Cl^-$ symport (r7). Water is presumed to be taken up largely through aquaporin-1 channels of the NPE cells (not shown). Solute and water can then proceed from the NPE to the PE cells through the gap junctions. Ions may be released into the stromal interstitial fluid through the Na^+–K^+ exchange pump (extrusion of 3 Na^+ in exchange for 2 K^+ taken up by the cell) (r8), and parallel K^+ (r9) and Cl^- channels (r10).

and concentration under some, but not all, experimental conditions (Macknight *et al.*, 1997; McLaughlin *et al.*, manuscript submitted).

2. Parallel Cl^-–HCO_3^- and NA^+–H^+ Antiports

Fluorometric measurements of intracellular pH have provided evidence for parallel Cl^-–HCO_3^- (Helbig *et al.*, 1988a, 1989) and Na^+–H^+ antiports (Helbig *et al.*, 1988b,c) in continuous lines of PE cells. Wiederholt *et al.* (1991) have suggested a role for bicarbonate reminiscent of that suggested for oxalate (Knickelbein *et al.*, 1986) and formate (Karniski and Aronson, 1987) in the renal proximal tubule. CO_2 is believed to enter the PE cells from the stroma by crossing the lipid bilayer, undergo carbonic anhydrase-catalyzed hydration, dissociate into bicarbonate and protons, and thereby stimulate cell uptake of Na^+ and Cl^- through the parallel antiports. The quantitative significance of this mechanism has been unclear until recently. Electron microprobe measurements have now documented that HCO_3^- indeed increases the content and concentration of Cl within the epithelial syncytium and that inhibition of carbonic anhydrase with acetazoleamide blocks this stimulation (Macknight *et al.*, 1997; McLaughlin *et al.*, manuscript submitted). Furthermore, the data suggest that bicarbonate-stimulated, acetazoleamide-inhibited Cl^- uptake is quantitatively more important than uptake through the Na^+–K^+–$2Cl^-$ symport of PE cells within the intact ciliary epithelium under physiologic conditions.

3. Cation Channels

Cation-nonselective channels have been detected in both PE (Stelling and Jacob, 1993) and NPE (Carré *et al.*, 1996a) cells. Stelling and Jacob (1993) have suggested that such channels may play a significant role in loading the PE cells with cation from the stroma. Voltage-gated, tetrodotoxin-blockable Na^+ channels have also been detected in cultured rabbit PE cells (Fain and Farahbakhsh, 1989). The role of these excitable Na^+ channels is unknown, but they may serve as a supplementary conduit for the Na^+ loading of PE cells.

4. Water Pores

It has long been appreciated from both thermodynamic considerations and measurements with black lipid membranes that specialized conduits are necessary for transmembrane movements of hydrophilic molecules and ions. The need for water channels has been less apparent. However, it has been known for more than 25 years that the equivalent rate constant for water exchange across erythrocyte membranes calculated from measurements of hydraulic conductivity is higher than that obtained from measuring diffusive water flow (reviewed by Solomon *et al.*, 1983). At about the same time that these early studies were being performed, mercurials were reported to inhibit transmembrane water flow (Macey and Farmer, 1970). Both sets of observations were considered to favor the possibility of water pores. In Chapter 5, Lee and coworkers review recent information documenting a superfamily of at least six such pores (aquaporins). As noted in Section IV,C, one aquaporin has been found in the NPE cells, but none of the known members of this family has been identified in the PE cells (Stamer *et al.,* 1994). Either an as-yet unidentified aquaporin subserves water transfer at the stromal surface or simple diffusive movement across the lipid bilayer of the PE cells is itself sufficiently rapid to support aqueous humor formation.

B. Transfer from PE to NPE Cells through Gap Junctions

Both structural (Reale and Spitznas, 1975; Raviola and Raviola, 1978; Coca-Prados *et al.,* 1992) and functional studies (Green *et al.,* 1985; Carré *et al.,* 1992; Edelman *et al.,* 1994; Oh *et al.,* 1994; Walker *et al.,* 1995; Bowler *et al.,* 1996; Mitchell and Civan, 1997) have unequivocally established that small ions and molecules can readily pass from the PE cells into the NPE cell layer through gap junctions. Coca-Prados *et al.* (1992) were the first to demonstrate that connexin 43 (Cx43) was an important component of the gap junctions in the ciliary epithelium. Wolosin and colleagues (1996) have presented data leading them to propose that the low-resistance pathways linking PE and NPE cells are novel heterotypic gap junctions, consisting of Cx43 in the PE and Cx50 in the NPE cells.

It is unclear to what extent the gap junctions are an important site of regulation of aqueous humor formation. The electron microprobe X-ray microanalyses of Bowler *et al.* (1996) have indicated that the Na, K, and Cl contents and concentrations of the PE and NPE cell layers are similar within intact rabbit ciliary epithelium, suggesting that the gap junctions are not rate limiting under baseline conditions. In contrast, Wolosin *et al.* (1997) have demonstrated that blocking the gap junctions by addition of 3 mM heptanol inhibits current through the transcellular pathway across rabbit ciliary epithelium. Heptanol in this concentration has recently been documented by whole-cell patch-clamp measurements to interrupt communication reversibly between PE–NPE cell couplets (Mitchell and Civan, 1997). Furthermore, Shi *et al.* (1996) have indicated that these intercellular communications can be modulated through α_1-adrenergic and cholinergic receptors of the PE cells. The potential importance of the gap junctions in regulating ciliary epithelial secretion is considered in depth in Chapter 6.

C. Transfer of Solute and Water by NPE Cells into Aqueous Humor

1. Na^+K^+-Activated ATPase

The formation of the aqueous humor is fundamentally dependent on active transport by the Na^+–K^+ exchange pump (Cole, 1960, 1977). Under physiologic conditions, 3 Na^+ ions are extruded and 2 K^+ ions accumulated at the expense of one adenosine triphosphate (ATP) molecule (Glynn, 1993). Molecular probes (Ghosh *et al.*, 1990, 1991), histochemical observations (Flügel and Lütjen-Drecoll, 1988), and functional measurements (Krupin *et al.*, 1984) have demonstrated that Na^+,K^+-ATPase is localized at the basolateral membranes of both the PE and NPE cells. It is presumed that the number of pump sites is greater on the aqueous surface than on the stromal surface, accounting for the vectorial direction of secretion.

Given the critical importance of the Na^+–K^+ exchange pump in vectorial transport, attention has been directed toward the possible functional significance of isozyme specificity and tissue topography of the pump sites. These issues are addressed in Chapter 2. Considerable effort has also been directed toward studying possible regulation of pump expression. Aldosterone is known to increase the rate of production of pump sites (Geering *et al.*, 1982), but efforts to detect hormonal regulation of pump kinetics have been less conclusive (Collins *et al.*, 1987). Delamere *et al.* (1990) and Delamere and King (1992) have reported that cyclic adenosine monophosphate (cAMP) inhibits Na^+,K^+-ATPase activity of rabbit ciliary epithelium, presumably by cAMP-activated phosphorylation of the pump (Aperia *et al.*, 1991; Delamere *et al.*, 1990; Bertorello *et al.*, 1991; Delamere & King, 1992) and of the protein-phosphase modulator DARPP-32 (Tsou *et al.*, 1993). Phosphorylation of DARPP-32 has been believed to prevent protein phosphatase 1 from dephosphorylating the pump (Aperia *et al.*, 1991; Snyder *et al.*, 1992). Carré and Civan (1995) have presented evidence suggesting that these inhibitory effects can be reversed by the second messenger cyclic guanidine monophosphate (cGMP), possibly by directly stimulating cAMP phosphodiesterase to lower the cAMP level (Mittag *et al.*, 1987) and indirectly by activating protein phosphatase 2A (through cGMP-activated kinase) to accelerate dephosphorylation of the pump and DARPP-32 (Tsou *et al.*, 1993).

2. K^+ Channels

As discussed elsewhere (Jacob and Civan, 1996), K^+ channels play two important functions. Like other vertebrate cells, the ciliary epithelial cells accumulate K^+ against an electrochemical gradient, so that K^+ channels certainly serve as a major conduit for K^+ release energetically downhill into the aqueous humor. The second function is to help fix the membrane potential of the ciliary epithelial syncytium at a large negative value to provide energy for driving Cl^- from the NPE cells into the aqueous humor. The intracellular Cl^- concentration is estimated to be ~45 mM (Bowler *et al.,* 1996) (two- to threefold lower than that of the aqueous humor) so that without the electrical driving force the concentration gradient *per se* would lead to reabsorption, not secretion. From electron probe X-ray microanalyses of the intact rabbit ciliary epithelium (Bowler *et al.,* 1996), the reversal potential for perfectly K^+-selective channels is about −85 mV. Operation of channels with so negative a reversal potential helps establish and maintain the membrane potential of the ciliary epithelial syncytium at very negative values [about −68 mV (Carré *et al.,* 1992)].

Evidence has been published for inward-rectifier (Cilluffo *et al.,* 1991; Gooch *et al.,* 1992), delayed-rectifier (Cilluffo *et al.,* 1991), and Ca^{2+}-activated (Barros *et al.,* 1991; Gooch *et al.,* 1992) K^+ channels in NPE cells. Which of these is dominant is as yet unknown (Jacob and Civan, 1996). As noted earlier, the ciliary epithelium falls into the category of leaky epithelia, whose stromal and luminal phases are electrically coupled through the paracellular pathway. Under these conditions, the K^+ channels of the PE cells also strongly contribute to the syncytial membrane potential. All three types of K^+ channels have also been identified in PE cells (Fain and Farahbakhsh, 1989; Jacob, 1991b; Stelling and Jacob, 1992).

Progress thus far has been limited in establishing the molecular identity of these K^+ channels in the ciliary epithelium. Because of the physiologic importance of ocular K^+ channels, a review of this area is provided by focusing on advances in our understanding of K^+ channels from another ocular epithelium, the lens (Chapter 4).

3. Cl^- Channels

Chloride is the major anion of the aqueous humor, and Cl^- channels are likely to be a major conduit for Cl^- transfer from the NPE cells to the aqueous (Fig. 1). However, it has been far more difficult to detect baseline NPE activity of Cl^- channels than of Na^+ pumps or K^+ channels in measurements of (1) transmural current across the iris-ciliary body, (2) cell-attached patch-clamping of the intact ciliary epithelium, and (3) patch-clamping of isolated NPE cells (Krupin and Civan, 1995; Jacob and Civan, 1996). These considerations have led to the hypothesis that activity of NPE Cl^- channels limits the rate of formation of the aqueous humor (Coca-Prados *et al.,* 1995).

The molecular basis of the Cl^- channels subserving Cl^- secretion by the NPE cells is unknown. However, certain functional characteristics of the NPE channels have provided clues to their possible identities. Two of the mechanisms best documented to stimulate NPE Cl^- channels are hypotonic swelling (Yantorno *et al.,* 1989; Edelman *et al.,* 1994; Wu *et al.,* 1996) and inhibition of endogenous protein kinase C (PKC) activity with staurosporine (Civan *et al.*, 1994; Coca-Prados *et al.*, 1995). An example of volume activation of Cl^- currents in a human NPE cell is presented in Fig. 3. Four families of nonsynaptic Cl^- channels of Cl^- channel regulators, consisting of at least 15 proteins, have been cloned and sequenced in other cells. Of these 15 proteins, only three [pI_{Cln} (Paulmichl *et al.*, 1992), P-glycoprotein (Valverde *et al.*, 1992), and ClC-2 (Gründer *et al.*, 1992)] have been reported to be activated by cell swelling, and only two [ClC-3 (Kawasaki *et al.*, 1994, 1995) and P-glycoprotein (Hardy *et al.*, 1995)] have been reported to be inhibited by PKC (Coca-Prados *et al.*, 1996). Coca-Prados *et al.* (1996) have suggested that the functional properties of NPE Cl^- channels could reflect operation of a PKC-inhibitable Cl^- conduit [possibly ClC-3 (Kawasaki *et*

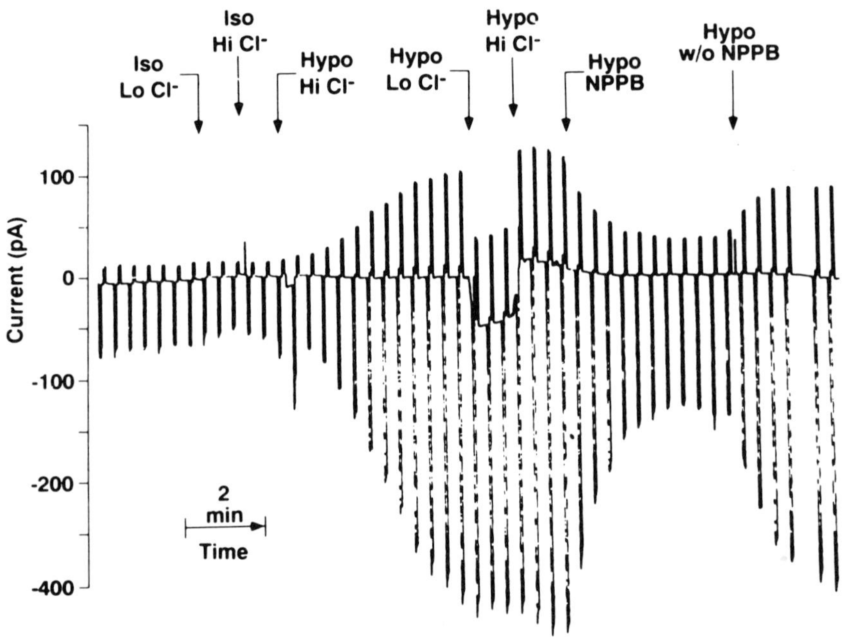

FIGURE 3 Activation of Cl^- channels by hypotonic swelling of cultured human NPE cell. The command voltage was held at −16 mV with periodic cycling to 0 and −82 mV during perforated-patch whole-cell recording. Reducing the osmolality from 315 to 204 mOsm by removing sucrose from the perfusate strongly activated currents at voltages displaced from the Cl^--reversal potential [−9 mV at the high external Cl^- concentration (85.5 m*M*)]. The volume-activated currents were reduced by partially replacing external Cl^- with methylsulfonate and were reversibly blocked by 100 μM NPPB [5-nitro-2-(3-phenylpropylamino)-benzoate]. [Reprinted from Anguíta *et al.* (1995) with permission from Academic Press.]

al., 1994, 1995)] providing the pathway for Cl^- release, which is regulated by a swelling-stimulated Cl^- channel modulator [possibly pl_{Cln} (Paulmichl *et al.*, 1992; Krapivinsky *et al.*, 1994)]. Both ClC-3 (Coca-Prados *et al.*, 1996) and pl_{Cln} (Coca-Prados *et al.*, 1995) are indeed expressed in human NPE cells. The hypothesis is consistent with several additional pharmacologic and electrophysiologic observations (Coca-Prados *et al.*, 1996) and is analogous to the documented action of the protein IsK in regulating I_{Ks} K^+ current through the K_vLQT1 conduits of the mammalian heart (Barhanin *et al.*, 1996; Sanguinetti *et al.*, 1996). P-glycoprotein may replace pl_{Cln} as a Cl^--channel regulator in bovine NPE cells (Wu *et al.*, 1996). Cl^- channels are more fully discussed in Chapters 2 and 3.

4. H^+-ATPase

Wax and his collaborators (Saito *et al.*, 1995; Wax *et al.*, 1997) have presented evidence that a bafilomycin-inhibitable vacuolar H^+-ATPase may play a significant role in regulating aqueous humor secretion. The precise mechanisms and full significance of these observations are not yet clear.

5. Water Pores

Aquaporin-1 (AQP1, initially called CHIP28) has been found to be plentifully distributed in the membranes of NPE cells, but not in PE cells (Stamer *et al.*, 1994). The clinical importance of the vasopressin-regulated aquaporin (AQP2) in the renal distal nephron is very well documented (Nielsen and Agre, 1995). The importance of the aquaporins in aqueous humor formation by the ciliary epithelium and in outflow through the trabecular meshwork is less clear. This issue is addressed in Chapter 5.

V. UNIDIRECTIONAL ABSORPTION

A. Uptake of Solute and Water at the Aqueous Surface by NPE Cells

The possible significance of vectorial transport in the opposite direction back to the stroma has been examined with the simplest possible model of aqueous humor reabsorption: the regulatory volume increase (RVI) of NPE cells (Civan *et al.*, 1996). With this approach, suspensions of cells are first hypotonically swollen to trigger a secondary release of KCl and water (the regulatory volume decrease or RVD) (Yantorno *et al.*, 1989, 1992; Civan *et al.*, 1994; Edelman *et al.*, 1994; Anguíta *et al.*, 1995; Botchkin and Matthews, 1995; Wu *et al.*, 1996). Sucrose is then added to restore isotonicity, shrinking the cells shrink to ~80% of their initial isotonic volumes. This sequence of events triggers a secondary regulatory response (the RVI) in many cells, in which solute and water are taken up by the cells. The RVI observed in human NPE cells displayed a volume-recovery rate of $0.144 \pm 0.007\%$/min (Civan *et al.,* 1996).

The reabsorption of solutes by human NPE cells reflects operation of at least four sets of transport mechanisms (Civan *et al.,* 1996): coupled Na^+–H^+ and Cl^-–HCO_3^- antiports, a hydrochlorthiadiazide-inhibitable Na^+–Cl^- symport, a Na^+–K^+–$2Cl^-$ symport, and an amiloride-sensitive Na^+ channel. Three of the four sets of mechanisms could be detected without elevating the K^+ concentration, but bumetanide-sensitive uptake through the Na^+–K^+–$2Cl^-$ symport was measurable only with an external K^+ concentration of 20 m*M* (Civan *et al.,* 1996). Independent evidence for the operation of a Na^+–K^+–$2Cl^-$ symport has been obtained from measurements of bumetanide-sensitive $^{86}Rb^+$ uptake by cultured human (Crook and Polansky, 1994; Crook and Riese, 1996) and rabbit (Dong and Delamere, 1994) NPE cells. Not only cell shrinkage, but also cAMP-activated protein kinase, appears to stimulate activity in the Na^+–K^+–$2Cl^-$ symport (Crook and Polansky, 1994; Crook and Riese, 1996). In contrast, activation of PKC inhibits activity of the Na^+–K^+–Cl^- symport in both NPE (Dong and Delamere, 1994) and PE cells (Von Brauchitsch and Crook, 1993).

Participation of an amiloride-sensitive epithelial Na^+ channel in the RVI is unusual in nonrenal cells, but has been noted in at least two other cell preparations (Okada and Hazama, 1989; Wehner *et al.,* 1995). Benzamil is a far more effective inhibitor of epithelial Na^+ channels than of Na^+–H^+ antiport exchange (Kleyman and Cragoe, 1988), and it inhibited the RVI of NPE cells at a very low concentration (1 μM) (Civan *et al.,* 1996). At the same concentration, benzamil had no effect on aliquots of the same cells in isosmotic suspension, suggesting that the sequence of hypotonic swelling followed by isotonic shrinkage activated the Na^+ channels (Civan *et al.,* 1997). Na^+ may also enter NPE cells from the aqueous humor through a cation-nonselective channel that has been detected by cell-attached patch-clamping of the intact rabbit ciliary epithelium (Carré *et al.,* 1996a).

Reabsorption of water is expected to follow the same route as for secretion, in part through the AQP1 channels (Stamer *et al.,* 1994).

B. Transfer from NPE to PE Cells through Gap Junctions

Reabsorption of solutes and water from NPE to PE cells should proceed through the gap junctions. Ionic movements can clearly flow in either direction across these junctions in isolated bovine NPE–PE cell couplets

(Wu *et al.,* 1996; Mitchell and Civan, 1997), but their possible rectifying properties have not yet been specifically characterized by patch-clamping.

C. Release of Solute and Water by PE Cells into Stroma

Extrusion of Na^+, K^+, and Cl^- from PE cells back into the stromal interstitial fluid is likely to proceed through the same classes of transporters as those subserving release in the opposite direction from the NPE cells into the aqueous humor. Na^+ will be pumped out of the cell through Na^+–K^+-activated ATPase, and K^+ and Cl^- will be released down their electrochemical gradients through parallel ion channels. However, the molecular basis and signaling pathways of these transport elements are likely different at the two membrane surfaces. As noted earlier, the isozyme components of Na^+–K^+-activated ATPase appear to be different in PE and NPE cells and to depend on location within the ciliary epithelium (Ghosh *et al.,* 1990, 1991) as well.

The same three types of K^+ channels noted in NPE cells have also been identified in PE cells (Fain and Farahbakhsh, 1989; Jacob, 1991b; Stelling and Jacob, 1992). It is unknown whether the molecular structures of the two sets of inward-rectifier, delayed-rectifier, and Ca^{2+}-activated K^+ channels are different. Recent advances in the molecular biology of ocular K^+ channels are discussed in Chapter 4.

Information is also available concerning Cl^- channels of PE cells, based on both volumetric and electrophysiologic measurements. A large-conductance Cl^- channel (~300 pS) has been observed in bovine PE cells under isotonic conditions (Mitchell *et al.,* 1997b). Hypotonic swelling of bovine PE cells activates a large-conductance (~100 pS) and a low-conductance (~9 pS) Cl^- channel (Zhang and Jacob, 1997). Recently, 10 μM tamoxifen has been found to accelerate the ATP-enhanced volume activation of Cl^- channels in cultured bovine PE cells (Mitchell *et al.,* 1997a). In contrast, the same concentration of tamoxifen blocks the volume-activated Cl^- channels of bovine NPE cells (Wu *et al.,* 1996; Mitchell *et al.,* 1997c). Clearly, one or more of the Cl^- channels must be different in the PE and NPE cells. The mechanism of tamoxifen's differential action on the two volume-activated Cl^- channels is not yet known. Tamoxifen has been widely used as an inhibitor of P-glycoprotein-associated Cl^- current (Valverde *et al.,* 1992), but at the same concentration acts as a calmodulin antagonist (Lam, 1984). It is unclear whether the different tamoxifen effects on the volume-activated Cl^- channels of PE and NPE cells arise from differences in the channels' molecular structure or in their regulation by calmodulin, P-glycoprotein, or another signaling cascade.

As for uptake of water from stroma to PE cell, we do not know whether water movement in the opposite direction (from PE cell to stroma) proceeds through as-yet unidentified water channels or through the bulk lipid phase.

VI. COORDINATED EFFECTS ON SECRETION AND ABSORPTION

The most striking evidence of endogenous regulation of the ciliary epithelial secretion is provided by the observation of the circadian rhythm. There is, as yet, no consensus on the basis for this striking two- to three-fold periodic change in secretory rate. Chapter 8 presents promising new information based on the strategies of molecular biology, and Chapter 9 considers clinical aspects of this important phenomenon.

A very wide range of regulatory pathways has been believed to modify the rate of aqueous humor formation, including the adrenergic system, arachidonic acid metabolites, melatonin, and corticosteroids. Recently, information has also become available concerning the effects of biologically active peptides (Carré and Civan, 1995; Crook *et al.*, 1994; Crook and Yabu, 1994) and purines (Carré *et al.*, 1996b, 1997a,b; Farahbakhsh and Cilluffo, 1997; Mitchell *et al.*, 1997a) on ciliary epithelial secretion. As noted in the introductory section, models of aqueous humor formation commonly equate changes in unidirectional secretion with changes in net secretion. Actually, it seems only reasonable to presume that modulators of net secretion could exert coordinated and opposite effects on unidirectional secretion and reabsorption. Without such coordination, increasing or decreasing the two antiparallel flows might leave the net formation of aqueous humor unchanged. At least two examples of such coordinated effects on the unidirectional flows have recently been described (Fig. 4).

One example has been provided by studying the effects of PKC on human NPE cells. As noted earlier, activation of PKC with a synthetic diacylglycerol reduces Cl^- channel activity (Civan *et al.*, 1994), and inhibition of PKC increases Cl^- channel activity of volume-activated NPE cells in suspension (Civan *et al.*, 1994) and of isotonically perfused cell-adherent preparations (Coca-Prados *et al.*, 1995). Activation of the NPE Cl^- channels favors unidirectional secretion (Fig. 1). In contrast, inhibiting PKC with staurosporine reduced reabsorption by NPE cells, measured as the RVI (Civan *et al.*, 1996). Thus, staurosporine is expected to increase net secretion by both actions, reducing unidirectional backflow and stimulating unidirectional secretion.

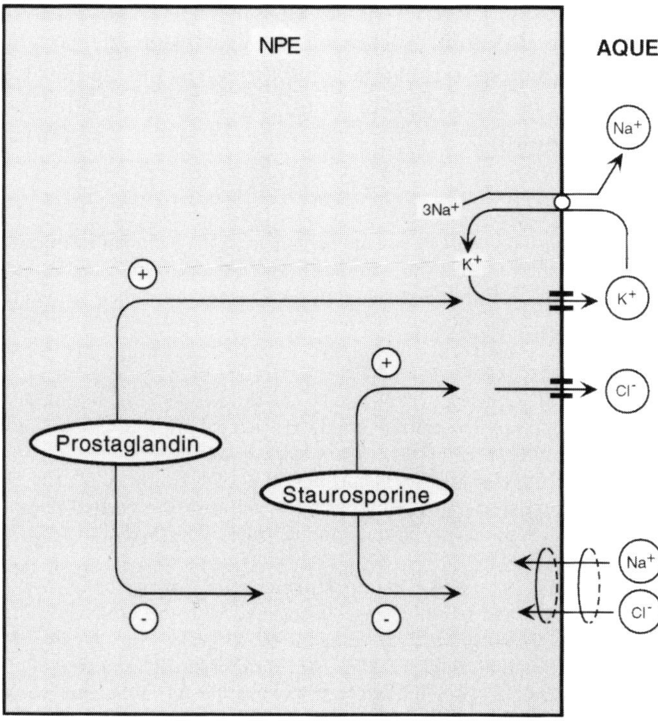

FIGURE 4 Regulation of net aqueous humor formation by coordinated effects on unidirectional secretion and reabsorption at the basolateral surface of NPE cells. [Modified from Civan *et al.* (1997), with permission of the *Journal of Experimental Zoology.*]

A second example is given by the actions of the arachidonic acid metabolite PGE_2 on coordinate modification of unidirectional secretion and reabsorption (Fig. 4). PGE_2 stimulates K^+ channel activity in human ODM NPE cells (Civan *et al.*, 1994), an action that should increase the electrical driving force for Cl^- secretion. The K^+ can recycle by being taken up by the Na^+–K^+ exchange pump. The net effect will be to stimulate unidirectional Cl^- secretion. Like staurosporine, PGE_2 reduces reabsorption by the human NPE cells, as measured by the RVI (Civan *et al.*, 1996). The combined actions of PGE_2 on the unidirectional fluxes will be to stimulate net Cl^- secretion into the aqueous humor. Consistent with these observations, prostaglandin $PGF_{2\alpha}$ has been reported to increase the short-circuit current across the ciliary epithelium (Candia *et al.*, 1989).

These observations indicate that second messenger cascades can trigger coordinate and opposing actions on unidirectional secretion and reabsorption. Net aqueous humor formation can be accelerated both by stimulating unidirectional secretion and by slowing unidirectional reabsorption. The converse is also expected, so that a novel approach to the medical treatment of glaucoma could be to accelerate unidirectional reabsorption to reduce net aqueous flow and IOP.

Acknowledgments

This work was supported in part by research grants from the National Institutes of Health [EY08343, EY10691, and EY01583 (for core facilities)]. I thank Drs. Claire H. Mitchell and Richard A. Stone, David A. Carré, and Kim Peterson-Yantorno for their helpful comments.

References

Anguíta, J., Chalfant, M. L., Civan, M. M., and Coca-Prados, M. (1995). Molecular cloning of the human volume-sensitive chloride conductance regulatory protein, pl_{Cln}, from ocular ciliary epithelium. *Biochem. Biophys. Res. Commun.* **208,** 89–95.

Aperia, A., Fryckstedt, J., Svensson, L., Hemmings, H. C., Jr., Nairn, A. C., and Greengard, P. (1991). Phosphorylated Mr 32,000 dopamine- and cAMP-regulated phosphoprotein inhibits Na^+,K^+-ATPase activity in renal tubule cells. *Proc. Natl. Acad. Sci. U.S.A.* **88,** 2798–2801.

Bairati, A., and Orzalesi, N. (1966). The ultrastructure of the ciliary body: A study of the junctional complexes and of the changes associated with the production of plasmoid aqueous humour. *Zeitschrift für Zellforschuung* **69,** 635–658.

Barhanin, J., Lesage, F., Guillemare, E., Fink, M., Lazdunski, M., and Romey, G. (1996). K_VLQT1 and IsK (minK) proteins associate to form the I_{Ks} cardiac potassium current. *Nature* **384,** 78–80.

Barros, F., Lôpez-Briones, L. G., Coca-Prados, M., and Belmonte, C. (1991). Detection and characterization of Ca^{2+}-activated K^+ channels in transformed cells of human nonpigmented ciliary epithelium. *Curr. Eye Res.* **10,** 731–738.

Bertorello, A. M., Aperia, A., Walaas, S. I., Nairn, A. C., and Greengard, P. (1991). Phosphorylation of the catalytic subunit of Na^+,K^+-ATPase inhibits the activity of the enzyme. *Proc. Natl. Acad. Sci. U.S.A.* **88,** 11359–11362.

Bill, A. (1973). The role of ciliary body blood flow and ultrafiltration in aqueous humor formation. *Exp. Eye Res.* **16,** 287–298.

Botchkin, L. M., and Matthews, G. (1995). Swelling activates chloride current and increases internal calcium in nonpigmented epithelial cells from the rabbit ciliary body. *J. Cell Physiol.* **164,** 286–294.

Bowler, J. M., Peart, D., Purves, R. D., Carré, D. A., Macknight, A. D. C., and Civan, M. M. (1996). Electron probe X-ray microanalysis of rabbit ciliary epithelium. *Exp. Eye Res.* **62,** 131–139.

Butler, G. A., Chen, M., Stegman, Z., and Wolosin, J. M. (1994). Na^+-Cl^-- and HCO_3^--dependent base uptake in the ciliary body pigment epithelium. *Exp. Eye Res.* **59,** 343–349.

Candia, O. A., Chu, T. C., and Alvarez, L. (1989). Prostaglandins and transepithelial ionic transport. *Prog. Clin. Biol. Res.* **312,** 149–154.

Carré, D. A., and Civan, M. M. (1995). cGMP modulates transport across the ciliary epithelium. *J. Membr. Biol.* **146,** 293–305.

Carré, D. A., Tang, C.-S. R., Krupin, T., and Civan, M. M. (1992). Effect of bicarbonate on intracellular potential of rabbit ciliary epithelium. *Curr. Eye Res.* **11,** 609–624.

Carré, D. A., Anguíta, J., Coca-Prados, M., and Civan, M. M. (1996a). Cell-attached patch clamping of the intact rabbit ciliary epithelium. *Curr. Eye. Res.* **15,** 193–201.

Carré, D. A., Mitchell, C. H., Peterson-Yantorno, K., and Civan, M. M. (1996b). Purinergic mechanisms in the ciliary epithelium. *Invest. Ophthalmol. Vis. Sci.* **37,** Suppl., S438.

Carré, D. A., Mitchell, C. H., Peterson-Yantorno, K., Coca-Prados, M., and Civan, M. M. (1997a). Adenosine activates Cl^- channels of NPE ciliary epithelial cells. *Invest. Ophthalmol. Vis. Sci.* **38,** Suppl., S1042.

Carré, D. A., Mitchell, C. H., Peterson-Yantorno, K., Coca-Prados, M., and Civan, M. M. (1997b). Adenosine stimulates Cl^- channels of nonpigmented ciliary epithelial cells. *Am. J. Physiol., Cell Physiol.,* in press.

Cilluffo, M. C., Cohen, B. N., and Fain, G. L. (1991). Nonpigmented cells of the ciliary body epithelium: Tissue culture and voltage gated currents. *Invest. Ophthalmol. Vis. Sci.* **32,** 1619–1629.

Civan, M. M. (1983). "Epithelial Ions and Transport: Application of Biophysical Techniques." Wiley-Interscience, New York.

Civan, M. M., Peterson-Yantorno, K., Coca-Prados, M., and Yantorno, R. E. (1992). Regulatory volume decrease in cultured non-pigmented ciliary epithelial cells. *Exp. Eye Res.* **54,** 181–191.

Civan, M. M., Coco-Prados, M., and Peterson-Yantorno, K. (1994). Pathways signalling the regulatory volume decrease of cultured non-pigmented ciliary epithelial cells. *Invest. Ophthalmol. Vis. Sci.* **35,** 2876–2886.

Civan, M. M., Coca-Prados, M., and Peterson-Yantorno, K. (1996). Regulatory volume increase of human non-pigmented ciliary epithelial cells. *Exp. Eye Res.* **62,** 627–240.

Civan, M. M., Peterson-Yantorno, K., Sánchez-Torres, J., and Coca-Prados, M. (1997). Potential contribution of epithelial Na^+ channel to net secretion of aqueous humor. *J. Exp. Zool.* in press.

Coca-Prados, M., Ghosh, S., Gigula, N. B., and Kumar, N. M. (1992). Expression and cellular distribution of the α1-gap junction gene product in the ocular pigmented ciliary epithelium. *Curr. Eye Res.* **11,** 113–122.

Coca-Prados, M., Anguíta, J., Chalfant, M. L., and Civan, M. M. (1995). PKC-sensitive Cl^- channels associated with ciliary epithelial homologue of pI_{Cln}. *Am. J. Physiol.* **268,** C572–C579.

Coca-Prados, M., Sánchez-Torres, J., Peterson-Yantorno, K., and Civan, M. M. (1996). Association of CIC-3 channel with Cl^- transport by human nonpigmented ciliary epithelial cells. *J. Membr. Biol.* **150,** 197–208.

Cole, D. F. (1960). Effects of some metabolic inhibitors upon the formation of the aqueous humour in rabbits. *Br. J. Ophthalmol.* **44,** 739–750.

Cole, D. F. (1977). Secretion of the aqueous humor. *Exp. Eye Res.* **25,** Suppl., 161–176.

Collins, S. A., Pon, D. J., and Sen, A. K. (1987). Phosphorylation of the alpha-subunit of (Na^+ + K^+)-ATPase by carbachol in tissue slices and the role of phosphoproteins in stimulus-secretion coupling. *Biochim. Biophys. Acta* **927,** 392–401.

Crook, R. B., and Polansky, J. R. (1994). Stimulation of Na^+, K^+, Cl^- cotransport by forskolin-activated adenylyl cyclase in fetal nonpigmented epithelial cells. *Invest. Ophthalmol. Vis. Sci.* **35,** 3374–3383.

Crook, R. B., and Riese, K. (1996). Beta-adrenergic stimulation of Na^+, K^+, Cl^- cotransport in fetal nonpigmented ciliary epithelial cells. *Invest. Ophthalmol. Vis. Sci.* **37,** 1047–1057.

Crook, R. B., and Yabu, J. M. (1994). Down-regulation of vasoactive intestinal peptide receptors by protein kinase C in fetal human non-pigmented ciliary epithelial cells. *Exp. Eye Res.* **59,** 31–39.

Crook, R. B., Lui, G. M., Alvarado, J. A., Fauss, D. J., and Polansky, J. R. (1994). High affinity vasoactive intestinal peptide receptors on fetal human nonpigmented ciliary epithelial cells. *Curr. Eye Res.* **13,** 271–279.

Delamere, N. A., and King, K. L. (1992). The influence of cyclic AMP upon Na,K-ATPase activity in rabbit ciliary epithelium. *Invest. Ophthalmol. Vis. Sci.* **33,** 430–435.

Delamere, N. A., Socci, R. R., and King, K. L. (1990). Alteration of sodium, potassium-adenosine triphosphatase activity in rabbit ciliary processes by cyclic adenosine monophosphate-dependent protein kinase. *Invest. Ophthalmol. Vis. Sci.* **31,** 2164–2170.

Dong, J., and Delamere, N. A. (1994). Protein kinase C inhibits Na^+-K^+-$2Cl^-$ cotransporter activity in cultured rabbit nonpigmented ciliary epithelium. *Am. J. Physiol.* **267,** C1553–C1560.

Edelman, J. L., Sachs, G., and Adorante, J. S. (1994). Ion transport asymmetry and functional coupling in bovine pigmented and nonpigmented ciliary epithelial cells. *Am. J. Physiol.* **266,** C1210–C1221.

Escribano, J., Ortego, J., and Coca-Prados, M. (1995). Isolation and characterization of cell-specific cDNA clones from a subtractive library of the ocular ciliary body of a single normal human donor: Transcription and synthesis of plasma proteins. *J. Biochem.* **118,** 921–931.

Fain, G. L., and Farahbakhsh, N. A. (1989). Voltage-activated currents recorded from rabbit pigmented ciliary body epithelial cells in culture. *J. Physiol.* (*London*) **417,** 83–103.

Farahbakhsh, N. A., and Cilluffo, M. C. (1997). Synergistic increase in Ca^{2+} produced by A_1 adenosine and muscarinic receptor activation via a pertussis-toxin-sensitive pathway in epithelial cells of the rabbit ciliary body. *Exp. Eye Res.* **64,** 173–179.

Farahbakhsh, N. A., Cilluffo, M. A., Chronis, C., and Fain, G. L. (1994). Dihydropyridine-sensitive Ca^{2+} spikes and Ca^{2+} currents in rabbit ciliary body epithelial cells. *Exp. Eye Res.* **58,** 197–206.

Flügel, C., and Lütjen-Drecoll, E. (1988). Presence and distribution of Na^+/K^+-ATPase in the ciliary epithelium of the rabbit. *Histochemistry* **88,** 613–621.

Frizzell, R. A., Field, M., and Schultz, S. G. (1979). Sodium-coupled chloride transport by epithelial tissues. *Am. J. Physiol.* **236,** F1–F8.

Frömter, E., and Diamond, J. M. (1972). Route of passive ion permeation in epithelia. *Nature New Biol.* **235,** 9–13.

Geck, P., Pietrzyk, C., Burckhardt, B.-C., Pfeiffer, B., and Heinz, E. (1980). Electrically silent cotransport of Na, K, and Cl in Ehrlich cells. *Biochim. Biophys. Acta* **600,** 432–447.

Geering, K., Girardet, M., Bron, C., Kraehenbuhl, J. P., and Rossier, B. C. (1982). Hormonal regulation of (Na^+,K^+)-ATPase biosynthesis in the toad bladder. Effect of aldosterone and 3,5,3′ triiodo-L-thyronine. *J. Biol. Chem.* **257,** 10338–10343.

Ghosh, S., Freitag, A. C., Martin-Vasallo, P., and Coca-Prados, M. (1990). Cellular distribution and differential gene expression of the three subunit isoforms of the Na,K-ATPase in the ocular ciliary epithelium. *J. Biol. Chem.* **265,** 2935–2940.

Ghosh, S., Hernando, N., Martin-Alonso, J. M., Martin-Vasallo, P., and Coca-Prados, M. (1991). Expression of multiple Na^+,K^+-ATPase genes reveals a gradient of isoforms along the nonpigmented ciliary epithelium: Functional implications in aqueous humor secretion. *J. Cell Physiol.* **149,** 184–194.

Glynn, I. M. (1993). Annual review prize lecture: All hands to the sodium pump. *J. Physiol.* (*London*) **462,** 1–30.

Gooch, A. J., Morgan, J., and Jacob, T. J. C. (1992). Adrenergic stimulation of bovine non-pigmented ciliary epithelial cells raises cAMP but has no effect on K^+ or Cl^- currents. *Curr. Eye Res.* **11,** 1019–1029.

Green, K., Bountra, C., Georgiou, P., and House, C. R. (1985). An electrophysiologic study of rabbit ciliary epithelial cells. *Invest. Ophthalmol. Vis. Sci.* **26,** 371–381.

Gründer, S., Thiemann, A., Pusch, M., and Jentsch, T. J. (1992). Regions involved in the opening of ClC-2 chloride channel by voltage and cell volume. *Nature* **360,** 759–762.

Hardy, S. P., Goodfellow, H. R., Valverde, M. A., Gill, D. R., Sepúlveda, F. V., and Higgins, C. F. (1995). Protein kinase C-mediated phosphorylation of the human multidrug resistance P-glycoprotein regulates cell volume-activated chloride channels. *EMBO J.* **14,** 68–75.

Hebert, S. C., Gamba, G., and Kaplan, M. (1996). The electroneutral Na^+-(K^+)-Cl^- cotransport family. *Kidney Int.* **49,** 1638–1641.

Helbig, H., Korbmacher, C., Kühner, D., Berweck, S., and Wiederholt, M. (1988a). Characterization of Cl^-/HCO_3^- exchange in cultured bovine pigmented ciliary epithelium. *Exp. Eye Res.* **47,** 515–523.

Helbig, H., Korbmacher, C., Berweck, S., Kuhner, D., and Wiederholt, M. (1988b). Kinetic properties of Na^+/H^+ exchange in cultured pigmented ciliary epithelial cells. *Pflügers Arch.* **412,** 80–85.

Helbig, H., Korbmacher, C., Stumpff, F., Coca-Prados, M., and Wiederholt, M. (1988c). Na^+/H^+ exchange regulates intracellular pH in a cell clone derived from bovine pigmented ciliary epithelium. *J. Cell Physiol.* **137,** 384–389.

Helbig, H., Korbmacher, C., Stumpff, F., Coca-Prados, M., and Wiederholt, M. (1989). Role of HCO_3^- in regulation of cytoplasmic pH in ciliary epithelial cells. *Am. J. Physiol.* **257,** C696–C705.

Jacob, T. J. C. (1991a). Identification of a low-threshold T-type calcium channel in bovine ciliary epithelial cells. *Am. J. Physiol.* **261,** C808–C813.

Jacob, T. J. C. (1991b). Two outward K^+ currents in bovine pigmented ciliary epithelial cells: $I_{K(Ca)}$ and $I_{K(V)}$. *Am. J. Physiol.* **261,** C1055–C1062.

Jacob, T. J. C., and Civan, M. M. (1996). The role of ion channels in aqueous humor formation. *Am. J. Physiol.* **271,** C703–C720.

Karniski, L. P., and Aronson, P. S. (1987). Anion exchange pathways for Cl^- transport in rabbit renal microvillus membranes. *Am. J. Physiol.* **253,** F513–F521.

Kawasaki, M., Uchida, S., Monkawa, T., Miyawaki, A., Mikoshiba, K., Marumo, F., and Saskai, S. (1994). Cloning and expression of a protein kinase C-regulated chloride channel abundantly expressed in rat brain neuronal cells. *Neuron* **12,** 597–604.

Kawasaki, M., Suzuki, M., Uchida, S., Sasaki, S., and Marumo, F. (1995). Stable and functional expression of the ClC-3 chloride channel in somatic cell lines. *Neuron* **14,** 12885–12891.

Kleyman, T. R., and Cragoe, E. J., Jr. (1988). Amiloride and its analogs as tools in the study of ion transport. *J. Membr. Biol.* **105,** 1–21.

Knickelbein, R. G., Aronson, P. S., and Dobbins, J. W. (1986). Oxalate transport by anion exchange across rabbit ileal brush border. *J. Clin. Invest.* **77,** 170–175.

Krapivinsky, G. B., Ackerman, M. J., Gordon, E. A., Krapivinsky, L. D., and Clapham, D. E. (1994). Molecular characterization of a swelling-induced chloride conductance regulatory protein, pl_{Cln}. *Cell* **76,** 439–448.

Krupin, T., and Civan, M. M. (1995). The physiologic basis of aqueous humor formation. *In* "The Glaucomas" (R. Ritch, M. B. Shields, and T. Krupin, eds.), 2nd ed., pp. 251–280. Mosby, St. Louis.

Krupin, T., Reinach, P. S., Candia, O. A., and Podos, S. M. (1984). Transepithelial electrical measurements of the isolated rabbit iris-ciliary body. *Exp. Eye Res.* **38,** 115–123.

Lam, H.-Y. P. (1984). Tamoxifen is a calmodulin antagonist in the activation of cAMP phosphodiesterase. *Biochem. Biophys. Res. Commun.* **118,** 27–32.

Loo, D. D. F., Zeuthen, T., Chandy, G., and Wright, E. M. (1996). Cotransport of water by the Na^+/glucose cotransporter. *Proc. Natl. Acad. Sci. U.S.A.* **93,** 13367–13370.

Macey, R. I., and Farmer, R. E. L. (1970). Inhibition of water and solute permeability in human red cells. *Biochim. Biophys. Acta* **211,** 104–106.

Macknight, A. D. C., McLaughlin, C. W., Peart, D., Purves, R. D., Carré, D. A., and Civan, M. M. (1997). Mechanism of mammalian aqueous humor production. *Abstr. XXXIII Int. Congr. Physiol. Sci.,* P013.04.

Mitchell, C. H., and Civan, M. M. (1997). Effects of uncoupling gap junctions between pairs of bovine NPE-PE ciliary epithelial cells of the eye. *FASEB J.* **11,** A301.

Mitchell, C. H., Peterson-Yantorno, K., Coca-Prados, M., and Civan, M. M. (1997a). Tamoxifen accelerates the ATP-activated regulatory volume decrease of bovine pigmented ciliary epithelial cells. *Invest. Ophthalmol. Vis. Sci.* **38,** S1042.

Mitchell, C. H., Wang, L. W., and Jacob, T. J. C. (1987b). A large-conductance chloride channel in pigmented ciliary epithelial cells activated by GTPγS. *J. Membr. Biol.* **158,** 167–176.

Mitchell, C. H., Zhang, J. J., Wang, L. W., and Jacob, T. J. C. (1997c). Volume-sensitive chloride current in pigmented ciliary epithelial cells: Role of phospholipases. *Am. J. Physiol.* **272,** C212–C222.

Mittag, T. W., Tormay, A., Ortega, M., and Severin, C. (1987). Atrial natriuretic peptide (ANP), guanylate cyclase, and intraocular pressure in the rabbit eye. *Curr. Eye Res.* **6,** 1189–1196.

Nielsen, S., and Agre, P. (1995). The aquaporin family of water channels in kidney. *Kidney Int.* **48,** 1057–1068.

Oh, J., Krupin, T., Tang, L. Q., Sveen, J., and Lahlum, R. A. (1994). Dye coupling of rabbit ciliary epithelial cells *in vitro. Invest. Ophthalmol. Vis. Sci.* **35,** 2509–2514.

Okada, Y., and Hazama, A. (1989). Volume-regulatory ion channels in epithelial cells. *NIPS* **4,** 238–242.

Paulmichl, M., Li, Y., Wickman, K., Ackerman, M., Peralta, E., and Clapham, D. (1992). New mammalian chloride channel identified by expression cloning. *Nature* **356,** 238–241.

Payne, J. A., and Forbush, B., III. (1995). Molecular characterization of the epithelial Na-K-Cl cotransporter isoforms. *Curr. Opin. Cell Biol.* **7,** 493–503.

Quigley, H. A. (1995). Ganglion cell death in glaucoma: Pathology recapitulates ontogeny. *Aust. N. Z. J. Ophthalmol.* **23,** 85–91.

Quigley, H. A., Nickells, R. W., Kerrigan, L. A., Pease, M. E., Thibault, D. J., and Zack, D. J. (1995). Retinal ganglion cell death in experimental glaucoma and after axotomy occurs by apoptosis. *Invest. Ophthalmol. Vis. Sci.* **36,** 774–786.

Raviola, G., and Raviola, E. (1978). Intercellular junctions in the ciliary epithelium. *Invest. Ophthalmol. Vis. Sci.* **17,** 958–981.

Reale, E., and Spitznas, M. (1975). Freeze-fracture analysis of junctional complexes in human ciliary epithelia. *Albrecht v. Graefes Arch. Klin. Exp. Ophthal.* **195,** 1–16.

Rose, R. C., and Schultz, S. C. (1971). Studies on the electrical potential profile across rabbit ileum. Effects of sugars and amino acids on transmural and transmucosal electrical potential differences. *J. Gen. Physiol.* **57,** 639–663.

Saito, I., Patil, R. V., and Wax, M. B. (1995). Characterization of vacuolar H^+-ATPase in ciliary epithelial cells. *Invest. Ophthalmol. Vis. Sci.* **36,** S591.

Sanguinetti, M. C., Curran, M. E., Zou, A., Shen, J., Spector, P. S., Atkinson, D. L., and Keating, M. T. (1996). Coassembly of K_VLQT1 and minK (IsK) proteins to form cardiac I_{Ks} potassium channel. *Nature* **384,** 80–83.

Sears, M. L. (1984). Autonomic nervous system: Adrenergic agonists. *In* "Pharmacology of the Eye" (M. L. Sears, ed.), pp. 193–248. Springer-Verlag, New York.

Shi, X.-P., Zamudio, A. C., Candia, O. A., and Wolosin, J. M. (1996). Adreno-cholinergic modulation of junctional communications between the pigmented and nonpigmented layers of the ciliary body epithelium. *Invest. Ophthalmol. Vis. Sci.* **37,** 1037–1046.

Snyder, S. H., Girault, J. A., Chen, J. Y., Czernik, A. J., Kebabian, J. W., Nathanson, J. A., and Greengard, P. A. (1992). Phosphorylation of DARPP-32 and protein phosphatase inhibitor-1 in rat choroid plexus: Regulation by factors other than dopamine. *J. Neurosci.* **12,** 3071–3083.

Solomon, A. K., Chasan, B., Dix, J. A., Lukacovic, M. F., Toon, M. R., and Verkman, A. S. (1983). The aqueous pore in the red cell membrane. *Ann. N. Y. Acad. Sci.* **414,** 97–124.

Stamer, W. D., Snyder, R. W., Smith, B. L., Agre, P., and Regan, J. W. (1994). Localization of aquaporin CHIP in the human eye: Implications in the pathogenesis of glaucoma and other disorders of ocular fluid balance. *Invest. Ophthalmol. Vis. Sci.* **35,** 3867–3872.

Stelling, J. W., and Jacob, T. J. C. (1992). The inward rectifier K^+ current underlies oscillatory membrane potential behavior in bovine pigmented ciliary epithelial cells. *J. Physiol. (London)* **458,** 439–456.

Stelling, J. W., and Jacob, T. J. C. (1993). Membrane potential oscillation from a novel combination of ion channels. *Am. J. Physiol.* **265,** C720–C727.

Stone, R. A., Laties, A. M., Hemmings, H. C., Jr., Ouimet, C. C., and Greengard, P. (1986). DARPP-32 in the ciliary epithelium of the eye: A neurotransmitter-regulated phosphoprotein of brain localizes to secretory cells. *J. Histochem. Cytochem.* **34,** 1465–1468.

Tsou, K., Snyder, G. L., and Greengard, P. (1993). Nitric oxide/cGMP pathway stimulates phosphorylation of DARPP-32, a dopamine- and cAMP-regulated phosphoprotein, in the substantia nigra. *Proc. Natl. Acad. Sci. U.S.A.* **90,** 3462–3465.

Valverde, M. A., Diaz, M., Sepúlveda, F. V., Gill, D. R., Hyde, S. C., and Higgins, C. F. (1992). Volume-regulated chloride channels associated with the human multidrug-resistance P-glycoprotein. *Nature* **355,** 830–833.

Von Brauchitsch, D. K., and Crook, R. B. (1993). Protein kinase C regulation of a Na^+, K^+, Cl^- cotransporter in fetal human pigmented ciliary epithelial cells. *Exp. Eye Res.* **57,** 699–708.

Vorwerk, C. K., Lipton, S. A., Zurakowski, D., Hyman, B. T., Sabel, B. A., and Dreyer, E. B. (1996). Chronic low-dose glutamate is toxic to retinal ganglion cells. Toxicity blocked by memantine. *Invest. Ophthalmol. Vis. Sci.* **37,** 1618–1624.

Walker, V. E., Miley, H. M., Camodeca, N., Stelling, J. W., Pollard, C. E., and Jacob, T. J. C. (1995). Cell volume, membrane potential and coupling in bovine ciliary epithelial cells. *Invest. Ophthalmol. Vis. Sci.* **36,** S590.

Wax, M. B., Saito, I., Tenkova, T., Krupin, T., Becker, B., Nelson, N., Brown, D., and Gluck, S. L. (1997). Vacuolar H^+-ATPase in ocular ciliary epithelium. *Proc. Natl. Acad. Sci. U.S.A.* in press.

Wehner, F., Sauer, H., and Kinne, R. K. H. (1995). Hypertonic stress increases the Na^+ conductance of rat hepatocytes in primary culture. *J. Gen. Physiol.* **105,** 507–535.

Weiderholt, M., and Zadunaisky, J. A. (1986). Membrane potentials and intracellular chloride activity in the ciliary body of the shark. *Pflügers Arch.* **407,** Suppl. 2, S112–S115.

Weiderholt, M., Helbig, H., and Korbmacher, C. (1991). Ion transport across the ciliary epithelium: Lessons from cultured cells and proposed role of the carbonic anhydrase. *In* "Carbonic Anhydrase" (F. Botre, G. Gross, and B. Storey, eds.), pp. 232–244. VCH Verlagsgesellschaft, Weinheim, Germany.

Wolosin, J. M., Chen, M., Gordon, R. E., Stegman, A., and Butler, G. A. D. (1993). Separation of the rabbit ciliary body epithelial layers in viable form: Identification of differences in bicarbonate transport. *Exp. Eye Res.* **56,** 401–409.

Wolosin, J. M., Schütte, M., and Chen, S. (1996). Connexin distribution in the rabbit and rat ciliary body; a case for heterotypic epithelial gap junctions. *Invest. Ophthalmol. Vis. Sci.* **38,** 341–348.

Wolosin, J. M., Candia, O. A., Peterson-Yantorno, K., Civan, M. M., and Shi, X.-P. (1997). Effect of heptanol on the short circuit currents of cornea and ciliary body demonstrates rate limiting role of heterocellular gap junctions in active ciliary body transport. *Exp. Eye Res.* in press.

Wu, J., Zhang, J. J., Koppel, H., and Jacob, T. J. (1996). P-glycoprotein regulates a volume-activated chloride current in bovine non-pigmented ciliary epithelial cells. *J. Physiol.* **491,** 743–755.

Yantorno, R. E., Coca-Prados, M., Krupin, T., Civan, M. M. (1989). Volume regulation of cultured, transformed, non-pigmented epithelial cells from human ciliary body. *Exp. Eye Res.* **49,** 423–437.

Yantorno, R. E., Carré, D. A., Coca-Prados, M., Krupin, T., and Civan, M. M. (1992). Whole-cell patch clamping of ciliary epithelial cells during anisosmotic swelling. *Am. J. Physiol.* **262,** C501–C509.

Zeuthen, T., Hamann, S., and la Cour, M. (1996). Cotransport of H^+, lactate and H_2O by membrane proteins in retinal pigment epithelium of bullfrog. *J. Physiol.* **497.1,** 3–17.

Zhang, G., Liu, Y., Ruoho, A. E., and Hurley, J. H. (1997). Structure of the adenylyl cyclase catalytic core. *Nature* **386,** 247–253.

Zhang, J. J., and Jacob, T. J. C. (1997). Three different Cl^- channels in the bovine ciliary epithelium activated by hypotonic stress. *J. Physiol.* **499.2,** 379–389.

CHAPTER 2

Molecular Approaches to the Study of the Na^+,K^+-ATPase and Chloride Channels in the Ocular Ciliary Epithelium

Miguel Coca-Prados and Juan Sánchez-Torres
Department of Ophthalmology and Visual Science, Yale University School of Medicine, New Haven, Connecticut 06520

I. INTRODUCTION

The ocular ciliary epithelium is a bilayer of neuroepithelial cells with a unique configuration in the mammalian eye, and possibly in the entire human body. The bilayer consists of two secretory polarized epithelial cells, the nonpigmented (NPE) and the pigmented (PE) ciliary epithelial cells, opposing each other at their apical plasma membranes. This unusual configuration is intriguing and has hampered progress in understanding how the ciliary epithelium regulates the secretion of aqueous humor into the posterior chamber of the eye. The generally accepted theory that the ciliary epithelium is the site of secretion of aqueous humor is based on anatomical and physiological evidence. The basal plasma membrane of the NPE cells

1063-5823/98 $25.00

faces the aqueous humor, and this layer of cells establishes a blood–aqueous humor barrier between the underlining PE cell layer and the posterior chamber of the eye. Thus, the NPE cell layer prevents the free passage of plasma proteins from the blood vessels into the posterior chamber, and from the posterior chamber to the stroma. The "tight" epithelial configuration of NPE cells contrasts with the "leaky" epithelia of the PE cells, which lack tight junctions. Both types of epithelial cells maintain an intimate cell-to-cell communication through numerous gap junctions located predominantly at their apical plasma membrane domains, and at the lateral plasma membrane domains among PE cells. Gap junction channels, which allow the passage of ions and molecules of low molecular weight, are preferentially of the $\alpha 1$ subtype in the ciliary epithelium (Coca-Prados *et al.,* 1992). Wiederholt *et al.* (1991) have suggested that the configuration of the ciliary epithelium might function physiologically as a syncytium.

Another important anatomical consideration of the ocular ciliary epithelium is its postsynaptic innervation by adrenergic and cholinergic fibers, which were observed in a few cases entering into the intercellular space between PE cells (Yamada, 1988). It has been suggested that the sympathetic and parasympathetic systems, acting on target cells of the ciliary body (i.e., ciliary epithelium, vascular endothelium and ciliary smooth muscle), may modulate physiological functions such as secretion of aqueous humor, accommodation, and intraocular pressure.

Along the ocular ciliary epithelium, several distinct anatomical regions have been defined, based on the morphological characteristics of the NPE and PE cells. The *pars plicata,* the most anterior region of the ciliary epithelium, is distinguished by numerous folding processes providing a large transport surface area. A middle region, or *pars plana,* is relatively flat in appearance and extends all the way to the *ora serrata,* which delineates the transition into the retina and retinal pigment epithelium. Although the length differs between species, these three anatomical regions confer anatomical demarcations to the ciliary epithelium. These regions are viewed as biochemical and functionally distinct, exhibiting the expression of genes in a differential fashion. One example of this is the Na^+–K^+ exchange pump and its multiple α- and β-subunit isoforms, which are differentially distributed within the ciliary epithelial cells. The Na^+,K^+-ATPase and isoforms result in an anterior to posterior gradient of expression along the NPE cell layer that coincides with the distinct anatomical regions within the ciliary epithelium (see Fig. 1). Another example is the expression of the cellular retinaldehyde-binding protein (CRALBP), and the neurotrophic factor, pigmented epithelium-derived factor (PEDF), which is restricted to distinct regions and cells within the ciliary epithelium (Martin-Alonso *et al.,* 1993; Ortego *et al.,* 1996). Overall, these data support the idea of

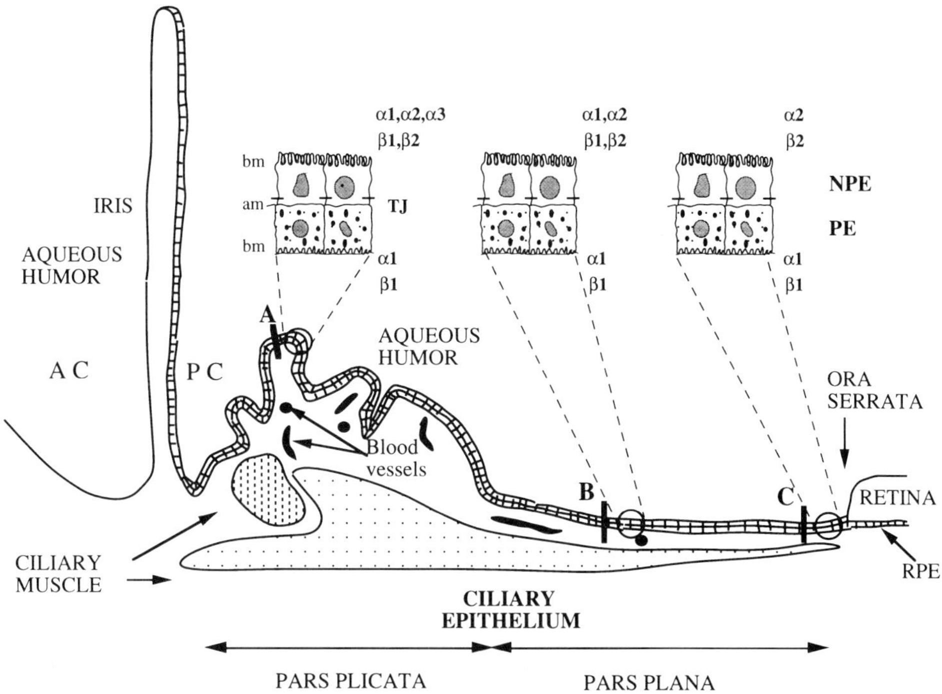

FIGURE 1 Schematic representation of the human ciliary epithelium and cellular distribution of the multiple α ($\alpha1$, $\alpha2$, $\alpha3$) and β ($\beta1$, $\beta2$) subunit isoforms of the Na^+,K^+-ATPase within three regional areas: (A) *pars plicata,* (B) *pars plana,* and (C) *ora serrata.* AC, Anterior chamber; PC, posterior chamber; NPE, nonpigmented ciliary epithelium; PE, pigmented ciliary epithelium; RPE, retinal pigment epithelium; TJ, tight junction; bm, basal plasma membrane; am, apical plasma membrane.

cell heterogeneity within NPE cells along the ciliary epithelium, and of differential mechanisms of gene expression between PE and NPE cells.

Analysis of the ionic composition of the aqueous humor has determined that the major cation transported by the ciliary epithelium is sodium (Na^+) by the Na^+,K^+-ATPase, and the major anion transported is chloride (Cl^-), by chloride channels. This chapter emphasizes the cellular and molecular biology approaches used to investigate the expression of the multiple α- and β-subunit isoforms of the Na^+,K^+-ATPase, the chloride channel ClC-3, and the putative channel regulator $pI_{Cln,}$ in the ocular ciliary epithelium. Finally, we refer to two increasingly useful techniques: the application of baculovirus vectors to express proteins, such as the Na,K-ATPase in insect cells; and the *two-hybrid system,* which is used to search for interacting proteins.

II. Na^+,K^+-ATPase

The Na^+,K^+-ATPase is an integral plasma membrane protein with enzymatic activity to transport Na^+ and K^+ in most eukaryotic cells. It uses the energy produced by the hydrolysis of adenosine triphosphate (ATP) to adenosine diphosphate (ADP) to transport actively three Na^+ ions outside the cell in exchange for two K^+ ions pumped into the cell. This transport produces both a chemical and an electrical ion gradient across the cell membrane that is essential for maintaining a number of vital cell functions. The Na^+ gradient is used by many transport systems to move phosphate, amino acids, or glucose into the cell, and to remove protons or Ca^{2+} from the cell. It plays a vital role in regulating osmotic balance and cell volume, in maintaining the resting membrane potential, and in establishing the ionic composition of fluids such as cerebrospinal fluid and aqueous humor (Cole, 1961, 1984).

The Na^+,K^+-ATPase consist of two subunits, α (110-kDa) and β (50 kDa), present in equimolar amounts. A third polypeptide, termed the γ-subunit (10 kDa) has also been demonstrated to be a structural component of the Na^+,K^+-ATPase, although is not essential for either ATP hydrolysis or transport (Mercer *et al.,* 1993). The $\alpha\beta$-unit forms the functional unit in pure soluble Na^+,K^+-ATPase. The α-subunit possesses the catalytic activity of the enzyme, and the binding sites for ATP, cations, and cardiac glycosides. The α-subunit is also phosphorylated at an aspartate residue by ATP and undergoes ligand-dependent conformational changes that accompany the binding, occlusion, and translocation of ions. The β-subunit is a glycoprotein with an unknown function. It may be necessary to the proper functional assembly of the α-subunit in the plasma membrane. Since the biochemical characterization of multiple α-subunit isoforms of the Na^+,K^+-ATPase (Sweadner, 1979), and the elucidation of the primary structure of the α-subunit of sheep and *Torpedo californica.* (Shull *et al.,* 1985; Kawakami *et al.,* 1985), recombinant DNA technology has led to the cloning and amino acid determination of multiple α ($\alpha1$, $\alpha2$, $\alpha3$, $\alpha4$) and β ($\beta1$, $\beta2$, $\beta3$) isoforms of the Na^+,K^+-ATPase (Lingrel *et al.,* 1990; Sweadner, 1991; Levenson, 1994). Each α and β isoform is encoded by a separate gene that has been mapped in different chromosomes. The multiple α and β isoforms display biochemical and molecular differences, including tissue specificity and developmental expression, sensitivity to cardiac glycosides, affinity for Na^+, and regulation by hormones (Lingrel *et al.,* 1990).

The $\alpha 1$ isoform has been found in all of the eukaryotic cells, whereas $\alpha 2$ and $\alpha 3$ are restricted to skeletal muscle, heart, retina and brain. The human isoforms $\alpha 1$, $\alpha 2$, and $\alpha 3$ exhibit 85% identity in their amino acid sequence. The new $\alpha 4$ isoform, found only in testis (Shamraj and Lingrel, 1994), exhibits 76–78% identity with the other α-subunits isoforms. The β-subunit isoforms ($\beta 1$, $\beta 2$, $\beta 3$) share less homology (32–37%) among themselves. The $\beta 1$ isoform is found in all eukaryotic cells, whereas $\beta 2$ and $\beta 3$ are found predominantly in brain and retina (Hernando *et al.,* 1994; Besirli *et al.,* 1996).

Information on the expression of multiple α- and β-subunit isoforms of the Na^+,K^+-ATPase isoforms in the ocular ciliary epithelium was obtained by applying several complementary approaches: (1) Northern blot hybridization analysis, (2) Western blot analysis, and (3) indirect immunofluorescence.

By Northern analysis, transcripts for five subunit isoforms ($\alpha 1$, $\alpha 2$, $\alpha 3$, $\beta 1$, and $\beta 2$) of the Na^+,K^+-ATPase were demonstrated in the ciliary epithelium of human and bovine eyes (Ghosh *et al.,* 1990, 1991; Coca-Prados *et al.,* 1995b). Figure 2A summarizes the patterns of mRNA hybridization of human $\alpha 1$, $\alpha 2$, $\alpha 3$, $\beta 1$, and $\beta 2$ cDNA probes in the ciliary epithelium and retina of a 51-year-old human eye donor (cadaver). The $\alpha 1$ cDNA probe hybridizes to a single mRNA size of 4.5 kb; the $\alpha 2$ hybridizes to two mRNAs of 4.5 and 6 kb; the $\alpha 3$ hybridizes to a single mRNA size of 4.5; the $\beta 1$ hybridizes to two mRNAs of 2 and 2.5 kb; and the $\beta 2$ probe hybridizes to a 3.6-kb mRNA. The high degree of nucleotide homology (85.8%) between $\alpha 2$ and $\alpha 3$ isoforms, requires highly stringent conditions to prevent cross-hybridization between the $\alpha 2$ probe and $\alpha 3$ RNA, or between the $\alpha 3$ probe and $\alpha 2$ mRNA. It has also been observed that the relative abundance of the two sizes of the $\alpha 2$ isoform is tissue specific, with the larger size transcript appearing predominantly in neural tissues, and the smaller size transcript predominantly in muscle (Mercer, 1993). The abundance of $\alpha 2$ transcripts is higher in the ciliary epithelium than in retina, whereas the expression of $\beta 2$ and $\alpha 3$ mRNA is higher in the retina than in the ciliary epithelium (Fig. 2A). Within the rat retina, it has been determined that $\alpha 3$ and $\beta 2$ are coexpressed preferentially in photoreceptor cells (Schneider and Kraig, 1990; Schneider *et al.,* 1991).

Although the relative abundance of transcripts for each of the α- and β-subunit isoforms has not been determined in NPE and PE cells separately, the isolation of pure cells by flow cytometric analysis (Ghosh *et al.,* 1990) provides a viable way to verify their level of expression by polymerase chain reaction (PCR). We tested and optimized conditions for PCR reactions to amplify DNA sequences of the human α and β isoforms from the ciliary epithelium (see Table I). Using DNA template from human cDNA libraries

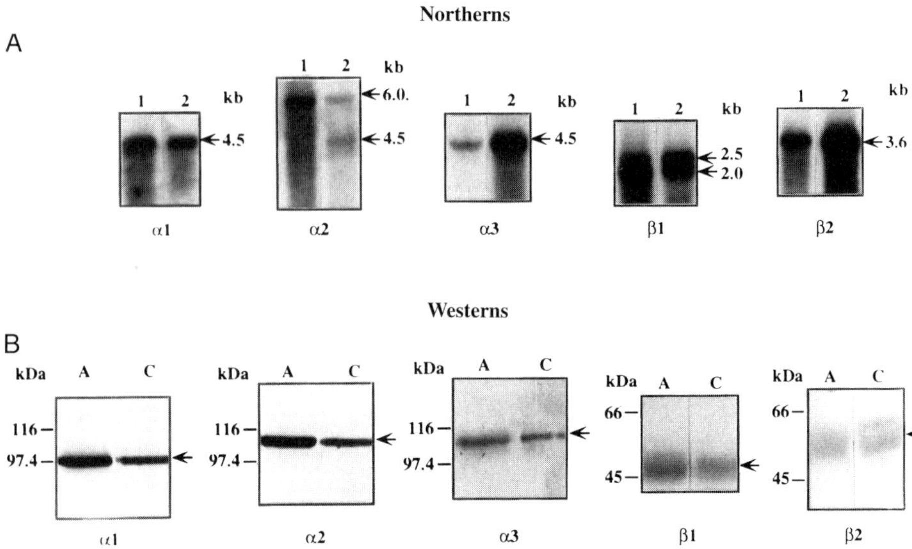

FIGURE 2 (A) Northern blot hybridization analysis of total RNA from the ciliary body (lane 1) and retina (lane 2) of a 51-year-old human eye donor. Five identical Northern blots were prepared with approximately 20 μg of total RNA in each lane and hybridized with human cDNA probes ($\alpha1$, $\alpha2$, $\alpha3$, $\beta1$, and $\beta2$) of the Na^+,K^+-ATPase. The arrows to the right of the gels indicate the size of the mRNAs hybridized to each of the α and β probes. (B) Immunoblotting analysis of Na^+,K^+-ATPase $\alpha1$-, $\alpha2$-, $\alpha3$-, $\beta1$- and $\beta2$-isoforms in the human ciliary epithelium and retina. Microsomes (60 μg/lane) were subjected to reduction–alkylation before separating the proteins in a 10% sodium dodecyl sulfate–polyacrylamide gel electrophoresis (SDS–PAGE) slab. Proteins were transferred to nitrocellulose filters and probed with isoform-specific antibodies to the Na^+,K^+-ATPase α- and β-subunit isoforms. At left is indicated the position of standard protein markers, and at right with an arrow the position of the labeled isoforms.

and the set of oligonucleotide primers shown in Table I, specific PCR products for the $\alpha1$, $\alpha3$, and $\beta2$ isoforms were obtained (Fig. 3).

The generation of isoform-specific antibodies, including monoclonal and polyclonal to the native protein (Urayama *et al.,* 1989; Ghosh *et al.,* 1991), recombinant proteins (Shyjan and Levenson, 1989), or synthetic peptides (Coca-Prados *et al.,* 1995b), has helped to determine its expression and cellular distribution in the ocular ciliary epithelium (Martin-Vasallo *et al.,* 1989; Dong *et al.,* 1994; Coca-Prados *et al.,* 1995). It has been determined that subjecting microsomal fractions from the ciliary epithelium to dithiothreitol (DTT) and iodoacetamide (reduction–alkylation) before separation by SDS–PAGE results in the resolution of $\alpha1$ from $\alpha2$ and $\alpha3$ isoforms (Ghosh *et al.,* 1990). Under these conditions, the $\alpha1$-subunit isoform is resolved as

TABLE I
Oligonucleotide Primers for Human Na^+, K^+-ATPase α and β Isoforms[a]

Gene/acc. no.	Forward/reverse	PCR product (bp)	Annealing temp (°C)
α1 (U16798)	5′>CTGGCCACTGTCACGGTCTGTCTG<3′ 5′>AGGTGTTGGGGCTCCGATGTGTT<3′	503	58.7
α3 (X129 10)	5′>ACCGCTGCTCCACCATCCTGCTAC<3′ 5′>CCGGGGGTTAACCTGGCTGACG<3′	428	62.7
β2 (M81 181)	5′>CCGCGTGGCCTTCAAACTC<3′ 5′>GTGGGCACTTCCGATAACTCTCAG<3′	379	59.3

[a] The oligonucleotide primers in this table have been designed and applied successfully on PCR reactions to amplify DNA sequences for the α1-, α3-, and β-subunit isoforms of the Na^+,K^+-ATPase from the human ciliary epithelium. The nucleotide sequence of each of the DNA products exhibited 100% homology with the corresponding nucleotide sequences reported respectively for α1, α3, and β2 human isoforms.

a 99-kDa protein, and the α2 and α3 resolve as a 105-kDa protein. The β1 and β2 isoforms resolve as broader bands due to their glycosylic properties (Fig. 2B).

Examination of the cellular distribution of the multiple α- and β-subunit isoforms within the ciliary epithelium has revealed marked differences between PE and NPE cells (Fig. 4) and within the NPE cell layer (Fig. 5). The Na^+,K^+-ATPase isoforms are restricted preferentially to the basolateral

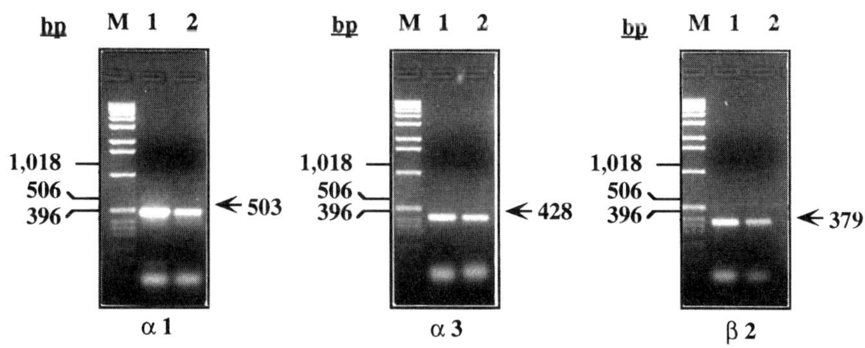

FIGURE 3 Polymerase chain reaction of Na,K-ATPase α and β isoforms from human ciliary epithelium (lane 1) and human retina (lane 2) libraries. The PCR method of Saiki *et al.* (1985) was used to anneal the oligonucleotide primers shown in Table I to DNA from 10^8 PFU of the λUni-ZAP XR cDNA libraries from human ciliary epithelium and retina. The PCR products were resolved on 1% agarose gels, and the expected PCR products, indicated with arrows, directly sequenced using a new method based on a Sequenase PCR sequencing kit (United States Biochemical, Cleveland, OH).

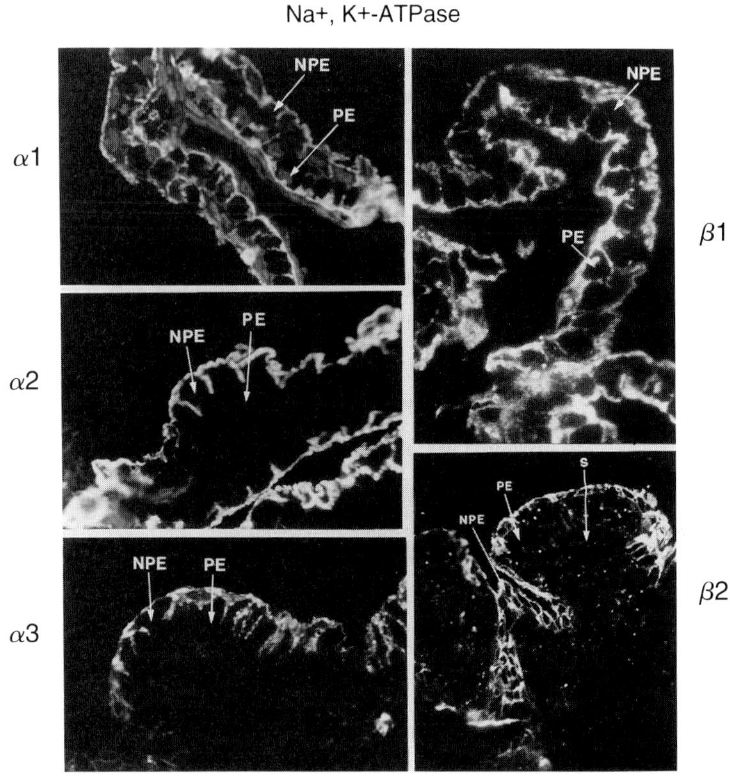

FIGURE 4 Differential cellular distribution of the Na^+,K^+-ATPase subunit isoforms $\alpha1$, $\alpha2$, $\alpha3$, $\beta1$, and $\beta2$ in NPE and PE cells at the anterior region of the ciliary epithelium. Semithin frozen sections (0.5 μm thick) were incubated with antisera against the different isoforms, and specific bound antibodies visualized with rhodamine-conjugated secondary antibodies as previously described (Ghosh *et al.,* 1990, 1991; Coca-Prados *et al.,* 1995b). Notice that antibodies to $\alpha2$, $\alpha3$, and $\beta2$ labeled selectively the basolateral plasma membrane domains of NPE cells, which contrasts with antisera to $\alpha1$ and $\beta1$, which labeled the basolateral plasma membrane domains of both NPE and PE cells. Arrows indicate NPE, nonpigmented epithelium; PE, pigmented epithelium; S, stroma. 500×. [Modified from *The Sodium Pump: Recent Developments—44th Annual Symposium—Part 2. Society of General Physiologists,* pp. 157–163, 1991, by copyright permission of The Rockefeller University Press.]

domains of NPE and PE cells, and the labeling within the NPE cell layer in particular follows a decreasing anterior to posterior gradient (Fig. 5). Along the NPE cell layer, this gradient parallels with a decreasing level of expression for $\alpha1$, $\alpha2$, and $\alpha3$ isoforms at the mRNA and protein level, and a decreasing Na^+,K^+-ATPase enzymatic activity (Ghosh *et al.,* 1991).

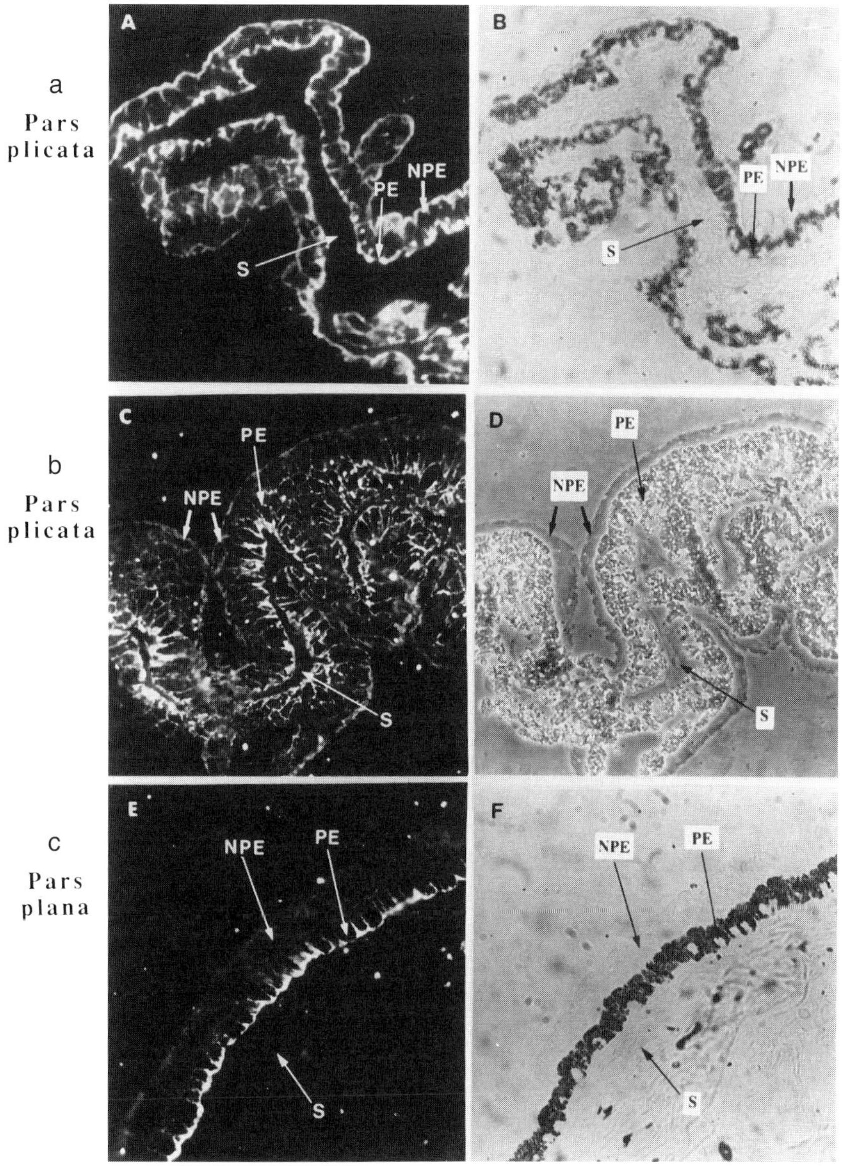

FIGURE 5 Immunostaining of the $\beta 1$-subunit isoform of the Na^+,K^+-ATPase determines an anterior to posterior decreasing gradient on the basolateral plasma membrane domain of NPE cells. The intense labeling at the basolateral plasma membrane of NPE cells in region (a) (panel A), decreases in region (b) (panel C) and disappears in region (c) (panel E). In contrast, the intense labeling of the basolateral plasma membrane of PE cells remains unchanged along the three regions. Panels B, D, and F are phase-contrast photographs of the fluorescence photographs at left (A, C, and E). Notice the presence of granules of melanin in PE layer. Arrows indicate NPE, nonpigmented epithelium; PE, pigmented epithelium; S, stroma. 500×. [Reprinted from Ghosh *et al.* (1991) *Journal of Cellular Physiology,* Copyright © 1991 John Wiley & Sons, Inc. Reprinted by permission of Wiley-Liss, Inc. a subsidiary of John Wiley & Sons, Inc.]

Another remarkable example of a differential displaying labeling among NPE cells along the ciliary epithelium has been documented with the monoclonal antibody (McB-X3.1 developed against the $\alpha3$ isoform (Urayama *et al.,* 1989). This antibody revealed two distinct NPE population of cells, one at the *pars plicata,* which positively labeled, and another at the *pars plana* region, which was not stained by McB-X3.1 (Ghosh *et al.,* 1990). Because transcripts and protein for the $\alpha3$ isoform are detected in these two regions of the ciliary epithelium, these findings were indicative of a possible conformational change of the epitope recognized by this antibody on NPE cells at the *pars plana* region (Ghosh *et al.,* 1990, 1991).

Based on the distinct patterns of subunit isoform immunolabeling within the ciliary epithelium, we suggested these NPE and PE cells may exercise distinct transport activities toward the posterior chamber (secretion) or toward the stroma (absorption). For example, NPE cells at the *pars plicata,* which exhibit labeling for $\alpha1/\alpha2/\alpha3/\beta1/\beta2$ isoforms, are coupled to PE cells with a restricted labeling for $\alpha1/\beta1$ and may favor a higher activity of Na^+ transport by NPE cells toward the aqueous humor side. The net result will be a driving force for secretion toward the posterior chamber at the *pars plicata* region. In contrast, NPE cells at the *pars plana* and *ora serrata* regions exhibiting a reduced expression for all isoforms, coupled to PE cells with an intense labeling for $\alpha1/\beta1$ isoforms, may favor Na^+ transport toward the stroma. This would indicate solute and water reabsorption toward the stroma in those regions. This possibility, although never tested physiologically in the ciliary epithelium, may find support when the substrate dependence properties of $\alpha1$, $\alpha2$, and $\alpha3$ ATPases for Na^+, K^+, and ATP are examined. In this regard, the most significant difference among the three α isoforms is that they possess distinct enzymatic activities under specific cellular conditions, becoming activated at different Na^+ or K^+ concentrations (Jewell *et al.,* 1992). For example, there is evidence that $\alpha1$ or $\alpha2$ isoforms exhibit similar affinity for Na^+. However, the $\alpha3$ isoform exhibits a two- to threefold higher affinity for Na^+ compared with that of $\alpha1$ or $\alpha2$. All three α isoforms have similar affinity for K^+. The ATP dependence of Na^+,K^+-ATPase activity is slightly higher in $\alpha3$ than in $\alpha1$ and $\alpha2$ (Levenson, 1994). Thus, the $\alpha3$-subunit isoform, which is immunodetected on NPE cells at the *pars plicata,* but not at the *pars plana* and *ora serrata* regions, may be more active under conditions of low intracellular Na^+ concentration than $\alpha1$ or $\alpha2$. Because there is no apparent change in the immunodetection of $\alpha1$ and $\beta1$ isoforms along the PE cell layer, we hypothesized that this gradient in expression along the NPE cell layer might be the result of variation in α-subunit and/or β-subunit isoform due to the need to express Na^+,K^+-ATPase with different enzymatic activities.

Additional evidence in support of the idea of functional differences along the ciliary epithelium is also found in studies with the carrier protein CRALBP, a likely component of the visual cycle (Martin-Alonso *et al.,* 1993), and the neurotrophic-survival factor PEDF, a new member of the serine protease inhibitor (serpin) superfamily (Ortego *et al.,* 1996). The cellular distribution of these proteins within the ciliary epithelium also reveals a differential expression between NPE and PE cells and among the regions. Diversity of Na^+,K^+-ATPase α- and β-subunit isoforms in different regions within the intestine (Giannela *et al.,* 1993) and the kidney (Tumlin *et al.,* 1994) have also been demonstrated. Finally, studies *in vitro* and *in vivo* have determined an asymmetric ion transport by PE and NPE cells, supporting the concept that PE and NPE display distinct transport properties leading to secretion or absorption (Edelman *et al.,* 1994).

Studies on the structure and function of the Na^+,K^+-ATPase have been facilitated by molecular approaches that led to the cloning of the genes coding for the α- and β-subunit polypeptides (Shull *et al.,* 1986; Mercer *et al.,* 1986). The Na^+,K^+-ATPase α-subunit is highly conserved among species, reaching approximately 98% between mammalians. The regions containing the phosphorylation site and ATP binding are very well conserved between species and related cation-transporting ATPases of the P-type (i.e., Ca-ATPase, H,K-ATPase). The use of gene transfer techniques, such as the transfer of rat ouabain-resistant enzyme in transfection experiments into primate cells, in conjunction with the use of site direct mutagenesis, and the chemical modification of amino acid residues with carboxyl-specific hydrophobic reagents have provided valuable information relative to which amino acid residues are involved in cation binding, interaction with cardiac glycosides, and ion translocation (Lingrel *et al.,* 1994; Lingrel and Kuntzweiler, 1994; Levenson, 1994). The generation of monoclonal antibodies and limited proteolytic digestion studies have also permitted epitope mapping of both the NH_2 terminus and COOH terminus (Felsenfeld and Sweadner, 1988; Ovchinnikov *et al.,* 1988). Although there is not yet a definitive model on the membrane topology of Na^+,K^+-ATPase, a "working model" proposes a 10 transmembrane model in which three main regions can be distinguished: the NH_2 terminus region containing four strongly hydrophobic transmembrane segments (H1–H4) crossing the membrane as α-helices; a middle region that has no hydrophobic stretches and is folded as a globular domain on the cytoplasmic surface; and a COOH terminus that proposes six putative transmembrane segments (H5–H10) (Lingrel and Kuntzweiler, 1994).

The expression of the multiple Na^+,K^+-ATPase isoforms in Sf9 insect cells, a cell line derived from the ovary of the fall armyworm *Spodoptera frugiperda* using recombinant baculoviruses, has proved to be very useful to study the function of an individual subunit isoform alone or in combination. These studies have revealed: (1) that the α-subunit of the Na^+,K^+-ATPase contains enzymatic activity in the absence of the β-subunit; (2) that the α-subunit assembles selectively with the β-subunit of the Na^+,K^+-ATPase but not with the related β-subunit of the H^+,K^+-ATPase; and (3) that when coexpressed, $\alpha3/\beta2$ displays a higher affinity for Na^+ than $\alpha3/\beta1$, but, is similar to the one exhibited by $\alpha1/\beta1$ (Blanco *et al.,* 1993, 1995).

A variety of genes has been expressed in insect cells with recombinant baculoviruses (O'Reilly *et al.,* 1992). An important advantage of the baculovirus-base system is that it allows the expression of cytoplasmic, secretory, and membrane-bound proteins, following processing steps similar to those in mammalian cells. It uses *Autographa californica,* a lytic nuclear polyhedrosis virus (AcNPV), which propagates to a high titer, and therefore a larger amount of recombinant protein in a soluble form can be obtained (up to 500 mg/liter). AcNPV has a large (130-kb) circular double-stranded DNA that replicates and assembles in viral particles in the nucleus of infected Sf9 cells. Two types of viral particles are formed during the life cycle of AcNPV, a nonoccluded virus that is released extracellularly from 12 to 18 hr postinfection, and an occluded virus or polyhedra-derived virus that accumulates in the nucleus from 18 to 72 hr postinfection. The occluded viral particles are embedded in a viral protein called *polyhedrin* (29-kDa), which is the major protein component of the occlusion bodies. Recombinant baculoviruses are constructed in a two-stage process: (1) The gene of interest is cloned into a plasmid transfer vector downstream from the baculovirus promoter and flanked at the 5′ and 3′ by viral-specific sequences, usually the polyhedrin gene. (2) This plasmid DNA is introduced into insect cells along with wild-type viral DNA. By homologous recombination, the foreign gene is inserted into the viral DNA, and the polyhedrin gene is excised. The gene of interest now is under the control of the polyhedrin promoter. Several biotechnology companies have developed viral AcNPV DNA containing deletions in portions of the gene that are essential for viral propagation, thus increasing the proportion of recombinant viruses from 1 to 99% in cotransfection experiments with the transfer vectors. Similarly, new baculovirus transfer vectors are now commercially available that contain the β-galactosidase gene, simplifying the identification of recombinant plaques and their purification.

Relatively limited information is available, on the function(s) of the $\beta1$- and $\beta2$-subunit isoforms of the Na^+,K^+-ATPase. Previous studies determined the $\beta1$-subunit to be essential for enzyme activity and the proper insertion of the α-subunit in the cell membrane (Geering, 1991). The hydropathy profiles of the $\beta1$ and $\beta2$ isoforms predict one single transmembrane-spanning region, a small cytoplasmic amino terminal tail, and a long extracellular carboxyl-terminal domain. In addition to several glycosylation sites near the carboxyl terminal of the protein, the $\beta1$ isoform contains several disulfide bonds interacting with cyteine residues, which may be important for the activity of the enzyme (Kawamura and Nagano, 1984). The $\beta2$ isoform was originally isolated from liver and found to be homologous to an adhesion molecule on glia (AMOG). These studies led to the hypothesis that the $\beta2$ isoform might be a recognition molecule to modulate several functions including: (1) Na^+,K^+-ATPase activity and (2) interaction with other molecules such as N-CAM. Interestingly, the $\beta2$-subunit isoform of the Na^+,K^+-ATPase is more homologous at the amino acid level (about 41%) with the β-subunit of the gastric Na^+,K^+-ATPase than with the $\beta1$ isoform (35%) (Shull, 1990). Although recent studies indicate that the gastric Na^+,K^+-ATPase is not expressed in the eye (Coca-Prados *et al.*, 1995a), the possibility exists that related α- or β-subunit of a yet to be identified nongastric Na^+,K^+-ATPase may be expressed in the eye. This hypothesis is supported by the observations of cross-reactivity of HK12.18 antibodies to epitope(s) on an antigen(s) on NPE cells distinct from the gastric Na^+,K^+-ATPase (Fain *et al.*, 1988) and the cloning of nongastric α- and β-subunits Na^+,K^+-ATPase(s) in different tissues (Crowson and Shull, 1992; Jaisser *et al.*, 1993a,b). The recent cloning of the human $\beta2$-subunit of the Na^+,K^+-ATPase from human retina (Hernando *et al.*, 1994) shows it to be widely expressed in the human eye, with the exception of the lens and perhaps adult cornea (Coca-Prados *et al.*, 1995b).

III. REGULATION OF Na^+,K^+-ATPase

A great deal of information is available about the regulation of the multiple α-subunit isoforms of the Na^+,K^+-ATPase in numerous tissues by hormones and effector molecules (Samuels *et al.*, 1988; Lingrel *et al.*, 1990), including mineralocorticoids, thyroid hormone, endothelin, prostaglandin E_2 (PGE_2), the nitric oxide. There is evidence that the Na,K-ATPase α isoforms are also under regulation during development (Orlowski and Lingrel, 1988). By analogy, the ocular ciliary epithelium may have some resemblance to the mammalian nephron or intestine in its segmental, functional, and biochemical differentiation. It is conceivable that the enzymatic activity and expression of different isoforms might be regulated in subsets of NPE cells along the ciliary epithelium by the extent to which the different isoforms respond to certain physiological demands (i.e., changes in Na or K balance), or to effector molecules. Recent studies have revealed that the $\alpha1$-subunit isoform of the Na^+,K^+-ATPase is phosphorylated by protein kinase A (PKA) and protein kinase C (PKC) *in vitro* and on stimulation of intact cells in culture with forskolin or phorbol esters, respectively (Beguin *et al.,* 1994). In contrast, these studies also revealed that $\alpha 2$- and $\alpha 3$-subunit isoforms are not substrates for PKC, suggesting that the $\alpha 1$-subunit isoforms might be differentially regulated by PKA phosphorylation and PKC (Feschenko and Sweadner, 1995).

An important area of investigation is the regulation of Na^+,K^+-ATPase isoforms in response to ionic stimuli (Pressley, 1988). At the level of the individual cell, regulation of Na^+,K^+-ATPase expression is key to maintaining low intracellular Na^+ and high intracellular K^+ concentrations. For example, an increase of $\alpha 1/\beta 1$ Na^+,K^+-ATPase expression of about twofold is seen in response to low K^+ incubation in renal cells (Lescale-Matys *et al.,* 1990; Tang and McDonough, 1992). In the kidney, while $\alpha 1$ is central to regulation of extracellular Na^+, $\alpha 2$ expression regulation is central to buffering against changes in extracellular K^+, and may have evolved to satisfy the needs of the organism for this capability (McDonough *et al.,* 1992). The expression of $\alpha 2$ isoform in glial cells has also been demonstrated to respond when extracellular K^+ concentrations are high (Watts *et al.,* 1991). Similarly, it has been determined that the $\alpha 3$-subunit isoform of the Na^+,K^+-ATPase in the pineal gland possesses a lower apparent Michaelis constant for Na^+ than does the $\alpha 1$ isoform, suggesting that it may be more efficient or active at reduced intracellular Na^+ concentrations (Shyjan *et al.,* 1990). This information could be of value in determining whether NPE cells from distinct regions along the ciliary epithelium might exhibit differential responses to ionic stimuli such as low or high extracellular K^+ or Na^+.

In the ocular ciliary epithelial cells, there is evidence that mineralocorticoid receptors are present, which makes it a likely target for mineralocorticoid hormone action (Starka *et al.,* 1977). The administration of aldosterone alters aqueous humor levels of Na^+ and K^+ (Starka *et al.,* 1986). These results are consistent with previous findings showing aldosterone as a potent stimulator of transepithelial Na^+ transport in a variety of target tissues, including the kidney cortical collecting tubules (Barlet-Bas and Doucet, 1988; Farman *et al.,* 1991; Farman, 1993), and as a regulator of Na^+,K^+-ATPase gene expression and enzymatic activity.

Another important effector that may be implicated in the regulation of aqueous humor formation by the ciliary epithelium is prostaglandin. Prostaglandins have been shown to induce diuretic as well as natriuretic effects in several experimental systems (Smith *et al.,* 1989). In the ocular ciliary epithelium there is substantial evidence for the presence of PGE_2-type receptors in the NPE cells (Csukas *et al.,* 1993). It has been suggested that the PGE_2-produced natriuretic effect may occur directly by inhibiting the Na^+,K^+-ATPase. Endothelin, which induces natriuresis *in vivo,* is one example of a hormone with activity that involves the regulation of Na^+,K^+-

ATPase in the kidney, through prostaglandin-dependent mechanisms. Interestingly, endothelin has been found to reduce Na^+,K^+-ATPase activity in a suspension of cells derived from the proximal tubule and inner medullary collecting duct (Zeidel *et al.,* 1989). Response to endothelin was blocked by ibuprofen, an inhibitor of cyclooxygenase, whereas the addition of PGE_2 resulted in an equivalent and nonadditive inhibition of ouabain-sensitive $^{86}Rb^+$ uptake. Based on these studies, it would be of interest to measure Na^+,K^+-ATPase activity in NPE and PE cells isolated from the different regions of the ciliary epithelium. These results could demonstrate whether PGE_2 inhibits Na^+,K^+-ATPase activity on these cells in a differential fashion.

Finally, another area of interest to be investigated is whether nitric oxide (NO) can affect Na^+,K^+-ATPase activity and expression of α- and β-subunit isoforms in the ciliary epithelial cells. Recent evidence suggests that NO, a major messenger molecule, may modulate Na^+,K^+-ATPase function in certain cells (Bachmann and Mundel, 1994).

IV. MOLECULAR CHARACTERIZATION OF THE CHLORIDE CHANNEL ClC-3 AND THE CHLORIDE CHANNEL REGULATOR, pI_{Cln}, IN THE OCULAR CILIARY EPITHELIUM

A. Structure of ClC-3

Chloride channels in general are a key component in ion transport and fluid in epithelial cells. Because chloride (Cl^-) is the main anion transported by the ciliary epithelium into the aqueous humor, and Cl channels have been observed by swelling nonpigmented ciliary epithelial cells *in vivo* and *in vitro* under anisosmotic conditions (Civan *et al.,* 1994), it has been postulated that the Cl^- channels are rate-determining the aqueous humor secretion during volume regulation (Coca-Prados *et al.,* 1995a; Anguita *et al.,* 1995). The molecular mechanisms underlying the process by which a swollen cell restores its original volume by secreting salt and water are not known.

Many functional properties are known concerning the chloride channels which may participate both in volume regulation and in aqueous humor formation (Civan *et al.,* 1994). To understand how these chloride channels are regulated in the ciliary epithelium at the molecular level, knowledge of their amino acid sequences is essential.

Three classes of chloride channels have been cloned from epithelial cells: (1) the ClC gene family; (2) the ABC (ATP-binding cassette) family, including the cystic fibrosis transmembrane conductance regulator (CFTR) and the multidrug resistant P-glycoprotein (*MDR1* gene product); and (3) pI_{Cln}. In this chapter, we emphasize the molecular evidence thus far accumulated, which indicates that the chloride channel ClC-3: (1) is a member of the multigene family of ClC chloride channels; (2) is the main chloride channel expressed in the human ocular ciliary epithelium; and (3) may be regulated by the putative chloride channel regulator pI_{Cln}. Within the multigene family of ClC channels, 10 related chloride channels have been cloned to date: ClC-0 through -7, ClC-Ka, and ClC-Kb. Although the physiological functions for most of these channels are still unknown, it has been suggested that they probably exhibit different properties in the tissues or cells in which they are expressed. Several recent review articles have addressed the molecular biology and molecular physiology of voltage-gated chloride channels of the ClC family (Jentsch, 1994; Pusch and Jentsch, 1994; Jentsch *et al.,* 1995).

As for many other cells, hypotonic swelling of human NPE cells activates both Cl^- and, to a lesser extent, K^+ channels (Civan *et al.,* 1994). The molecular identity of the K^+ channels has not yet been established, and only three proteins (P-glycoprotein, ClC-2, and pI_{Cln}) have been proposed to be involved in volume-regulated Cl^- currents of other cells. Both functional studies (Civan *et al.,* 1994; Coca-Prados *et al.,* 1995a,b) and PCR amplifications of human cDNA libraries (Coca-Prados, Sánchez-Torres, Chalfant, Peterson-Yantorno, and Civan, unpublished manuscript, 1997) render it unlikely that either P-glycoprotein or ClC-2 is involved in volume regulation of human NPE cells. However, two candidate genes for participation in the volume-activated chloride channel activity of human NPE cells have been identified by molecular cloning studies (Coca-Prados *et al.,* 1995; Anguita *et al.,* 1995). One cDNA clone was homologous to the nonsynaptic ClC-3 chloride channel (Coca-Prados *et al.,* 1996). The other cloned gene coded for the human homolog of pI_{Cln}, a putative regulator of a swelling-induced chloride conductance in human NPE cells (Coca-Prados *et al.,* 1995a; Anguita *et al.,* 1995). Civan and Coca-Prados have proposed a working model suggesting that pI_{Cln} is the volume-sensitive regulator of the chloride channel activated by hypotonic swelling of human NPE cells (Coca-Prados *et al.,* 1996). The possibility that P-glycoprotein may play an analogous role in regulating the swelling-activated chloride currents of bovine NPE cells (Wu *et al.,* 1996) is considered in Chapter 3 of this volume.

The ClC-3 chloride channel was originally cloned from a rat kidney cDNA library (Kawasaki *et al.,* 1994a), human retina, and mouse thymus (Borsani *et al.,* 1995). The ClC-3 gene is located on human chromosome 4 (Borsani *et al.,* 1995). ClC-3 was found to be expressed in abundant levels in neuronal cells in the hyppocampus, and in Purkinje cell in the cerebellum. In the ocular ciliary epithelium, the cloning of ClC-3 was carried out by screening a human ciliary body cDNA library with a partial ClC-3 DNA probe, obtained by PCR (Coca-Prados *et al.*, 1996) (Fig. 6).

The clone (JST1.5) has a cDNA insert of approximately 1.9-kb long, which contains the entire coding region of ClC-3. Partial nucleotide sequences of the 5′ and 3′ ends of JST1.5 confirm that it shares 100% homology with the nucleotide sequence for the ClC-3 cDNA isolated from human retina (Borsani *et al.*, 1995). The human ClC-3 differs by only two amino acid residues from the protein encoded by the rat ClC-3 (Kawasaki *et al.*, 1994a), suggesting it is a conserved protein between these two species. Based on the amino acid sequence of the coding region, the predicted molecular mass of ClC-3 is 84 kDa, sharing high similarity (77–87%) with the recently identified ClC-4 and ClC-5, and much lower similarity (18–22.5%) with the rest of the ClC members. Accordingly, ClC-3 is a closer relative to ClC-4 and ClC-5 than to the other members. The ClC-3 chloride

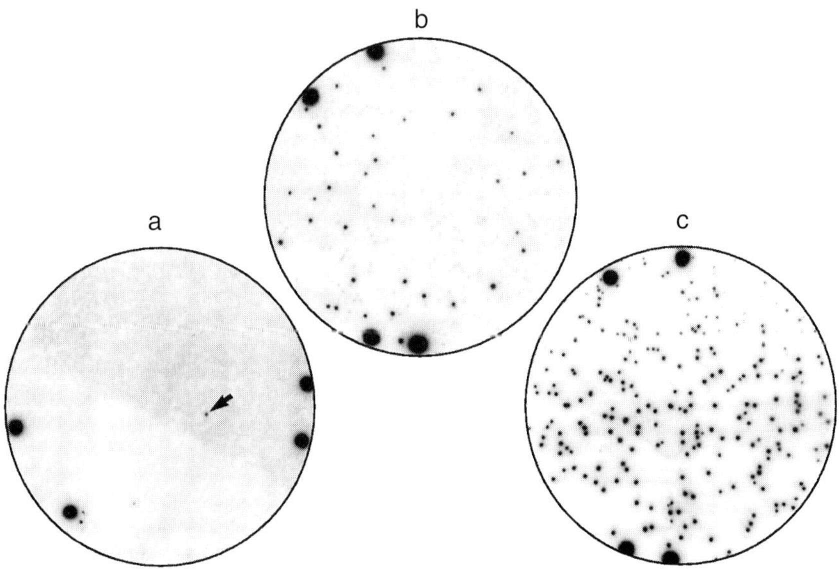

FIGURE 6 Screening and purification of a ClC-3 clone from a human ciliary epithelium cDNA library. After screening approximately 10^5 recombinant clones from a human ciliary epithelial cDNA library with an [α-^{32}P]-labeled ClC-3 DNA (0.5-kb) probe (Coca-Prados *et al.*, 1995a) to high stringency conditions (Sambrook *et al.*, 1989), one positive clone (arrow in panel a) was isolated and purified after three rounds of screening (panels b and c). Partial nucleotide sequence of this clone exhibited complete homology with the nucleotide sequence reported for human ClC-3 (Borsani *et al.*, 1995). The four dense spots in the periphery of a, b, and c are markers used to match the orientation of the filters with the corresponding agar plates.

channel has been successfully expressed in *Xenopus* oocytes, where the elicited chloride current can be blocked by phorbol ester (Kawasaki *et al.,* 1994b).

Figure 7A shows the hydropathy profile of the human ClC-3 chloride channel and its secondary structure based on the one proposed for ClC-0 (Pusch *et al.,* 1995; Jentsch *et al.,* 1995). Hydrophobic regions are plotted above the *x* axis and hydrophilic regions below. The hydropathy and secondary structure analysis predicts 12 to 13 transmembrane-spanning domains. Using the Prosite program of DNAstar consensus sequences on ClC-3 the following posttranslational modifications were revealed: (1) *N*-glycosylation sites, (2) protein phosphorylation by cAMP-dependent protein kinase; (3) protein phosphorylation by PKC; (4) two consensus sequences for protein phosphorylation by phosphorylation by casein-kinase II; and (5) numerous consensus sequences for myristylation modifications.

The secondary structure of ClC-3 in Fig. 7b is based on the topological model proposed for ClC-0 chloride channel (Pusch *et al.,* 1995). In this model the amino and carboxyl termini are placed in the cytoplasm and predict up to 13 transmembrane domains (D1 to D13). Domains D4 and D13 are predicted to not cross the membrane (Gründer *et al.,* 1992). Domains D9 to D12 consist of a broad region that is hydrophobic. Although there is no definitive configuration for any of the members of the ClC chloride channels, a quaternary structure for ClC-0 made of two identical subunits, with each subunit making a chloride conduction pore, has recently been proposed (Middleton *et al.,* 1996; Ludewig *et al.,* 1996). At present, it is not known whether this unique structure (homodimer) can be extended to the rest of the members of the ClC family. The ongoing preparation of polyclonal antibodies to recombinant ClC-3 protein and synthetic peptides will help to determine its cellular distribution in the intact ciliary epithelium. These probes will be very helpful in investigating aspects of the cellular biology of this chloride channel in NPE cells, such as the rate of its synthesis, processing, and delivery to the plasma membrane. Topographic mapping of the structure of ClC-3 protein, using epitope-directed antibodies, and mutagenesis experiments will also provide valuable information about its function and regulation. Thus far, no genetic diseases have been found to be associated with mutations in ClC-3.

B. Molecular Cloning of pI_{Cln}, a Putative Cl^- Channel Regulator, from the Ocular Ciliary Epithelium

Civan's laboratory was the first to describe a chloride current in NPE cells that is activated by cell swelling (Yantorno *et al.,* 1992). The electro-

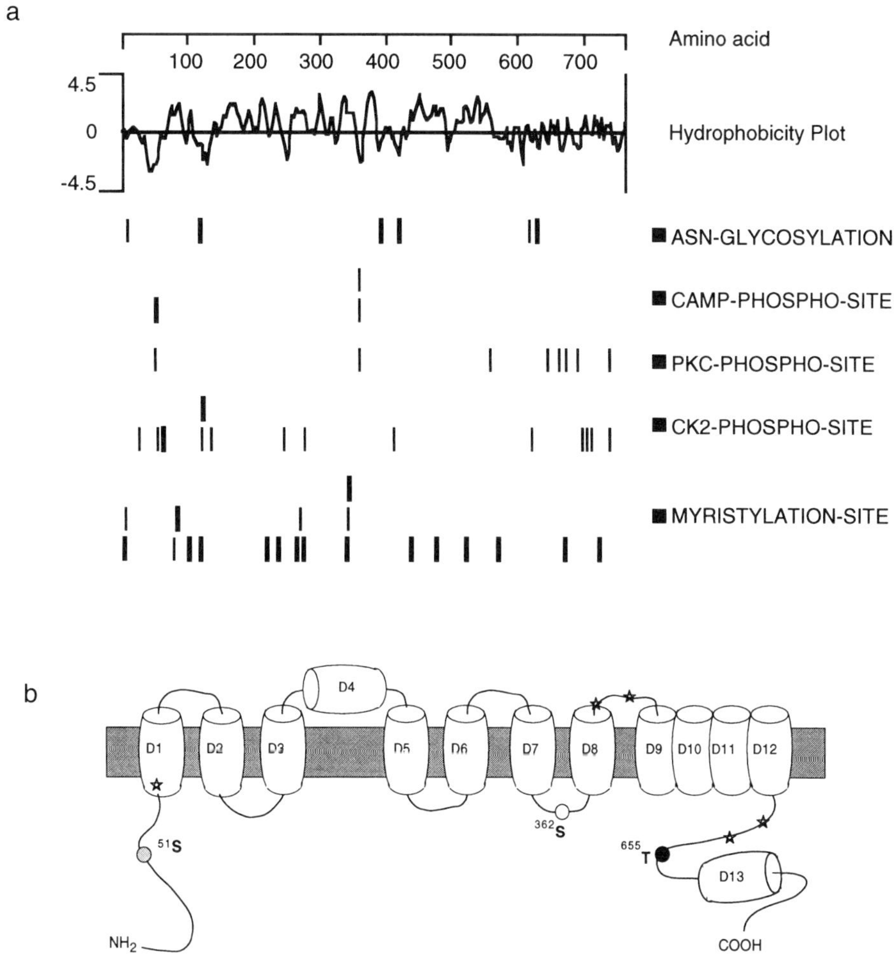

FIGURE 7 7ClC-3 structure. (a) Hydropathy plot of the human ClC-3 generated by the method of Kyte and Doolitle. Hydrophobic regions are above the *x*-axis and hydrophilic regions below. The Prosite program (DNAstar, Madison, WI) revealed the following putative consensus sites (indicated by solid boxes) for posttranslational modifications: (1) glycosylation; (2) phosphorylation by cAMP-dependent PKA; (3) phosphorylation by PKC; (4) phosphorylation by casein-kinase II; and (5) myristylation. (b) Transmembrane orientation of ClC-3 based on the model proposed for ClC-0, the prototype of chloride channels of the ClC family (Jentsch *et al.*, 1995). Stars denote glycosylated residues, and small circles indicate putative phosphorylated residues.

physiological and pharmacological characteristics of the chloride current resembled cell-swelling activated chloride current ($I_{Cl.swell}$) present in many eukaryotic cells and *Xenopus* oocytes. Several studies propose a link between this chloride current $I_{Cl.swell}$ with the pI_{Cln} protein. It has been suggested both that this protein behaves like a chloride channel (Paulmichl *et al.,* 1992) and that it is similar to a cytosolic protein that regulates $I_{Cl.swell}$ current (Krapivinsky *et al.,* 1994). The cDNA encoding the chloride channel/chloride channel regulator was originally isolated from MDCK epithelial cells (Paulmichl *et al.,* 1992), and later from kidney, heart, and ciliary epithelium (Krapivinsky *et al.,* 1994; Okada *et al.,* 1995; Anguita *et al.,* 1995).

We briefly review the work that led to the cloning of pI_{Cln} from the human ciliary epithelium. The human homolog of the pI_{Cln} cDNA was obtained by PCR screening of a cDNA library of the human ciliary epithelium (Anguita *et al.,* 1995). Figure 8 shows an alignment of the deduced amino acid sequences for the chloride channel/chloride channel regulator pI_{Cln} protein among different species. The nucleotide sequence of the human pI_{Cln} cDNA exhibited 89% similarity with the rat pI_{Cln} (Krapivinsky *et al.,* 1994; Abe *et al.,* 1993), 91.9% with the canine pI_{Cln} (Paulmichl *et al.,* 1992), 90.4% with the rabbit pI_{Cln} (Okada *et al.,* 1995), and 67.2% with the *Xenopus laevis* pI_{Cln} (Krapivinsky *et al.,* 1994), indicating that the human, rat, dog, and rabbit cDNA clones encode the same protein with small species differences.

Recently a prokaryotic expression vector, pET (Novagen), was used to express the entire coding region of the human pI_{Cln} in bacteria. The entire open reading frame of pI_{Cln} ($237\alpha\alpha$) was cloned in the expression vector $pET20b^+$, and expressed as a recombinant protein in *Escherichia coli* BL21(DE3)pLysS (Fig. 9). The recombinant pI_{Cln} protein purified from bacteria protein by chromatography using the polyhistidine tag tail at the C-terminus of the recombinant protein has been used to raise polyclonal antibodies (Sánchez-Torres and Coca-Prados, 1996).

Although the predicted molecular mass of pI_{Cln} is 27 kDa, the purified recombinant fusion protein displayed a molecular mass of 37 kDa under denaturing SDS–PAGE conditions. This molecular weight protein was unsuspected since the predicted size for pI_{Cln}, based on its amino acid composition, was 27 kDa. Why the molecular mass was higher than the one predicted by SDS–PAGE is not known. The lack of potential glycosylation sites of pI_{Cln} eliminates this as a possible cause of the abnormal electrophoretic mobility. Perhaps the abundance of negatively charged residues, with a predicted $pI = 3.8$ is involved.

When pI_{Cln} mRNA is translated *in vitro,* the 37-kDa protein is the main product obtained and recognized by antibodies to pI_{Cln} (Krapivinsky *et al.,* 1994). Another puzzling observation is the fact that pI_{Cln} is associated

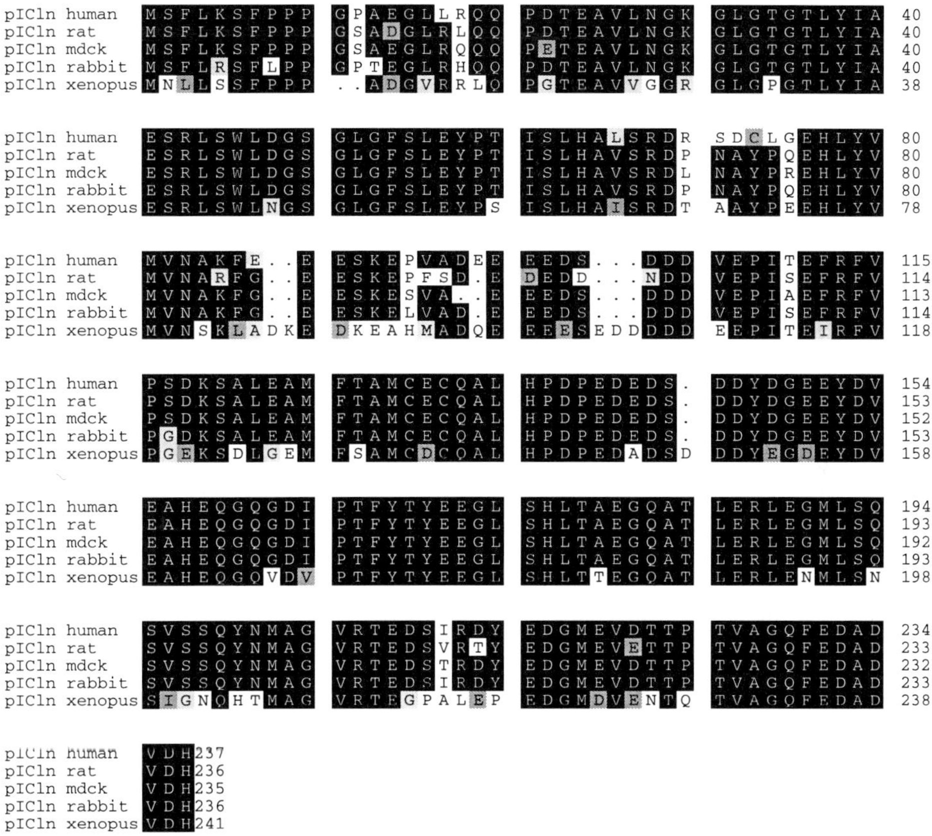

FIGURE 8 Comparison of the predicted pI_{Cln} amino acid sequences from human, rat, dog, rabbit, and *Xenopus.* The sequence similarity of protein pI_{Cln} sequences was determined with the program Pileup of the University of Wisconsin Genetics Computer Group sequence analysis package. Residues conserved are indicated in black boxes. Dots denote gaps introduced into the alignment to allow best fit. Amino acid substitutions are in gray or unboxed. Amino acids are numbered at right. The amino acid sequences were obtained from GenBank/EMBL, accession numbers: U17899 (human); X65450 (MDCK); L26450 (rat); D26076 (rabbit), and L26449 (*X. laevis*).

between 70–80% with the cytosolic fraction, 15–20% with the nuclei fraction, and less than 5% with the microsomal fraction. These observations suggest that pI_{Cln} may form multiple oligomeric complexes with cytosolic proteins, cytoskeleton proteins, or membrane-associated proteins. If pI_{Cln} is, in fact, a swelling-induced protein rather than a channel protein, then the

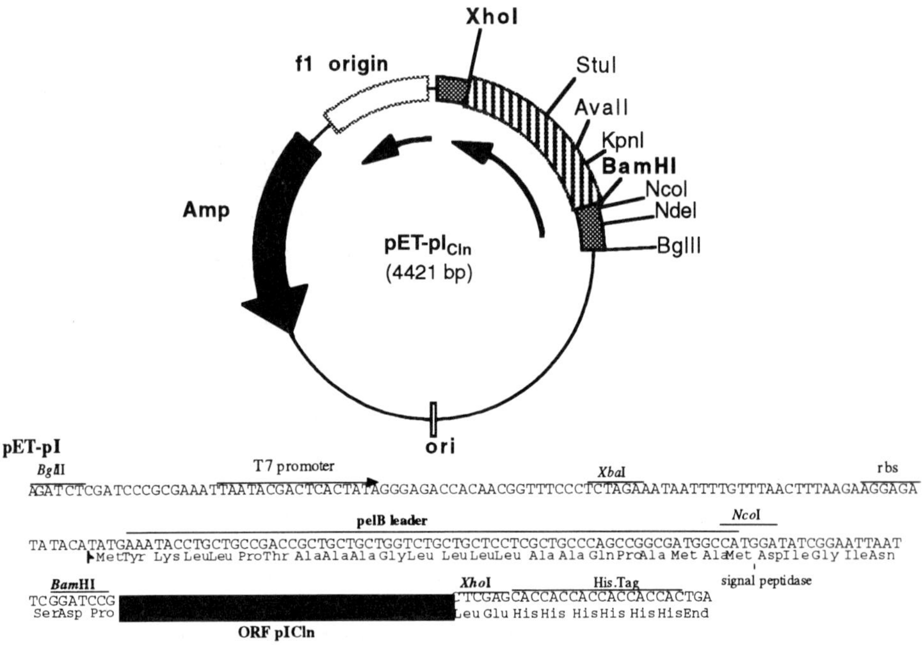

FIGURE 9 Expression of the human pI_{Cln} as recombinant protein in E. coli. The coding region of pI_{Cln} ($237\alpha\alpha$) was subcloned in frame, using *Bam*H1 site in the 5′ end and *Xho*-I in the 3′ end, in the $pET20b^+$ vector. The open reading frame does not include the natural stop codon (TGA) of pI_{Cln}. The recombinant protein of pI_{Cln} was induced in *E. coli* BL21 (DE3)pLysS cells with isopropyl-β-D-thiogalactopyranoside. The recombinant protein contained a tail of six histidines (His-Tag) at the carboxyl end and was purified by chromatography (Sánchez-Torres and Coca-Prados, 1996).

molecular identity of the hypotonicity-activated chloride channels remains unknown (Krapivinsky *et al.,* 1994).

The quaternary structure of pI_{Cln} is not yet known. However, several structure models have been proposed based on its hydrobophicity profile. One model suggests no potential transmembrane helices in pI_{Cln}, but instead assumes that pI_{Cln} forms a channel by dimerization of two identical subunits built of four antiparallel β sheets within each individual subunit (Paulmichl *et al.,* 1992). Site-directed mutagenesis experiments helped to determine important effects of pI_{Cln} on the chloride activities. For example, a single mutation of glycine 49 for alanine within the nucleotide-binding domain (residues 49–53) reduces the sensitivity to extracellular cAMP, while multiple mutations of all three glycines completely abolished its inhibitory effect. In addition, antisense oligonucleotides against pI_{Cln} suppressed the volume-activated chloride currents in 3T3 cells (Gschwentner *et al.*, 1994), and antibodies directed against pI_{Cln} into *Xenopus* oocytes suppressed their swelling-activated chloride (Krapivinsky *et al.*, 1994).

A model of the function of pI_{Cln} in NPE cells was recently proposed by Civan (Coca-Prados *et al.*, 1996). In this working model, pI_{Cln} interacts with cytoplasmic proteins to modulate the function of the chloride channel ClC-3 (Coca-Prados *et al.*, 1996). Based on studies by Krapivinsky *et al.* (1994), one of the putative proteins with which pI_{Cln} may interact has been identified as actin. However, other unknown proteins have been suggested to interact with pI_{Cln}. A powerful technique recently developed and of increasing importance in molecular biology, called the *two-hybrid system* (Bartel *et al.*, 1993), could be used to test whether pI_{Cln} interacts with ClC-3, and to identify other putative interacting proteins with pI_{Cln} and ClC-3. The two-hybrid system enables the identification of interacting proteins in the native conformation, by isolating the cDNA clones for these proteins. If the proteins in question are known, because their amino acid sequences were previously determined (e.g., pI_{Cln} and ClC-3), further analysis can be carried out to determine the domains of interaction between the two proteins. The two-hybrid system is based on the idea that many eukaryotic transcriptional activators, such as the yeast GAL4, are comprised of two domains: (1) a target-specific DNA-binding domain (DNA-BD) and (2) a target-independent activation domain (AD). Although the two domains are normally part of the same protein, a functional activator can be assembled *in vivo* from separated domains of the transcription factor. The two-hybrid system contains sequences encoding the two functional domains of the GAL4 transcriptional activator in two different cloning vectors: (1) GAL4 DNA-BD vector and (2) GAL4 AD vector. The GAL4 DNA-BD vector is used to generate a fusion protein of the DNA-binding domain and a target protein (e.g., pI_{Cln}), and the GAL4 AD vector is used to generate a fusion protein and a target protein (e.g., actin). As an alternative to the GAL4 AD vector containing one target protein, an entire library of hybrids with the activation domain could be used. When interaction occurs between the target protein (i.e., pI_{Cln}) and a candidate protein (known or unknown), the two GAL4 functional domains, responsible for DNA-binding and activation, interact with each other, resulting in the functional restoration of transcriptional activation. The two cloning hybrid vectors, containing the cloned genes for target protein and interacting proteins, are cotransfected into a yeast host strain carrying reporter genes (i.e, *HIS3* and *lacZ*) having upstream GAL4 binding sites. The expression of the reporter genes indicates interaction between a candidate gene and the target protein. Today, commercial kits are available for identifying and investigating protein–protein interactions using the GAL4-based two-hybrid system.

V. ADDITIONAL TRANSPORTER GENES IDENTIFIED IN THE OCULAR CILIARY EPITHELIUM

Since the generation of cDNA libraries representative of the intact ciliary body and ciliary epithelial cells (NPE and PE), a growing number of genes coding for known transporter proteins as well as novel transporter proteins have been found to be expressed within the ciliary epithelium. Although their role in aqueous humor secretion is not yet completely understood, their expression within the ciliary epithelium demonstrates a potential role in aqueous humor secretion. The identification of these transporter proteins has been determined by PCR screening of ciliary epithelial libraries, followed by nucleotide sequencing of the cDNA clone; or alternatively by Northern blot hybridization analysis. Among these genes are the following: (1) a novel water channel (CHIP 29) (Patil *et al.,* 1994); (2) a cGMP-activated gated channel homologous to rod photoreceptor cGMP-gated cation channel (Carré *et al.,* 1996); (3) an amiloride-sensitive sodium channel (Civan *et al.,* 1997); (4) a prostaglandin transporter (Schuster *et al.,* 1997); and (5) a glucose transporter with homology to glucose transporter-like protein III (Escribano *et al.,* 1995).

Acknowledgments

Supported by research grants from the National Institutes of Health (EY 08672 and EY 04873). Additional support was from the Connecticut Lions Foundation and the Glaucoma Foundation. J. S.-T. was supported in part by an ARVO/Alcon Laboratories research fellowship grant. We are grateful to Dr. Mortimer M. Civan, University of Pennsylvania, for his constant encouragement and support.

References

Abe, T., Takeuchi, K., Ishii K., and Abe, K. (1993). Molecular cloning and expression of a rat cDNA encoding MDCK-type chloride channel. *Biochim. Biophys. Acta* **1173,** 353–356.

Anguita, J., Chalfant, M. L., Civan, M. M., and Coca-Prados, M. (1995). Molecular cloning of the human volume-sensitive chloride conductance regulatory protein, pICln, from ocular ciliary epithelium. *Biochem. Biophys. Res. Commun.* **208,** 89–95.

Bachmann, S., and Mundel, P. (1994). Nitric oxide in the kidney: Synthesis, localization, and function. *Am. J. Kidney Dis.* **24**(1), 112–129.

Barlet-Bas, C., and Doucet, A. (1988). Aldosterone and sodium induce kidney Na-K-ATPase *in vitro* by two different mechanisms. *Prog. Clin. Biol. Res.* **268B,** 339–344.

Bartel, P., Chien, C. T., Sternglanz, R., and Fields, S. (1993). Elimination of false positives that arise in using the two-hybrid system. *Biotechniques* **14**(6), 920–924.

Bartel, P. L., and Fields, S. (1995). Analyzing protein–protein interactions using two-hybrid system. *Methods Enzymol.* **254,** 241–263.

Beguin, P., Beggah, A. T., Chibalin, A. V., Burgener-Kairuz, P., Jaisser, F., Mathews, P. M., Rossier, B. C., Cotecchia, S., and Geering, K. (1994). Phosphorylation of the Na,K-ATPase alpha-subunit by protein kinase A and C *in vitro* and in intact cells. Identification of a novel motif for PKC-mediated phosphorylation. *J. Biol. Chem.* **269,** 24437–24445.

Besirli, C., Gong, T.-W. L., and Lomax, M. I. (1997). Novel β3 isoform of the Na,K-ATPase β subunit from mouse retina. *Biochem. Biophys. Acta.* **1350**(1), 21–26.

Blanco, G., Xie, Z. J., and Mercer, R. W. (1993). Functional expression of the alpha 2 and alpha 3 isoforms of the Na,K-ATPase in baculovirus-infected insect cells. *Proc. Natl. Acad. Sci. U.S.A.* **90,** 1824–1828.

Blanco, G., Sanchez, G., and Mercer, R. W. (1995). Comparison of the enzymatic properties of the Na,K-ATPase α3 β1 and α3 β2 isozymes. *Biochemistry* **34,** 9897–9903.

Borsani, G., Rugarli, E. I., Taglialatela, M., Wong, C., and Ballabio, A. (1995). Characterization of a human and murine gene (CLCN3) sharing similarities to voltage-gated chloride channels and to a yeast integral membrane protein. *Genomics* **27,** 131–141.

Carré, D. A., Anguíta, J., Coca-Prados, M., and Civan, M. M. (1996). Cell-attached patch clamping of the intact rabbit ciliary epithelium. *Curr. Eye Res.* **15,** 193–201.

Civan, M. M., Coca-Prados, M., and Peterson-Yantorno, K. (1994). Pathways signaling the regulatory volume decrease of cultured nonpigmented ciliary epithelial cells. *Invest. Ophthalmol. Vis. Sci.* **35,** 2876–2886.

Civan, M. M., Peterson-Yantorno, R., and Coca-Prados, M. (1997). Potential contribution of epithelial Na^+ Channel to net secretion of aqueous humor. *J. Exp. Zool.* in press.

Coca-Prados, M., Martin-Vasallo, P., Hernando-Sobrino, N., and Ghosh, S. (1991). Cellular distribution and differential expression of the Na,K-ATPase α isoform (α1, α2, α3), β1, and β2/AMOG genes in the ocular ciliary epithelium. *In* "The Sodium Pump: Recent Developments" (J. H. Kaplan and P. De Weer, Eds.), pp. 157–163. The Rockefeller University Press, New York.

Coca-Prados, M., Ghosh, S., Gilula, N. B., and Kumar, N. M. (1992). Expression and cellular distribution of the α1 gap junction gene product in the ocular pigmented ciliary epithelium. *Curr. Eye Res.* **11,** 113–122.

Coca-Prados, M., Anguíta, J., Chalfant, M. L., and Civan, M. M. (1995a). PKC-sensitive Cl^- channels associated with ciliary epithelial homologue of pI_{Cln}. *Am. J. Physiol: Cell* **268,** C572–C579.

Coca-Prados, M., Fernandez-Cabezudo, M. J., Sánchez-Torres, J., Crabb, J. W., and Ghosh, S. (1995b). Cell-specific expression of the human Na^+,K^+-ATPase β2 subunit isoform in the nonpigmented ciliary epithelium. *Invest. Ophthalmol. Vis. Sci.* **36,** 2717–2728.

Coca-Prados, M., Sánchez-Torres, J., Peterson-Yantorno, K., and Civan, M. M. (1996). Association of ClC-3 channel with Cl^- transport by human nonpigmented ciliary epithelial cells. *J. Membr. Biol.* **150,** 197–208.

Cole, D. F. (1961). Electrochemical changes associated with the formation of the aqueous humor. *Br. J. Ophthalmol.* **45,** 202–217.

Cole, D. F. (1984). *In* "The Eye" (H. Davson, Ed.), pp. 230–269. Academic Press, San Diego.

Crowson, M. S., and Shull, G. E. (1992). Isolation and characterization of a cDNA encoding the putative distal colon Na^+,K^+-ATPase. Similarity of deduced amino acid sequence to gastric Na^+,K^+-ATPase and Na^+,K^+-ATPase and mRNA expression of distal colon, kidney, and uterus. *J. Biol. Chem.* **267,** 13740–13748.

Csukas, S., Bhattacherjee, P., Rhodes, L., and Paterson, C. A. (1993). Prostaglandin E2 and F2 alpha binding sites in the bovine iris ciliary body. *Invest. Ophthalmol. Vis. Sci.* **34,** 2237–2245.

Dong, J., Delamere, N. A., and Coca-Prados, M. (1994). Na^+,K^+-ATPase inhibition activates $Na^+/K^+/2Cl^-$ cotransporter activity in cultured rabbit nonpigmented ciliary epithelium. *Am. J. Physiol.* **266,** C198–C205.

Edelman, J. L., Sachs, G., and Adorante, J. (1994). Ion transport asymmetry and functional coupling in bovine pigmented and nonpigmented ciliary epithelial cells. *Am. J. Physiol.* **266,** C1210–C1221.

Escribano, J., Ortego J., and Coca-Prados, M. (1995). Isolation and characterization of cell-specific cDNA clones from a subtractive library of the ocular ciliary body of a single normal human donor: Transcription and synthesis of plasma proteins. *J. Biochem.* (*Tokyo*) **118,** 921–931.

Fain, G. L., Smolka, A., Cilluffo, M. C., Fain, M. J., Lee, D. A., Brecha, N. C., and Sachs, G. (1988). Monoclonal antibodies to the H^+-K^+ ATPase of gastric mucosa selectively stain the non-pigmented cells of the rabbit ciliary body epithelium. *Invest. Ophthalmol. Vis. Sci.* **29,** 785–794.

Farman, N. (1993). Recent aspects of the action of aldosterone. *Rev. Nephrol.* **14,** 163–169.

Farman, N., Oblin, M. E., Lombes, M., Delahaye, F., Westphal, H. M., Bonvalet, J. P., and Gasc, J. M. (1991). Immunolocalization of gluco- and mineralocorticoid receptors in rabbit kidney. *Am. J. Physiol.* **260,** C226–C233.

Felsenfeld, D. P., and Sweadner, K. J. (1988). Fine specificity mapping and topography of an isozyme-specific epitope of the Na,K-ATPase catalytic subunit. *J. Biol. Chem.* **263,** 10932–10942.

Feschenko, M. S., and Sweadner, K. J. (1995). Structural basis for species-specific differences in the phosphorylation of Na,K-ATPase by protein kinase C. *J. Biol. Chem.* **270,** 14072–14077.

Geering, K. (1991). The functional role of the beta-subunit in the maturation and intracellular transport of Na, K-ATPase. *FEBS Lett.* **285,** 189–193.

Ghosh, S., Freitag, A. C., Martin-Vasallo, P., and Coca-Prados, M. (1990). Cellular distribution and differential gene expression of the three α subunit isoforms of the Na^+,K^+-ATPase in the ocular ciliary epithelium. *J. Biol. Chem.* **265,** 2935–2940.

Ghosh, S., Hernando, N., Martín-Alonso, J. M., Martin-Vasallo, P., and Coca-Prados, M. (1991). Expression of multiple Na^+,K^+-ATPase genes reveals a gradient of isoforms along the nonpigmented ciliary epithelium: Functional implications in aqueous humor secretion. *J. Cell. Physiol.* **149,** 184–194.

Giannella, R. A., Orlowski, J., Jump, M. L., and Lingrel, J. B. (1993). Na^+,K^+-ATPase gene expression in rat intestine and Caco-2 cells: Response to thyroid hormone. *Am. J. Physiol.* **265,** G775–G782.

Gründer, S., Thiemann, A., Pusch, M., and Jentsch, T. J. (1992). Regions involved in the opening of ClC-2 chloride channel by voltage and cell volume. *Nature* **360,** 759–762.

Gschwentner, M., Nagl, U. O., Schmarda, A., Woll, E., Ritter, M., Waitz, W., Deetjen, P., and Paulmichl, M. (1994). Structure–function relation of a cloned epithelial chloride channel. *Renal Physiol. Biochem.* **17,** 148–152.

Hernando, N., Martín-Vasallo, P., Ghosh, S., Ghosh, P., Swaroop, A., and Coca-Prados, M. (1994). Nucleotide sequence of a cDNA for the $\beta 2$ subunit isoform of Na^+,K^+-ATPase from human retina. *Biochim. Biophys. Acta Bio-Membr.* **1189,** 109–111.

Jaisser, F., Coutry, N., Farman, N., Binder, H. J., and Rossier, B. C. (1993a). A putative H^+,K^+-ATPase is selectively expressed in surface epithelial cells of rat distal colon. *Am. J. Physiol.* **265,** C1080–C1089.

Jaisser, F., Horisberger, J. D., Geering, K., and Rossier, B. C. (1993b). Mechanisms of urinary K^+ and H^+ excretion: Primary structure and functional expression of a novel H,K-ATPase. *J. Cell Biol.* **123,** 1421–1429.

Jentsch, T. J. (1994). Molecular biology of voltage-gated chloride channels. *In* "Current Topics in Membranes," Vol. 42, Chap. 2, pp. 35–57. Academic Press, San Diego.

Jentsch, T. J., Günther, W., Pusch, M., and Schwappach, B. (1995). Properties of voltage-gated chloride channels of the ClC gene family. [Review] *J. Physiol.* **482,** 19S–25S.

Jewell, E. A., Shamraj, O. I., and Lingrel, J. B. (1992). Isoforms of the α subunit of Na,K-ATPase and their significance. *Acta Physiol. Scand.* **146,** 161–169.

Kawakami, K., Noguchi, S., Noda, M., Takahashi, H., Ohta, T., Kawamura, M., Nojima, H., Nagano, K., Hirose, T., Inayama, S., Hayashida, H., Miyata, T., and Numa, S. (1985). Primary structure of the α-subunit of *Torpedo californica* (Na$^+$,K$^+$)ATPase deduced from cDNA sequence. *Nature* **316,** 733–736.

Kawamura, M., and Nagano, K. (1984). Evidence for essential disulfide bonds in the beta-subunit of (Na+, K+)-ATPase. *Biochem. Biophys. Acta* **774,** 188–192.

Kawasaki, M., Uchida, S., Monkawa, T., Miyawaki, A., Mikoshiba, K., Marumo, F., and Sasaki, S. (1994a). Cloning and expression of a protein kinase C-regulated chloride channel abundantly expressed in rat brain neuronal cells. *Neuron* **12,** 597–604.

Kawasaki, M., Uchida, S., Sasaki, S., and Marumo, F. (1994b). Single-channel currents of cloned ClC-3 channels in stably transfected cells. *J. Am. Soc. Nephrol.* **5,** 288.

Krapivinsky, G. B., Ackerman, M. J., Gordon, E. A., Krapivinsky, L. D., and Clapham, D. E. (1994). Molecular characterization of a swelling-induced chloride conductance regulatory protein, pICln. *Cell* **76,** 439–448.

Lescale-Matys, L., Hensley, C. B., Crnkovic-Markovic, R., Putnam, D. S., and McDonough, A. A. (1990). Low K$^+$ increases Na$^+$,K$^+$-ATPase abundance in LLC-PK1/C14 cells by differentially increasing β, and not α, subunit mRNA. *J. Biol. Chem.* **265,** 17935–17940.

Levenson, R. (1994). Isoforms of the Na,K-ATPase: Family membranes in search of function. *Rev. Physiol. Pharmacol.* **123,** 1–45.

Lingrel, J. B., and Kuntzweler, T. (1994). Na,K,-ATPase. *J. Biol. Chem.* **269,** 19659–19662.

Lingrel, J. B., Orlowski, J., Shull, M. M., and Price, E. M. (1990). Molecular genetics of Na,K-ATPase. *In* "Prog. Nucl. Acid. Res. Mol. Biol.," Vol. 38, pp. 37–89. Academic Press, San Diego.

Lingrel, J. B., Huysse, J. V., O'Brien, W., Jewell-Motz, E., Askew, R., and Schultheis, P. (1994). Structure–function studies of the Na,K-ATPase. *Kidney Int.* **45,** S32–S39.

Ludewig, U., Pusch, M., and Jentsch, T. J. (1996). Two physically distinct pores in the dimeric ClC-0 chloride channel. *Nature* **383,** 340–343.

Martín-Alonso, J. M., Ghosh, S., Hernando, N., Crabb, J. W., and Coca-Prados, M. (1993). Differential expression of the cellular retinaldehyde-binding protein in bovine ciliary epithelium. *Exp. Eye Res.* **56,** 659–669.

Martin-Vasallo, P., Ghosh, S., and Coca-Prados, M. (1989). Expression of Na$^+$,K$^+$-ATPase alpha subunit isoforms in the human ciliary body and cultured ciliary epithelial cells. *J. Cell. Physiol.* **141,** 243–252.

McDonough, A. A., Azuma, K. K., Lescale-Matys, L., Tang, M. J., Nakhoul, F., Hensley, C. B., and Komatsu, Y. (1992). Physiologic rationale for multiple sodium pump isoforms. Differential regulation of $\alpha 1$ versus $\alpha 2$ by ionic stimuli. *Ann. N.Y. Acad. Sci.* **671,** 156–169.

Mercer, R. W. (1993). Structure of the Na,K-ATPase. [Review] *Int. Rev. Cytol.* **137C,** 139–168.

Mercer, R. W., Schneider, J. W., Savitz, A., Emanuel, J., Benz, E. J., Jr., and Levenson, R. (1986). Rat-brain Na,K-ATPase beta-chain gene: Primary structure, tissue-specific expression, and amplification in ouabain-resistant HeLa C$^+$ cells. *Mol. Cell. Biol.* **6,** 3884–3890.

Mercer, R. W., Biemesderfer, D., Bliss, D. P., Jr., Collins, J. H., and Forbush, B. III, (1993). Molecular cloning and immunological characterization of the γ polypeptide, a small protein associated with the Na,K-ATPase. *J. Cell Biol.* **121,** 579–586.

Middleton, R. E., Pheasant, D. J., and Miller, C. (1996). Homodimeric architecture of a ClC-type chloride ion channel. *Nature* **383,** 337–340.

Okada, H., Ishii, K., Nunoki, K., and Taira, N. (1995). Cloning of a swelling-induced chloride current related protein from rabbit heart. *Biochim. Biophys. Acta* **1234,** 145–148.

O'Reilly, D. R., Miller, L. K., and Luckow, V. A. (1992). "Baculovirus Expression Vectors: A laboratory Manual," pp. 1–347. W. H. Freeman, New York.

Orlowski, J., and Lingrel, J. B. (1988). Differential expression of the Na,K-ATPase $\alpha 1$ and $\alpha 2$ subunit genes in a murine myogenic cell line. Induction of the $\alpha 2$ isozyme during myocyte differentiation. *J. Biol. Chem.* **263**(33), 17817–17821.

Ortego, J., Escribano, J., Patricia Becerra, S., and Coca-Prados, M. (1996). Gene expression of the neurotrophic factor PEDF in the human ciliary epithelium. *Invest. Ophthalmol. Vis. Sci.* **37,** 2759–2767.

Ovchinnikov, Yu. A., Luneva, N. M., Arystarkhova, E. A., Gevondyan, N. M., Arzamazova, N. M., Kozhich, A. T., Nesmeyanov, V. A., and Modyanov, N. N. (1988). Topology of Na^+,K^+-ATPase. Identification of the extra- and intracellular hydrophilic loops of the catalytic subunit by specific antibodies. *FEBS Lett.* **227,** 230–234.

Patil, R. V., Yang, X., Saito, J., Coca-Prados, M., and Wax, M. B. (1994). Cloning of a novel cDNA homologous to CHIP28 water channel from ocular ciliary epithelium. *Biochem. Biophys. Res. Commun.* **204**(2), 861–866.

Paulmichl, M., Li, Y., Wickman, K., Ackerman, M., Peralta, E., and Clapham, D. (1992). New mammalian chloride channel identified by expression cloning. *Nature* **356,** 238–241.

Pressley, T. A. (1988). Ion concentration dependent regulation of Na,K-pump abundance. *J. Membr. Biol.* **105,** 187–195.

Pusch, M., and Jentsch, T. J. (1994). Molecular physiology of voltage-gated chloride channels. [Review] *Physiol. Rev.* **74,** 813–827.

Pusch, M., Ludewig, V., Rehfeldt, A., and Jentsch, T. J. (1995). Gating of the voltage-dependent chloride channel ClC–0 by the permeant anion. *Nature* **373**(6514), 527–531.

Saiki, R. K., Scharf, S., Faloona, F., Mullis, K. B., Horn, G. T., Erlich, H. A., and Arnheim, N. (1985). Enzymatic amplification of β-globin genomic sequences and restriction site analysis for diagnosis of sickle cell anemia. *Science* **230,** 1350–1354.

Sambrook, J., Fritsch, E. F., and Maniatis, T. (1989). "Molecular Cloning: A Laboratory Manual." Cold Spring Harbor Laboratory Press, Cold Spring Harbor, NY.

Samuels, H. H., Forman, B. M., Horowitz, Z. D., and Ye, Z.-S. (1988). Regulation of gene expression by thyroid hormone. *J. Clin. Invest.* **81,** 957–967.

Sánchez-Torres, J., and Coca-Prados, M. (1996). Structural and functional studies of the swelling-induced chloride channel/channel regulator, pICln, with the aid of specific antibodies and antisense oligonucleotides. *Invest. Ophthalmol. Vis. Sci.* **37,** S439.

Sanger, F., Nicklen, S., and Coulson, A. R. (1977). DNA sequencing with chain terminating inhibitors. *Proc. Natl. Acad. Sci. U.S.A.* **74,** 5463–5467.

Schneider, B. G., and Kraig, E. (1990). Na^+,K^+-ATPase of the photoreceptor: Selective expression of $\alpha 3$ and $\beta 2$ isoforms. *Exp. Eye Res.* **51,** 553–564.

Schneider, B. G., Shyjan, A. W., and Levenson, R. (1991). Co-localization and polarized distribution of Na,K-ATPase $\alpha 3$ and $\beta 2$ subunits in photoreceptor cells. *J. Histochem. Cytochem.* **39,** 507–517.

Schuster, V. L., Lu, R., and Coca-Prados, M. (1997). The prostaglandin transporter "PGT" is widely-expressed in ocular tissues. *Surv. Ophthalmol.* **41**(Suppl. 2), S41–S45.

Shamraj, O. I., and Lingrel, J. B. (1994). A putative fourth Na,K-ATPase alpha-subunit gene is expressed in testis. *Proc. Natl. Acad. Sci. U.S.A.* **91,** 12952–12956.

Shull, G. E. (1990). cDNA cloning of the rat gastric H,K-ATPase. *J. Biol. Chem.* **265,** 12123–12126.

Shull, G. E., Schwartz, A., and Lingrel, J. B. (1985). Amino-acid sequence of the catalytic subunit of the (Na^+,K^+)ATPase deduced from a complementary DNA. *Nature* **316,** 691–695.

Shull, G. E., Greeb, J., and Lingrel, J. B. (1986). Molecular cloning of three distinct forms of the Na^+,K^+-ATPase α-subunit from rat brain. *Biochemistry* **25,** 8125–8132.

Shyjan, A. W., and Levenson, R. (1989). Antisera specific for the $\alpha1$, $\alpha2$, $\alpha3$, and β subunits of the Na,K-ATPase: Differential expression of α and β subunit in rat tissue membranes. *Biochemistry* **28,** 4531–4535.

Shyjan, A. W., Cena, V., Klein, D. C., and Levenson, R. (1990). Differential expression and enzymatic properties of the Na^+,K^+-ATPase $\alpha3$ isoenzyme in rat pineal glands. *Proc. Natl. Acad. Sci. U.S.A.* **87,** 1178–1182.

Smith, W. L., Sonnenburg, W. K., Allen, M. L., Watanabe, T., Zhu, J., and El-Harith, E. A. (1989). The biosynthesis and actions of prostaglandins in the renal collecting tubule and thick ascending limb. *Adv. Exp. Biol.* **259,** 131–147.

Starka, L., Hampl, R., and Obenberger, J. (1977). Aldosterone binding in bovine ciliary body. *Endocrinol. Exper.* **11,** 203–208.

Starka, L., Hampl, R., Obenberger, J., and Doskocil, M. (1986). The role of corticosteroids in the homeostasis of the eye. *J. Steroid Biochem.* **24,** 199–205.

Sweadner, K. (1979). Two molecular forms of (Na^+,K^+)-stimulated ATPase in rat brain: Separation and difference in affinity for strophanthidin. *J. Biol. Chem.* **254,** 6060–6067.

Sweadner, K. J. (1991). Overview: Subunit diversity in the Na,K-ATPase. [Review] *Soc. Gen. Physiol. Series.* **46,** 63–76.

Tang, M. J., and McDonough, A. A. (1992). Low K^+ increases Na^+,K^+-ATPase α- and β-subunit mRNA and protein abundance in cultured renal proximal tubule cells. *Am. J. Physiol.* **263,** C436–C442.

Tumlin, J. A., Hoban, C. A., Medford, R. M., and Sands, J. M. (1994). Expression of Na^+,K^+-ATPase α- and β-subunit mRNA and protein isoforms in the rat nephron. *Am. J. Physiol.* **266,** F240–F245.

Urayama, O., Shutt, H., and Sweadner, K. J. (1989). Identification of three isozyme proteins of the catalytic subunit of the Na,K-ATPase in rat brain. *J. Biol. Chem.* **264,** 8271–8280.

Watts, A. G., Sánchez-Watts, G., Emanuel, J. R., and Levenson, R. (1991). Cell-specific expression of mRNAs encoding Na^+,K^+-ATPase α- and β-subunit isoforms within the rat central nervous system. *Proc. Natl. Acad. Sci. U.S.A.* **88,** 7425–7429.

Wiederholt, M., Helbig, H., and Korbmacher, C. (1991). Ion transport across the ciliary epithelium: Lessons from cultured cells and proposed role of the carbonic anhydrase. *In* "Carbonic Anhydrase" (F. Botre, G. Gross, and B. T. Storey, Eds.), pp. 232–244.

Wiederholt, M., Sturm, A., and Lepple-Wienhues, A. (1994). Relaxation of trabecular meshwork and ciliary muscle by release of nitric oxide. *Invest. Ophthalmol. Vis. Sci.* **35,** 2515–2520.

Wu, J., Zhang, J. J., Koppel, H., and Jacob, T. J. (1996). P-glycoprotein regulates a volume-activated chloride current in bovine non-pigmented ciliary epithelial cells. *J. Physiol.* **491**(3), 743–755.

Yamada E. (1988). Intraepithelial nerve fibers in the rabbit ocular ciliary epithelium. *Arch. Histol. Cytol.* **51,** 43–51.

Yantorno, R. E., Carré, D. A., Coca-Prados, M., Krupin, T., and Civan, M. M. (1992). Whole-cell patch clamping of ciliary epithelial cells during anisosmotic swelling. *Am. J. Physiol.* **262,** C501–C509.

Zeidel, M. L., Brady, H. R., Kone, B. C., Gullans, S. R., and Brenner, B. M. (1989). Endothelin, a peptide inhibitor of Na(+)-K(+)-ATPase in intact renal tubular epithelial cells. *Am. J. Physiol.* **257,** C1101–C1107.

CHAPTER 3

Chloride Channels in the Ciliary Epithelium

Tim J. C. Jacob
School of Molecular and Medical Biosciences, University of Wales, Cardiff CF1 3US, United Kingdom

I. INTRODUCTION

Why should one devote an entire chapter to chloride channels in a book titled *The Eye's Aqueous Humor: From Secretion to Glaucoma*? Current evidence indicates that chloride may be the major, or lead, ion in the transport of fluid from the blood into the eye.

It has long been known that the ciliary transepithelial potential is negative (inside of the eye with respect to the serosal side; Watanabe and Saito, 1978; Burstein *et al.*, 1984) and this was linked to the transport of chloride ions by Chu *et al.* (1987) when they found that the chloride channel blocker 4,4′-diisothiocyanostilbene-2,2′-disulphonic acid (DIDS), applied to the

Current Topics in Membranes, Volume 45

1063-5823/98 $25.00

aqueous side of the tissue, reduced the short-circuit current (Fig. 1). This finding, of a contribution from chloride to the short-circuit current, has recently been confirmed by Mok and To (1996) and Carre *et al.* (1996a). Furthermore, it has been suggested that the same mechanisms that underlie volume regulation may also be involved in fluid secretion (Schultz, 1981; Yantorno *et al.,* 1992; Farahbakhsh and Fain, 1987), and chloride channels are implicated in regulatory volume decrease (RVD). With this in mind, the finding that the adrenergic agonist isoprenaline induces a chloride inward current and a decrease in cell volume in nonpigmented cells (Chen *et al.,* 1994) acquires a special significance.

Chloride channels form a wide and molecularly heterogeneous group of membrane proteins. In their review, Jacob and Civan (1996) listed 11 nonsynaptic Cl^- channels or Cl^--channel regulators. Cl^- channels do not lend themselves to systematic classification on the basis, say, of inhibitor sensitivity nor do they divide neatly into molecular families, so this author is going to divide them into rather arbitrary groups to facilitate a discussion of their characteristics and possible function: volume-activated Cl^- channels, agonist-activated Cl^- channels, anion-selective channels, and nonselective channels.

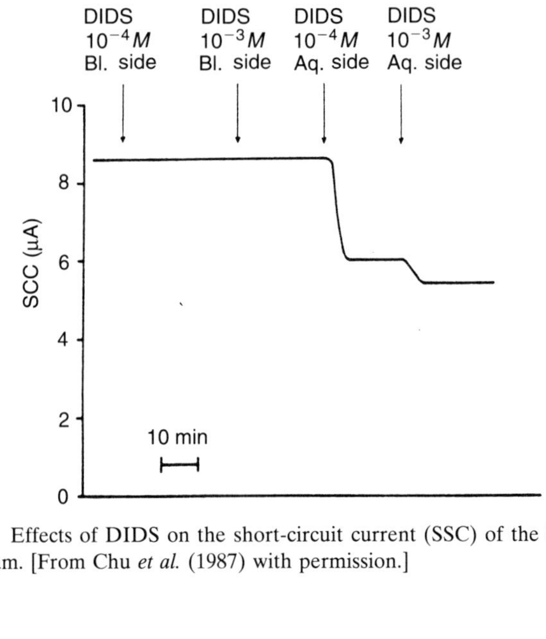

FIGURE 1 Effects of DIDS on the short-circuit current (SSC) of the isolated monkey ciliary epithelium. [From Chu *et al.* (1987) with permission.]

II. VOLUME-ACTIVATED CHLORIDE CHANNELS

A. Whole-Cell Patch-Clamp Studies

1. Nonpigmented Cells

Using the whole-cell recording variant of the patch-clamp technique, a hypotonic-activated chloride current was first reported by Yantorno *et al.* (1992) in the cell line derived from human nonpigmented ciliary epithelial (NPE) cells, ODM-2. Later this current was also found in native bovine (Wu and Jacob, 1994) and rabbit cells (Botchkin and Matthews, 1995). Anguita *et al.* (1995) demonstrated that the same current could be activated under conditions where the cell's interior was not dialyzed, by the less invasive technique of perforated patch whole-cell recording. Yantorno *et al.* (1992) found that, under resting isotonic conditions, chloride conductance appeared to be of little significance, most of the conductance being provided by potassium. However, on exposure to hypotonic solution, most of the current activated was inhibitable by quinidine and 5-nitro-2(3-phenylpropylamino)benzoic acid (NPPB) and they concluded that it was carried by chloride. This they confirmed by conducting ion substitution experiments. The physiological importance of this chloride current was demonstrated by the observation that, in 50% hypotonic solution, RVD in the human nonpigmented cell line (ODM-2) was inhibited by (1) prior chloride substitution and (2) the chloride channel inhibitor DIDS (Civan *et al.,* 1992). This was later confirmed by the demonstration that a range of chloride channel blockers inhibited RVD in freshly isolated bovine NPE cells (Miley *et al.,* 1995b). However, the same range of chloride (and potassium) channel blockers that inhibited RVD in NPE cells had no effect in pigmented ciliary epithelial (PE) cells (Miley *et al.,* 1995b), an observation that we come back to later.

The nature of this volume-activated chloride current in NPE cells has been investigated further using pharmacological, molecular, and immunocytochemical techniques. In rabbit NPE cells, Botchkin and Matthews (1995) found that increases in cell volume increased intracellular calcium and activated a chloride channel, although there was no evidence that calcium or cyclic adenosine monophosphate (cAMP) was involved in the regulation of the chloride channel. The chloride current was inhibited by DIDS, 4-acetamido-4′-isothiocyanatostilbene-2,2′-disulphonic acid (SITS), NPPB, and [(dihydro-indenyl)oxy]alkanoic acid (DIOA). Wu *et al.* (1996) found that the volume-activated chloride current in fresh bovine NPE cells was adenosine triphosphate (ATP_i) dependent, calcium independent, and they concluded from a pharmacological study that the spectrum of inhibitors (NPPB, SITS, DIDS, verapamil, quinidine, tamoxifen, dideoxyforskolin) that blocked the current identified it as the chloride channel associated with the multidrug resistance gene product, P-glycoprotein (P-gp). They followed up this observation by demonstrating the presence of P-gp with the monoclonal antibodies C219 and JSB-1, and by inhibiting volume activation of the chloride current by inclusion of C219 in the patch pipette. C219 recognizes the ATP-binding domain and prevents the binding of ATP, a prerequisite to channel activation. Coca-Prados *et al.* (1995a), working on the cell line ODM-2, found that the protein kinase C (PKC) inhibitor, staurosporine, upregulated a chloride current the kinetics of which were similar to the volume-activated chloride current in these cells. They narrowed down the possible candidates to cystic fibrosis transmembrane conductance regulator (CFTR) or the chloride channel associated with pI_{Cln}. Using the polymerase chain reaction (PCR) method and Northern blotting, they demonstrated the presence of pI_{Cln} transcripts in both the ODM-2 cells and human ciliary body tissue and concluded that their findings were consistent with the presence of a volume-activated, PKC-sensitive chloride channel regulated by pI_{Cln}. The more conventional second messenger signaling pathway involving arachidonic acid metabolism has also been demonstrated to couple cell swelling to channel activation (Civan *et al.,* 1994).

Civan's group have found evidence consistent with an association of the chloride channel ClC-3, with the volume- and staurosporine-activated chloride transport in ODM-2 cells (Coca-Prados *et al.,* 1996). In a volumetric study, they found inhibition of RVD by a range of inhibitors. In support of the involvement of pI_{Cln} was inhibition by NPPB and ATP. In support of the involvement of P-gp was inhibition by verapamil and dideoxyforskolin and possession of consensus phosphorylation sites. The failure of forskolin to inhibit RVD was taken as evidence against the involvement of P-gp. However, there is no evidence in favor of a strong inhibition by forskolin of the P-gp-associated volume-activated chloride channel. Against the involvement of ClC-3 alone is the fact that it has not been shown to be volume activated, so that Coca-Prados *et al.* (1996) have suggested that ClC-3 is regulated by the volume-sensitive pI_{Cln}. Polymerase chain reaction demonstrated the presence of ClC-3 in the human ciliary body and ODM-2 cells.

There are interesting ramifications to these studies, and they need not be mutually exclusive as we see later. More work is necessary to clarify the situation.

2. Pigmented Cells

Before the description of volume-activated chloride channels in the pigmented cells is begun, the author wants to make a case for the much neglected pigmented cell. It was consigned to a marginal role by Okisaka *et al.* (1974), who selectively destroyed the pigmented cell layer and demonstrated that, after 3–6 weeks, the intraocular pressure returned to near normal levels. This study reveals nothing about secretion. The result *cannot* therefore be taken as evidence that the pigmented cells are not involved in the process of aqueous humor secretion.

Pigmented cells, when challenged with hypotonic solution, respond with an increased chloride current. However, as mentioned earlier, this current has a completely different pharmacological sensitivity from that of the NPE cells. Mitchell *et al.* (1996) found that the current was blocked by NPPB, SITS, and DIDS in a voltage-dependent manner but that, unlike the NPE cells, tamoxifen, dideoxyforskolin, and quinidine did not affect it. The current was not dependent on external or internal calcium but was abolished by the phospholipase C inhibitor neomycin. That no chloride channel blockers inhibited the process of RVD when measured volumetrically (Miley *et al.*, 1995b) can be explained by the voltage dependence of the block recorded electrophysiologically. At the resting potential (≈ -70 mV) there is little inhibition. The involvement of G proteins in the volume-activated current was demonstrated, and by selective inhibition of the phospholipases A_2 and C, it was concluded that both were involved in the transduction of the swelling stimulus into a membrane current, the key being the level of the arachidonic acid pool (Mitchell *et al.*, 1996).

B. Single-Channel Patch-Clamp Studies

The whole-cell studies have revealed volume-activated chloride currents and a variety of possible candidates for the role of channel and/or channel regulator (ClC-3, P-gp, pI_{Cln}). The single channel studies reveal that there are, in addition to this multiplicity of candidates, a number of different volume-activated chloride channels and that the channels activated in NPE and PE cells are different.

Zhang and Jacob (1996,1997) found a number of chloride channels that were activated by cell swelling in cell-attached patch-clamp experiments. The NPE cells possessed a 7.3- and a 19-pS chloride channel and the PE cells possessed an 8.6- and a 105-pS chloride channel (Fig. 2), all of which activated, and inactivated, within the duration of a 5–10 min hypotonic challenge. All channels were inhibited by NPPB but only the intermediate conductance (19-pS) channel of the NPE cells was inhibited by tamoxifen. Zhang and Jacob (1997) suggested that the 19-pS, tamoxifen-sensitive channel had similarities with the volume-sensitive organic osmolyte and anion channel (VSOAC) described by Jackson *et al.* (1995). It has been suggested that VSOAC is pI_{Cln} (Strange *et al.*, 1996).

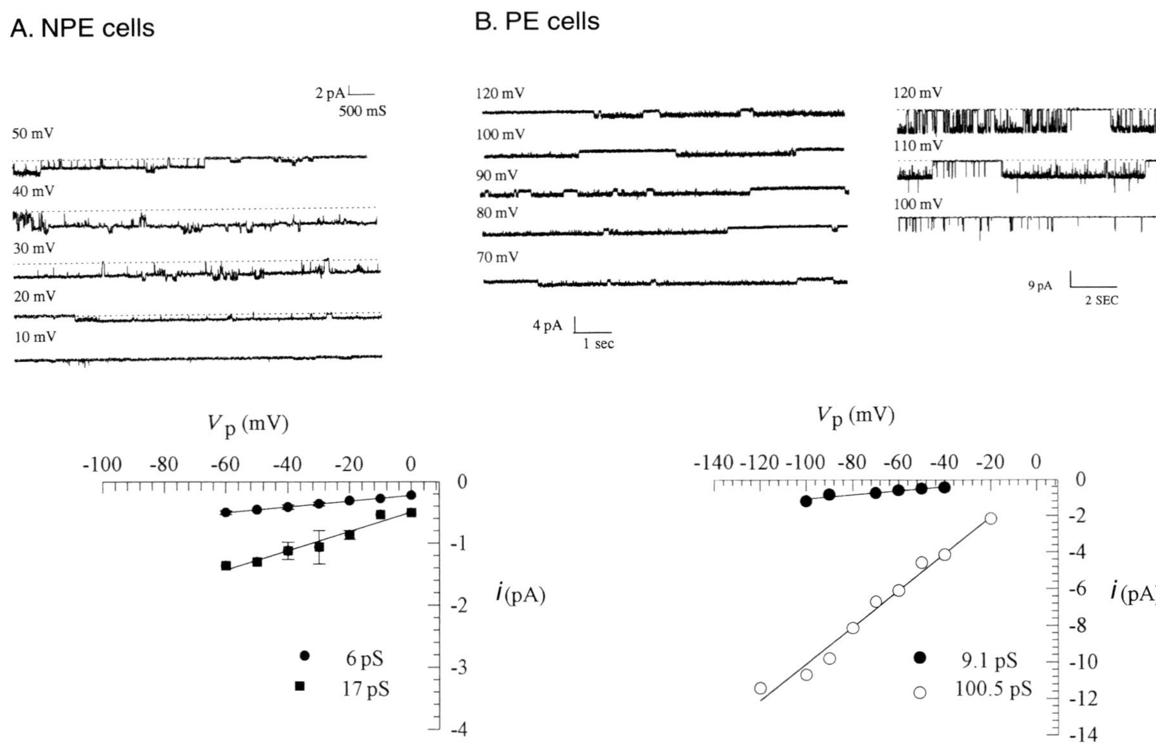

FIGURE 2 Volume-activated single chloride channel activity recorded by cell-attached patch-clamp in (A) nonpigmented and (B) pigmented ciliary epithelial cells. A low-conductance chloride channel (~8 pS) was found in both cell types. An intermediate conductance chloride channel (17 pS) was found in the NPE cells and a large conductance chloride channel (100 pS) was found in the PE cells. The patch pipette contained NMDG-Cl and only hyperpolarizing potentials were used. Downward deflections indicate inward (chloride) currents. [Modified from Zhang and Jacob (1997).]

The task ahead is to assign regulator to channel and to design experiments that will distinguish between the roles of channel or channel regulator for the increasing number of cloned candidate molecules involved in chloride transport.

III. AGONIST-ACTIVATED CHLORIDE CHANNELS

A. *Whole-Cell Studies*

The β-adrenoceptor antagonist, timolol, has been used for many years in the treatment of glaucoma. It was satisfactory, if not entirely unanticipated, therefore, when Chen *et al.* (1994) reported that isoprenaline, a forskolin derivative, and cyclic adenosine monophosphate (cAMP) activated a chloride conductance in dog NPE cells and concomitant with this activation there was a decrease in cell volume (Fig. 3). The current was inhibited by DIDS. However, in a study by Gooch *et al.* (1992), bovine NPE cells failed to produce any chloride current in response to either isoprenaline, even though intracellular levels of cAMP were raised, or to membrane permeant analogues of cAMP.

B. *Patch-Clamp Studies*

Consistent with the former, but not the latter of the preceding two studies, Edelman *et al.* (1995) have described a chloride channel with a conductance of 21 pS on the basolateral membrane of bovine NPE cells that was reported to be activated by dibutyryl-cAMP. This channel may play a role in the agonist-induced efflux of chloride and may therefore participate in aqueous humor secretion.

C. *Efflux Studies*

An increase in chloride efflux from PE cells was observed during cell swelling, and maneuvers that caused elevation of intracellular calcium or cAMP. Activation of P_{2u} receptors increased $[Ca^{2+}]_i$ and isoprenaline caused a significant increase in $cAMP_i$ (Miley *et al.,* 1995a).

IV. ANION-SELECTIVE CHANNELS

An anion channel with a large conductance ($\approx$200 pS) and variable selectivity has been described in a number of tissues. This channel has a

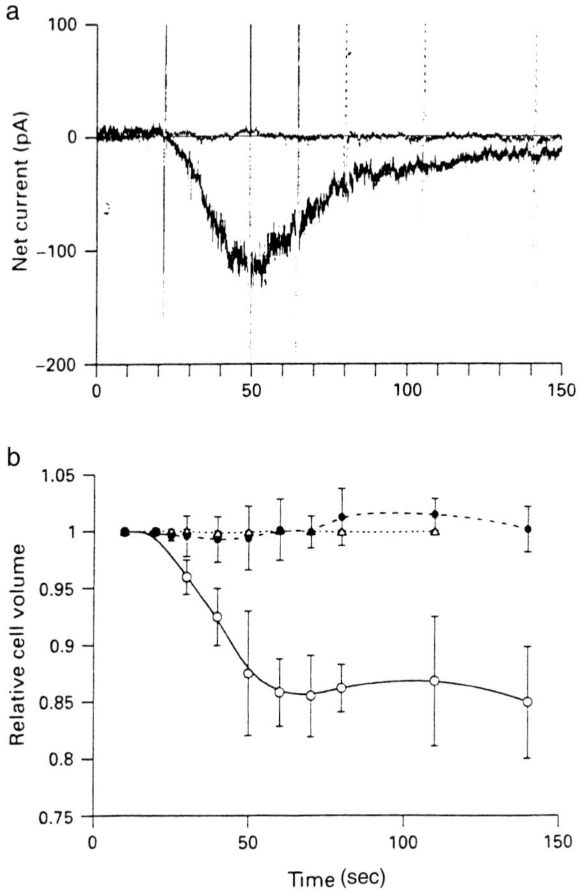

FIGURE 3 Time course of the isoprenaline-induced inward current and cell volume changes. (a) Typical records of isoprenaline-induced inward currents in the absence (lower curve) and presence (upper curve) of 10 μM DIDS, with nystatin-perforated recording (KCl in the pipette). Isoprenaline (20 μM) was introduced into the bath at time zero. The holding potential was −70 mV. (b) The time-dependent profile of cell volume change in the absence (open circles) and presence (filled circles) of 10 μM DIDS or of 1 μM timolol (open triangles). Isoprenaline was applied at time zero. [From Chen *et al.* (1994) with permission.]

P_{Cl}/P_{Na} ratio ranging from 33:1 in B lymphocytes (McCann *et al.*, 1989) and 6:1 in rabbit colonic smooth muscle (Sun *et al.*, 1992) to 0.67 in T84 cells (Vaca and Kunze, 1992). This channel, if indeed it represents a single group and this is by no means certain, is variously referred to as the *high-conductance anion channel* or the *maxi-Cl⁻* channel. A large conductance anion channel has been found in the PE cells with a conductance of 293 pS in excised inside-out patches when bathed with symmetrical 130 m*M* NaCl and 209 pS when the cytoplasmic NaCl concentration was reduced to 33 m*M* (Fig. 4; Mitchell and Jacob, 1994,1996). The channel is highly selective for chloride with $P_{Cl}/P_{Na} = 24$. Both SITS and DIDS blocked the channel and it was activated by guanosine thiotriphosphate (GTPγS) applied to the cytoplasmic membrane surface. The physiological relevance of this channel is uncertain and it is rarely seen in the intact cell. Mitchell and Jacob (1996) observed the channel in only 2 of 90 cell-attached patches that subsequently were shown to contain the channel when excised. Recently, using cell-attached recording, Zhang and Jacob (1997) have found a channel with a conductance of 105 pS activated by cell swelling in PE cells. It was inhibited by NPPB identifying it with other maxi-Cl^- channels and, if this channel proves to be the high-conductance anion channel, it would suggest a role in volume regulation in PE cells. It was not observed in NPE cells under the same conditions.

V. NONSELECTIVE CHANNELS

One final class of channel has been described in the PE cells that discriminates poorly between anions and cations, but, because it would conduct chloride ions, it is included in this chapter. A unique feature of this channel is that it would conduct chloride *into* the cell. The mechanism for multi-ion permeation that would allow anions and cations to move as mixed complexes through the channel has been discussed by Franciolini and Nonner (1987,1994). Such a channel has been described by Mitchell and Jacob (1996). It has a conductance of 285 pS in excised patches bathed in symmetrical 130 m*M* NaCl solutions and a P_{Cl}/P_{Na} ratio of 1.3 : 1 (Fig. 5). This selectivity increased to 3.7 : 1 when the inner face of the patch was bathed with a concentrated NaCl solution. The channel was found in both excised and cell-attached patches and was activated by strong hyperpolarization and inhibited by gadolinium (Mitchell and Jacob, 1996). Since it was found in PE cells, it is believed that this channel may act as an uptake pathway for ions during the first stage of aqueous humor production. Stelling and Jacob (1992,1993) have reported a large nonselective current, activated by hyperpolarization, that reverses at 0 mV, in whole-cell recording. It was responsible for the repolarizing phase (repolarization from hyperpolarized potentials back to the resting potential) of potential oscillations found in PE, but not NPE, cells. This current may represent the activity of the large nonselective conductance in the PE cells.

VI. ROLE OF CHLORIDE CHANNELS

The role played by the variety of chloride channels described previously depends to a large extent on their localization in the ciliary body. Those

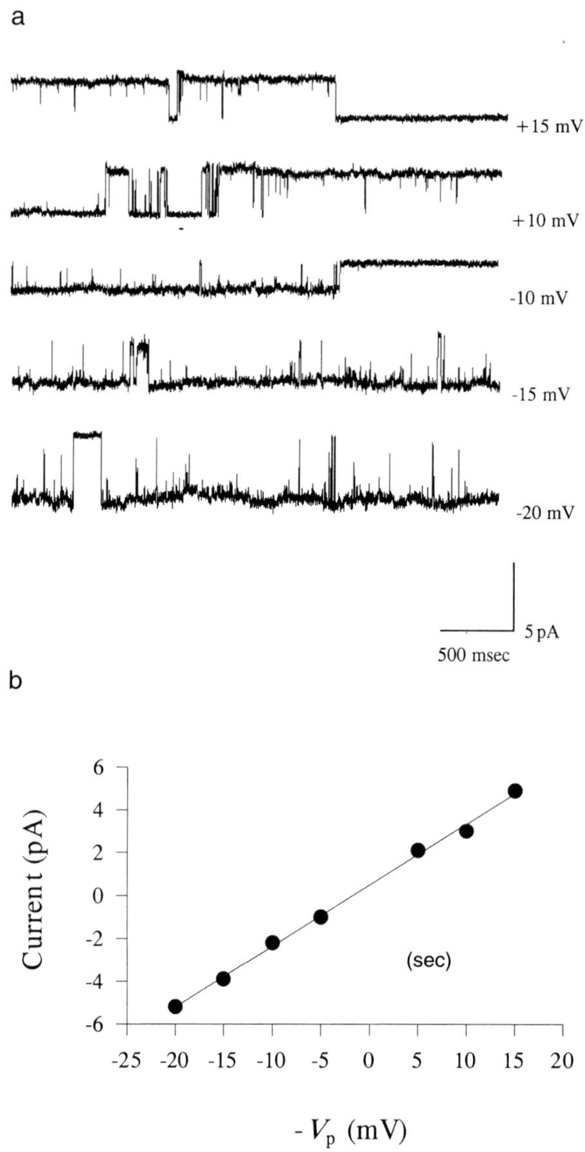

FIGURE 4 Maxi-chloride channel. (a) These current traces were obtained from an inside-out patch from a PE cell bathed in symmetrical 130 mM NaCl solutions. The patch was polarized to the potentials shown to the right of each trace. Upward deflections from this baseline indicate outward current through open channels. (b) Current–voltage plot of the record in part (a). The conductance was 295 pS and the channel reversed at −2 mV. [From Mitchell and Jacob (1996) with permission.]

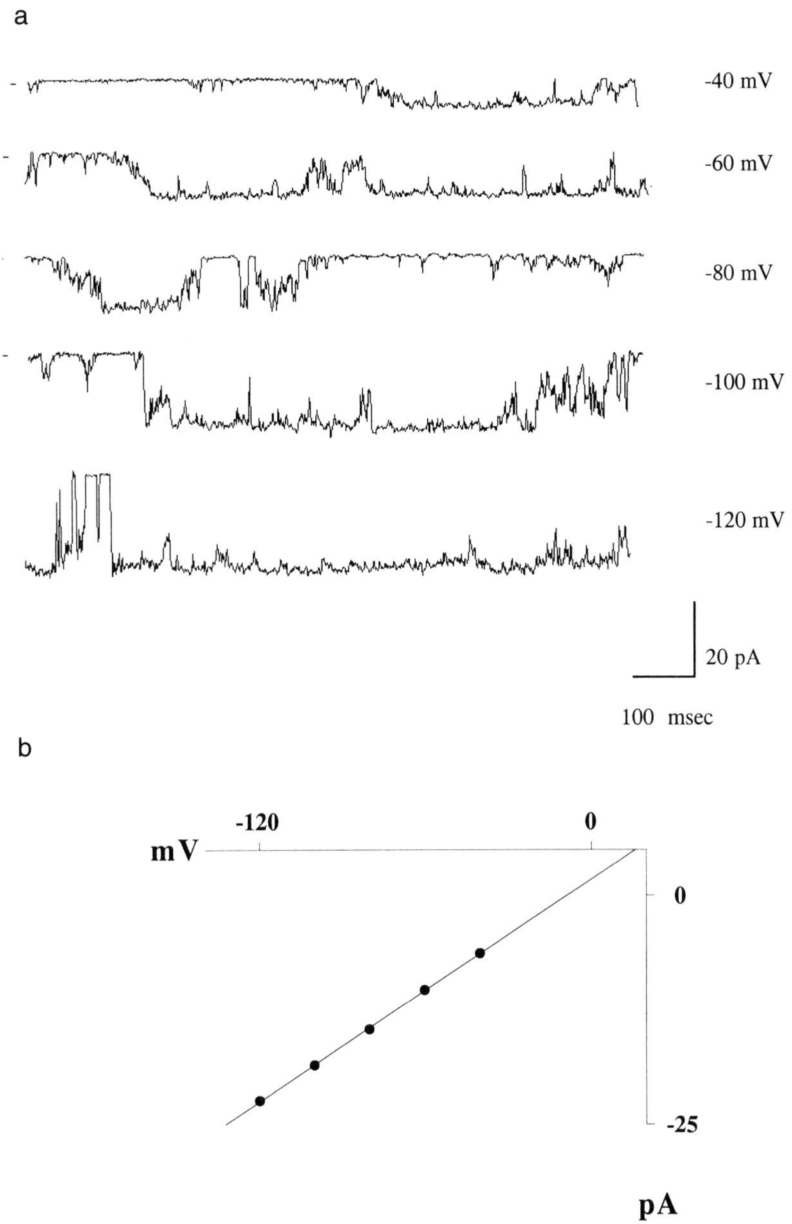

FIGURE 5 Single-channel current from an excised inside-out patch from a PE cell bathed in symmetrical sodium gluconate. (a) The negative of the pipette potential is shown to the right of each trace, and the closed channel level is indicated by a dash to the left of each trace. (b) Downward deflections indicate inward current through open channels. The conductance of this channel was 215 pS. [From Mitchell *et al.* (1997) with permission.]

channels situated on the basolateral membrane of the PE cells, that is, the membrane facing the blood, may well be involved in the first phase of aqueous humor secretion, the loading of ions into the ciliary epithelium. The large nonselective channel (Mitchell and Jacob, 1996) may be involved in this operation. Those channels situated on the basolateral membrane of the NPE cells, that is, the membrane facing the inside of the eye, may well be involved in the final phase of aqueous humor secretion, the efflux into the eye. For example, the volume-activated chloride channels of the NPE cells described by Yantorno *et al.* (1992), Wu *et al.* (1996), and Botchkin and Matthews (1995) may be involved in constitutive secretion, whereas the cAMP-activated chloride channel described by Edelman *et al.* (1995) may allow an efflux of chloride into the eye controllable by adrenergic agonists. In relation to the former, constitutive secretion of aqueous humor, Stelling and Jacob (1993) have proposed that this might be driven by potential oscillations observed in the pigmented cells, resulting in a pulsatile secretion.

Some channels will undoubtedly serve a housekeeping role as the primary defense against cell swelling but, as pointed out by Yantorno *et al.* (1992), these volume regulatory mechanisms may well be recruited for the process of aqueous humor secretion. There are a variety of channels in the two cell types, modulated by a number of channels regulators (e.g., P-gp, pI_{Cln}), which could contribute to the major function of the tissue—aqueous humor secretion.

In closing, it might be of interest to mention those chloride channels that have not been found in the ciliary epithelium. To date, the calcium-activated chloride channel has not been found in either cell type; in fact, Stelling and Jacob (1996) provided evidence against its presence in PE cells, and patients suffering from cystic fibrosis, and therefore with defective CFTR Cl^- channels, do not appear to have an altered rate of aqueous humor secretion (McCannel *et al.*, 1992), possibly ruling out a major role for CFTR in aqueous secretion.

Research into the nature and regulation of chloride channels in the ciliary body is providing insight into the secretion of aqueous humor and offering interesting new avenues for the development of antiglaucoma therapies.

References

Anguita, J., Chalfant, M. L., Civan, M. M., and Coca-Prados, M. (1995). Molecular cloning of the human volume-sensitive chloride conductance regulatory protein, pI_{Cln}, from ocular ciliary epithelium. *Biochem. Biophys. Res. Comm.* **208,** 89–95.

Botchkin, L. M., and Matthews, G. (1995). Swelling activates chloride current and increases internal calcium in nonpigmented epithelia cells from the rabbit ciliary body. *J. Cell. Physiol.* **164,** 286–294.

Burstein, N. L., Fischberg, J., Liebovitch, L., and Cole, D. F. (1984). Electrical potential, resistance, and fluid secretion across isolated ciliary body. *Exp. Eye Res.* **39,** 771–779.

Carre, D. A., Mitchell, C. H., Peterson-Yantorno, K., Coca-Prados, M., and Civan, M. M. (1996a). Adenosine stimulates Cl^- channels of nonpigmented ciliary epithelial cells. *Am. J. Physiol.* in press.

Carre, D. A., Mitchell, C. H., Peterson-Yantorno, K., Coca-Prados, M., and Civan, M. M. (1996b). Purinergic mechanisms in the ciliary epithelium. *Invest. Ophthalmol. Vis. Sci.* **37,** S438.

Chen, S., Inoue, R., Inomata, H., and Ito, Y. (1994). Role of cyclic AMP-induced Cl conductance in aqueous humour formation by the dog ciliary epithelium. *Br. J. Pharmacol.* **112,** 1137–1145.

Chu, T. C., Candia, O. A., and Podos, S. M. (1987). Electrical parameters of the isolated monkey ciliary epithelium and effects of pharmacological agents. *Invest. Ophthalmol. Vis. Sci.* **28,** 1644–1648.

Civan, M. M., Peterson-Yantorno, K., Coca-Prados, M., and Yantorno, R. E. (1992). Regulatory volume decrease by cultured non-pigmented ciliary epithelial cells. *Exp. Eye Res.* **54,** 181–191.

Civan, M. M., Coca-Prados, M., and Peterson-Yantorno, K. (1994). Pathways signaling the regulatory volume decrease of cultured nonpigmented ciliary epithelial cells. *Invest. Ophthalmol. Vis. Sci.* **35,** 2876–2886.

Coca-Prados, M., Anguita, J., Chalfant, M. L., and Civan, M. M. (1995a). PKC-sensitive Cl^- channels associated with ciliary epithelial homologue of pI_{Cln}. *Am. J. Physiol.* **268,** C572–C579.

Coca-Prados, M., Ghosh, S., Gilula, N. B., and Kumar, N. M. (1995b). Expression and cellular distribution of the $\alpha 1$ gap junction gene product in the ocular pigmented ciliary epithelium. *Curr. Eye Res.* **11,** 113–122.

Coca-Prados, M., Sanchez-Torres, J., Peterson-Yantorno, K., and Civan, M. M. (1996). *J. Membr. Biol.* **150,** 197–208.

Edelman, J. L., Sachs, G., and Adorante, J. S. (1994). Ion transport asymmetry and functional coupling in bovine pigmented and nonpigmented ciliary epithelial cells. *Am. J. Physiol.* **266,** C1210–C1221.

Edelman, J. L., Loo, D. D. F., and Sachs, G. (1995). Characterization of potassium and chloride channels in the basolateral membrane of bovine nonpigmented ciliary epithelial cells. *Invest. Ophthalmol. Vis. Sci.* **36,** 2706–2716.

Farahbakhsh, N. A., and Fain, G. L. (1987). Volume regulation of non-pigmented cells from ciliary epithelium. *Invest. Ophthalmol. Vis. Sci.* **28,** 934–944.

Franciolini, F., and Nonner, W. (1987). Anion and cation permeability of a chloride channel in rat hippocampal neurones. *J. Gen. Physiol.* **90,** 453–478.

Franciolini, F., and Nonner, W. (1994). Anion–cation interactions in the pore of neuronal background chloride channels. *J. Gen. Physiol.* **104,** 711–723.

Gooch, A. J., Morgan, J., and Jacob, T. J. C. (1992). Adrenergic stimulation of bovine nonpigmented ciliary epithelial cells raises cAMP but has no effect on K^+ or Cl^- currents. *Curr. Eye Res.* **11,** 1019–1029.

Jackson, P. S., Morrison, R., and Strange, K. (1995). The volume-sensitive organic osmolyte-anion channel VSOAC is regulated by nonhydrolytic ATP binding. *Am. J. Physiol.* **267,** C1203–C1209.

Jacob, T. J. C., and Civan, M. M. (1996). Role of ion channels in aqueous humor formation. *Am. J. Physiol.* **271,** C703–C720.

Jacob, T. J. C., and Mitchell, C. H. (1995). Chloride current activated by hypotonic shock in pigmented ciliary epithelial cells. *Invest. Ophthalmol. Vis. Sci.* **36,** S586.

McCann, F. V., McCarthy, D. C., Keller, T. M., and Noelle, R. J. (1989). Characterization of a large conductance non-selective channel in B lymphocytes. *Cell. Signal.* **1,** 31–44.

McCannel, C. A., Scanlon, P. D., Thibodeau, S., and Brubaker, R. F. (1992). A study of aqueous humor formation in patients with cystic fibrosis. *Invest. Ophthalmol. Vis. Sci.* **33,** 160–164.

Miley, H. E., Alexander, P. D., Li, S. W., Jacob, T. J. C., and Pollard, C. E. (1995a). Chloride conductance in freshly dissociated bovine pigmented ciliary epithelial cells. *J. Physiol. (London)* **489,** 97P.

Miley, H. E., Walker, V. E., Pollard, C. E., and Jacob, T. J. C. (1995b). Regulatory volume increase in ciliary epithelial cells. *Invest. Ophthalmol. Vis. Sci.* **36,** Suppl., 586.

Mitchell, C. H., and Jacob, T. J. C. (1994). Two maxi-conductance channel types in bovine pigmented ciliary epithelial cells. *Invest. Ophthalmol. Vis. Sci.* **35,** Suppl., 1455.

Mitchell, C. H., and Jacob, T. J. C. (1996). Properties of a non-selective high conductance channel in bovine pigmented ciliary epithelial cells. *J. Membr. Biol.* **150,** 105–111.

Mitchell, C. H., Zhang, J. J., Wang, L., and Jacob, T. J. C. (1996). Volume-sensitive chloride current in pigmented ciliary epithelial cells: The role of phospholipases. *Am. J. Physiol.* **272,** C212–C222.

Mitchell, C. H., Wang, L., and Jacob, T. J. C. (1997). A large-conductance chloride channel in pigmented ciliary epithelial cells activated by GTPγS. *J. Membr. Biol.* **158.**

Mok, K. H., and To, C. H. (1996). Chloride and bicarbonate transport across isolated bovine iris-ciliary body (ICB). *Invest. Ophthalmol. Vis. Sci.* **37,** Suppl., S1108.

Okisaka, S., Kuwabara, T., and Rapoport, S. I. (1974). Selective destruction of the pigmented epithelium in the ciliary body of the eye. *Science* **184,** 1298–1299.

Schultz, S. G. (1981). Homocellular regulatory mechanisms in sodium-transporting epithelia: Avoidance of extinction by "flush-through." *Am. J. Physiol.* **241,** F579–F590.

Stelling, J. W., and Jacob, T. J. C. (1992). The inward rectifier K^+ current underlies oscillatory membrane potential behaviour in bovine pigmented ciliary epithelial cells. *J. Physiol. (London)* **458,** 439–456.

Stelling, J. W., and Jacob, T. J. C. (1993). Membrane potential oscillations from a novel combination of ion channels. *Am. J. Physiol.* **265,** C720–C727.

Stelling, J. W., and Jacob, T. J. C. (1996). Transient activation of K^+ channels by carbachol in bovine pigmented ciliary body epithelial cells. *Am. J. Physiol.* **271,** C203–C209.

Strange, K., Emma, F., and Jackson, P. S. (1996). Cellular and molecular physiology of volume-activated anion channels. *Am. J. Physiol.* **270,** C711–C730.

Sun, X. P., Supplisson, S., Torres, R., Sachs, G., and Mayer, E. (1992). Characterization of large-conductance chloride channels in rabbit colonic smooth muscle. *J. Physiol. (London)* **448,** 355–382.

Vaca, L., and Kunze, D. L. (1992). Anion and cation permeability of a large-conductance anion channel in T84 human colonic cell line. *J. Membr. Biol.* **130,** 241–249.

Watanabe, T., and Saito, Y. (1978). Characteristics of ion transport across the isolated epithelium of the toad as studied by electrical measurements. *Exp. Eye Res.* **27,** 215–226.

Wu, J. Q., and Jacob, T. J. C. (1994). Volume-activated chloride currents in isolated, nonpigmented ciliary epithelial cells of the bovine eye. *J. Physiol. (London)* **475,** 108P.

Wu, J., Zhang, J. J., Koppel, H., and Jacob, T. J. C. (1996). P-glycoprotein regulates a volume-activated chloride current in bovine non-pigmented ciliary epithelial cells. *J. Physiol. (London)* **491.3,** 743–755.

Yantorno, R. E., Carre, D. A., Coca-Prados, M., Krupin, T., and Civan, M. M. (1992). Whole cell patch clamping of ciliary epithelial cells during anisosmotic swelling. *Am. J. Physiol.* **262,** C501–C509.

Zhang, J. J., and Jacob, T. J. C. (1996). Swelling-induced channels in on-cell patches of bovine pigmented and non-pigmented ciliary epithelial cells. *J. Physiol. (London)* **493.P,** 79P.

Zhang, J. J., and Jacob, T. J. C. (1997). Three different Cl^- channels in the ciliary epithelium activated by hypotonic stress. *J. Physiol. (London)* **499.2,** 379–389.

CHAPTER 4

Identification of Potassium Channels in Human Lens Epithelium

James L. Rae and Allan R. Shepard
Departments of Physiology and Biophysics and Ophthalmology, Mayo Foundation, Rochester, Minnesota 55905

I. INTRODUCTION

Potassium channels are vital to the functioning of almost every cell type. Their general role is to set the resting membrane potential of the cell although they are never the only transporters involved in this important function. If they were to dominate in resting membrane potential control, the transmembrane voltage would sit very near the potassium equilibrium potential, some −90 mV in most cells. Voltages this negative hardly ever occur in cells under normal conditions, so it is clear that other processes

1063-5823/98 $25.00

and transporters are involved in regulation of the cell's voltage. Potassium channels sequenced to date usually show several phosphorylation consensus sequences and so have substantial potential to be regulated by cellular second messenger cascades following receptor binding of agonists or via other cell signaling cascades. In addition, potassium channels have substantial differences in the voltage dependence of their gating, making their role and time course of action very complicated during times when the transmembrane voltage is changing. This voltage-regulating function is important in offsetting the natural swelling tendency of cells due to the Donnan equilibrium associated with the large impermeant anion concentration present in all cells. By causing the cell to have large, negative resting potentials, potassium channels inhibit this potential cell swelling although they are by no means the only important players in this scheme. Because most electrogenic transporters are voltage dependent in their transport rates, potassium channels can affect indirectly the function of these transporters through the channels' role in transmembrane voltage regulation. In some instances, unlike transmembrane voltage regulation, the net flux through potassium channels can be large (Peterson and Maruyama, 1984) and can play a vital role in the transcellular movement of potassium and even in whole-body potassium regulation.

It is important to identify in molecular detail the particular kinds of potassium channels that function in the particular cells under investigation. This is because the particular channels involved may have already been well studied in the literature and thus the investigator of a particular cell type may be able to start with a wealth of knowledge about the channels without having to do all of the experimental work required to gain that knowledge. If the sequence is known, the existence of regulatory consensus sequences can be found and thus give the investigator some notion about which regulatory pathways to look for in the particular cell type. In the event that the channel is unique and unreported, that also directs the investigator to required studies for functional characterization of the channel.

Identification of potassium channels can be done both by electrophysiological approaches and molecular biology techniques. Channels can usually be identified by the voltage and time dependence of their gating, their single-channel conductance, and their sensitivity to blocking and regulatory substances. They can also be identifed by the primary amino acid sequences and number of their subunits. Eventually, their static and dynamic crystalline structures will be important to their classification and properties.

In this chapter, a brief overview of both electrophysiological characterization and some molecular biology characterizations of several potassium channels commonly found in human lens epithelium is presented. The channels identified are unlikely to be the only channels involved in the function of these cells. In fact, they are probably not even the only potassium channels involved. Not only are the channels identified but the kinds of approaches that can lead to their identification and characterization are also described. Because of the substantial sequence homology of human lens potassium channels with other published potassium channel sequences, much more progress with the structure–function relationships of human lens channels has been made than with those from other ocular epithelia.

II. ELECTROPHYSIOLOGICAL CHARACTERIZATION

A. Introduction

Unambiguous electrophysiological identification of the potassium channels contained in a single whole-cell type can be extremely difficult. This is largely because of the large number of different potassium channels that have been found to date and because many different potassium channels have very similar electrophysiological and pharmacological properties. In addition, most cell types have currents other than potassium currents and so whole-cell currents are almost always a complex mixture of channel types with different selectivities, different voltage dependences, and different time dependences to their gating kinetics. Often, channel types end up with operational definitions such as those passing current in the presence of Ba or those passing current in the presence of tetraethylammonium, or those remaining at a holding voltage of zero millivolts, etc. These operational definitions can often give rise to ambiguities if, for example, the effects of Ba or Cs are very voltage dependent. At some voltages, the blockade by these substances might be complete but at others incomplete, so difference currents might well be due to variable voltage contributions from more than one channel type. Some of these kinds of problems often can be clarified if currents are recorded from a large enough number of different cells from the same preparation. If, as is often the case, not every cell contains every conductance characteristic of the whole preparation, then some cells will lack some of the confounding conductances and in ideal cases may contain only one or two channel types. In those cells, one may be able to record the currents with little or no ambiguity but even there it is hard to rule out the existence of two or more channel types with similar properties. In some cases, help may come from cloning the cDNA from one channel type into an expression vector, expressing the channels, and then characterizing the currents in the "idealized" expression system. One then hopes that the expressed current has identical properties to that found in the native preparation. Even that approach is fraught with difficulty since one cannot *a priori* be certain that the expressed channel does not require another homologous or heterologous subunit for full function or that the channels are, for example, phosphorylated differently in the expression system than in the native preparation. One might believe that if single-channel currents could be recorded, then the ambiguity about properties might be resolved. However, even then, one cannot be assured that the gating kinetics seen in a patch are identical to those from channel in "unperturbed" membrane. So at present, ambiguities, either real or potential, remain.

B. Techniques

Whole-cell currents in small epithelial cells are best studied by perforated patch recordings (Horn and Marty, 1988; Rae *et al.,* 1990a; Rae and Fernandez, 1991; Tajima *et al.,* 1996). With this procedure (Fig. 1), the patch pipette is filled with an intracellular-like salt solution to which has been added a perforating agent such as amphotericin, nystatin, or gramicidin. Following gigohm seal formation, the perforating agent diffuses to the membrane patch, partitions into the membrane lipid, and forms myriad tiny channels between the patch pipette lumen and the interior of the cell. These channels allow the movement of small molecules such as monovalent cations and in some cases monovalent anions, while disallowing the movement of larger substances such as Ca ions, glucose, second messengers, and cytoplasmic proteins. This approach allows electrical access to the cell's interior without allowing loss of cellular constituents as is the case with standard whole-cell patch recording.

A similar approach allows the recording of single-channel currents in perforated outside-out vesicles (Fig. 2). Here, either following whole-cell perforated patch recording or essentially immediately after gigohm sealing, the patch pipette is slowly withdrawn from the cell. What usually happens is that a tube of cell membrane still attached to the cell follows the tip until the tube becomes so stretched that it ruptures and forms a tiny enclosed vesicle (Rae *et al.,* 1990a; Levitan and Kramer, 1990). Often, cytoplasmic elements, even mitochondria, are trapped in the vesicle making the vesicle a tiny cell attached to the electrode tip with the outside membrane being that which was outside facing in the original cell. The vesicle is so small that it has very little capacitance and so exhibits only a small amount of electrical noise from the series combination of the access resistance into the vesicle and the vesicle capacitance. The overall noise levels are such that picoamp-level single-channel currents are easily recorded under somewhat

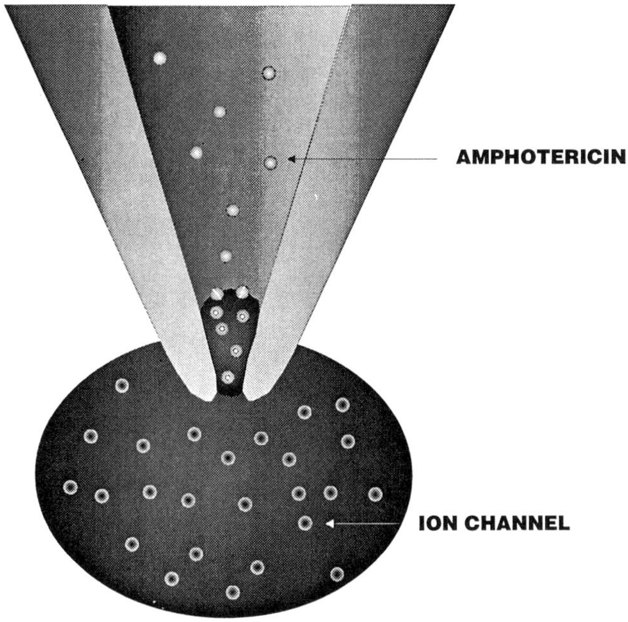

FIGURE 1 Schematic representation of a perforated patch whole-cell recording configuration. Amphotericin in the pipette filling solution partitions into the membrane patch in the pipette tip and makes channels permeable to monovalent cations and anions. These channels provide electrical access to the cell interior so that the transmembrane voltage can be voltage clamped but do not allow proteins, Ca^{2+}, or other small molecules to exit the cell.

physiological conditions. At least the conditions are more physiological than those associated with standard inside-out or outside-out patches.

These electrical measurement techniques can be used either with freshly dissociated or cultured cells from human lens epithelium or from mammalian cell culture expression systems such as HEK 293 cells. Using proper instrumentation, it is possible to apply computer-generated complex voltage protocols to measure both conductance and gating parameters from the channels of interest.

C. Currents Identified

1. Inward Rectifiers

Inward rectifier potassium currents are easily identified from either cultured or freshly dissociated human lens epithelial cells by measuring whole-cell currents. When the cells are bathed in a 3–4 m*M* K^+ containing normal

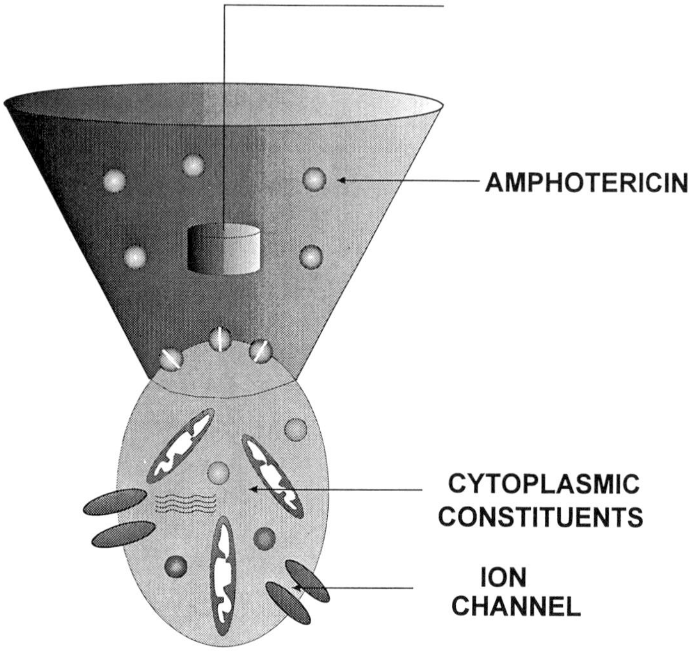

FIGURE 2 Schematic representation of a perforated outside-out vesicle recording configuration. Following gigohm sealing and partitioning of amphotericin from the pipette filling solution into the plasma membrane patch, the electrode is withdrawn slowly from the cell to produce an enclosed vesicle whose interior communicates with the pipette filling solution through the open pores created by the amphotericin molecules. Useful for single-channel recording.

Ringer's solution, inward currents are barely perceptible (Rae *et al.*, 1996). This is because the conductance of the channel is small when small concentrations of potassium are carrying the currents. Inward currents can be resolved under similar conditions in, for example, guinea-pig cardiac muscle cells (Saigusa and Matsuda, 1988). In this preparation, the channel density is enormous so that even though the current through any one channel is as small as it is in lens, the macroscopic current is large because of the large total number of channels. Enriching the bath's potassium concentration increases the conductance of the channel and so inward currents can then be seen even in lens epithelial cells (Fig. 3). With 150 mM K^+ in the patch pipette, outward currents cannot be seen in lens epithelial cells since the particular inward rectifiers found there (designated as IRK1) are blocked by internal Mg^+ and spermidine (VandenBerg, 1987; Stanfield *et al.*, 1994; Matsuda, 1991; Lu and MacKinnon, 1994; Loptain *et al.*, 1994; Ficker *et al.*, 1994; Fakler *et al.*, 1994, 1995; Tagliatatela *et al.*, 1994; Van Dongen *et al.*, 1990; Wible *et al.*, 1994; Yang

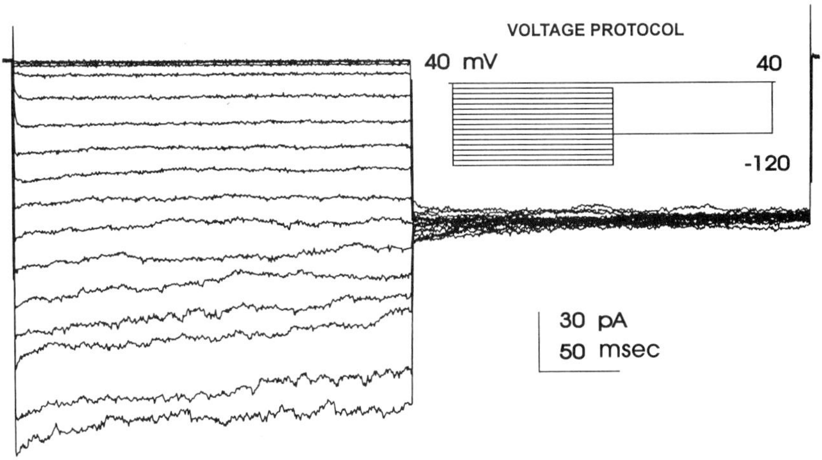

FIGURE 3 Whole-cell currents measured from a cultured human lens epithelial cell bathed in a Ringer's solution containing 2.5 m*M* $CaCl_2$, 5 m*M* HEPES, and 150 m*M* KCl (Rae and Shepard, unpublished). Measurements come from a perforated patch whole-cell recording configuration where the pipette contained 120 μg/ml amphotericin B, 15 m*M* NaCl, 5 m*M* KCl, 130 m*M* $KMeSO_4$, 5 m*M* HEPES, and 2 m*M* EGTA. The voltage protocol used is that shown in the inset. The current is typical of an inward rectifier.

et al., 1995). Therefore, currents that would normally be outward are rapidly blocked. In addition, even in the absence of blockers, the channels close quickly at voltages where the current would be outward. The net result is that in lens epithelium, outward currents cannot normally be seen (Cooper *et al.,* 1991). Again, transient outward currents can be seen in heart cells where the channel density is very high. A channel that carriers no inward or outward current in normal Ringer's but has large inward currents with no outward currents in an elevated potassium bathing solution is probably an inward rectifier. In addition, inward currents through IRK1 are blocked by submillimolar concentrations of external Cs^+ or Ba^{2+} (Cooper *et al.,* 1991). When all of these properties are found electrophysiologically, the currents are IRK1-like.

2. Delayed Rectifiers

The most common current seen in human lens epithelial cells comes from channels that are largely closed at voltages more negative than −40 mV (Cooper *et al.,* 1990; Rae and Rae, 1992). When recording currents in a normal Ringer's solution with a patch pipette containing 130–150 m*M* K^+, outward currents begin to appear when the voltage is depolarized to −40 mV (Fig. 4a). At voltages between −40 and +30 mV, the outward currents continue to get larger both because the driving force ($E_m - E_k$) gets

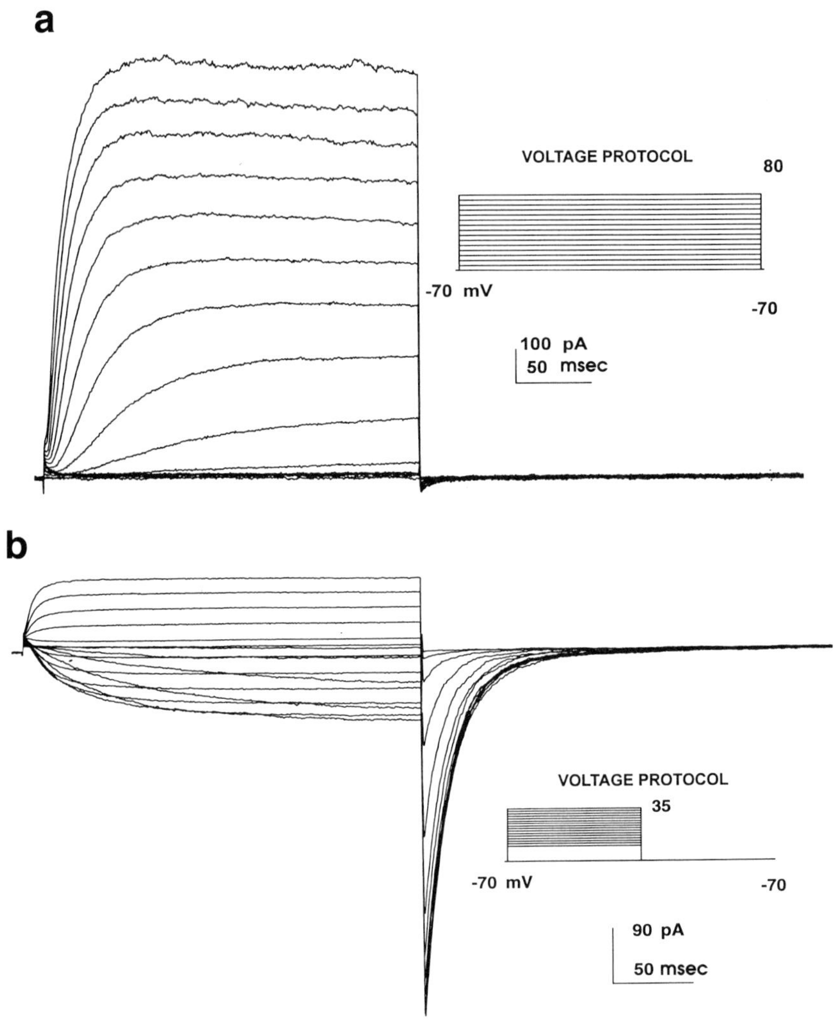

FIGURE 4 Whole-cell currents measured from a cultured human lens epithelial cell bathed in a Ringer's solution containing 2.5 m*M* $CaCl_2$, 5 m*M* HEPES (Rae and Shepard, unpublished), and (a) 4 m*M* KCl and 150 m*M* NaCl or (b) 150 m*M* KCl. Measurements come from a perforated patch whole-cell recording configuration where the pipette contained amphotericin B, 15 m*M* NaCl, 5 m*M* KCl, 130 m*M* $KMeSO_4$, 5 m*M* HEPES, and 2 m*M* EGTA. The voltage protocol used is that shown in the inset. The currents come from a delayed rectifier.

larger and because the open probability increases. At voltages depolarized to +30 mV, the currents get larger because of a larger driving force but not because of an open probability (P_o) increase since P_o is maximum around +30 mV. These outward currents are blocked by tetraethylammonium (TEA) used in the bathing medium at concentrations greater than 10 m*M* (Kirsch *et al.*, 1991; Heginbotham and MacKinnon, 1992; Tagliatatela *et al.*, 1991; Yellen *et al.*, 1991; MacKinnon and Yellen, 1990). When the bath is enriched for potassium, the outward currents look similar to those in Ringer's although the reversal potential shifts in accordance with E_K. Now however, clear inward currents can be seen starting at a voltage of about −40 mV where the channels begin to open, and when the voltage is jumped back from the test voltage to a holding voltage (usually −70 mV), large tail currents can be seen that deactivate sufficiently slowly that their time course can be easily resolved (Fig. 4b). If the cell is held at depolarized voltages (e.g., 0 mV), the channels almost totally inactivate so that brief voltage steps of 100 msec or so produce neither inward nor outward currents through the channel. A final signature of this channel is that when the voltage is positive to −40 mV, the channels open with a delay. When all of these properties are found, one is dealing with a delayed rectifier-like channel. Unfortunately, many studies have shown that there are many outwardly rectifying potassium channels that have properties similar to these and so it might not be possible from whole-cell electrophysiological measurements alone to determine exactly which potassium channels the cell might contain (Rae, 1994; Pongs, 1992; Stühmer *et al.*, 1989). In human lens epithelium, we hypothesized that the channel involved was h-DRK1 because of the similarity of its electrophysiological properties with those of previously recorded h-DRK1 channel currents. Molecular biological results have supported this initial guess.

3. Calcium-Activated Potassium Channels

Both cultured and freshly dissociated human lens epithelia produce currents from large single-channel conductance calcium-activated potassium channels (Cooper *et al.*, 1990). As reported earlier, these channels have a conductance of some 270 pS in 150 m*M* symmetrical potassium (Latorre *et al.*, 1989). The whole-cell currents associated with these channels are quite distinctive and easily recognized. The open probability of these channels is dependent on both voltage and the internal calcium concentration. At the 200 n*M* or less amount of internal calcium found in most cells, rather extreme depolarizations are required to activate the channels. In human lens epithelium, opening of the channels begins at about +50 mV and the channels are not yet fully open at +200 mV (Fig. 5a). These transmembrane voltages are largely impossible in lens epithelium and so it is imperative that, if these channels are to be important physiologically, their opening must be associated with a transient rise in intracellular calcium. Given such

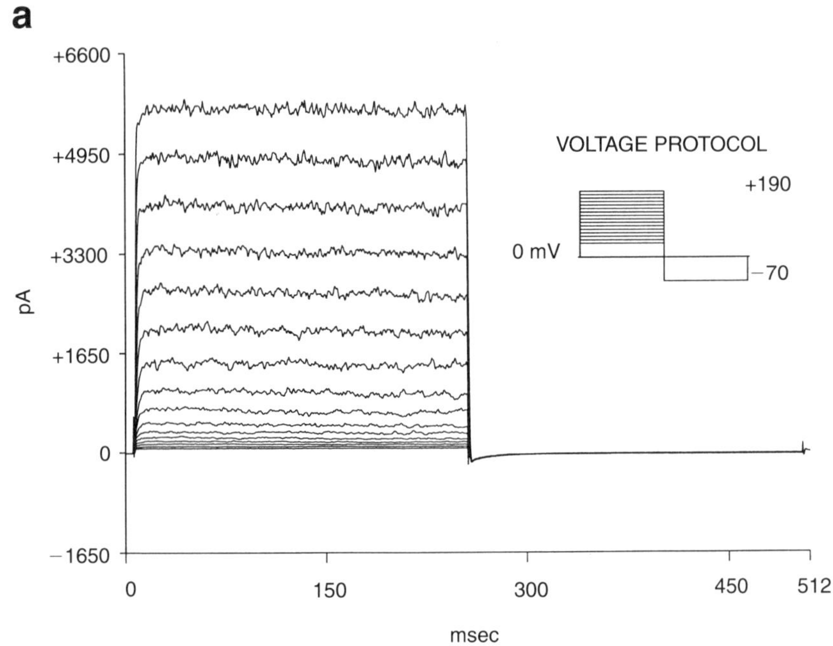

FIGURE 5 Whole-cell currents measured from a cultured human lens epithelial cell bathed in a Ringer's solution containing 2.5 m*M* $CaCl_2$, 5 m*M* HEPES, and 150 m*M* KCl (Rae and Shepard, unpublished). Measurements come from a perforated patch whole-cell recording configuration where the pipette contained amphotericin B, 15 m*M* NaCl, 5 m*M* KCl, 130 m*M* $KMeSO_4$, 5 m*M* HEPES, and 2 m*M* EGTA. The voltage protocols used are shown in the inset. The currents come from calcium-activated potassium channels. Note the absence in both (a) and (b) of an inward tail current following a voltage step applied at the end of each activating voltage step. In (b) we can see that the reason for the lack of tail current is simply that the channels deactivate rapidly following the step to generate tail currents. The deactivation is increasingly rapid as the voltage is made more negative. In fact, by the time the tail current generating voltage is near 0 mV where the currents should become inward, the deactivation is so fast that no inward channel currents can be seen.

a calcium rise, the voltage dependence of gating shifts substantially to more negative potentials. If, for example, intracellular calcium in the region of the channel were to rise to 1–2 μM, these channels would begin gating near the resting voltage of the cell. Therefore, in cells with normal intracellular calcium concentrations, a potassium current that first becomes evident at voltages substantially positive to 0 mV is likely a calcium-activated potassium channel. Because the single-channel conductance is so large, the currents are likely to appear noisy although this property will not exist if the total current comes from a large number of these channels. When the bathing solution around the cell is enriched in potassium, the anticipated inward steady-state currents and tail currents are rarely seen. This is because

b

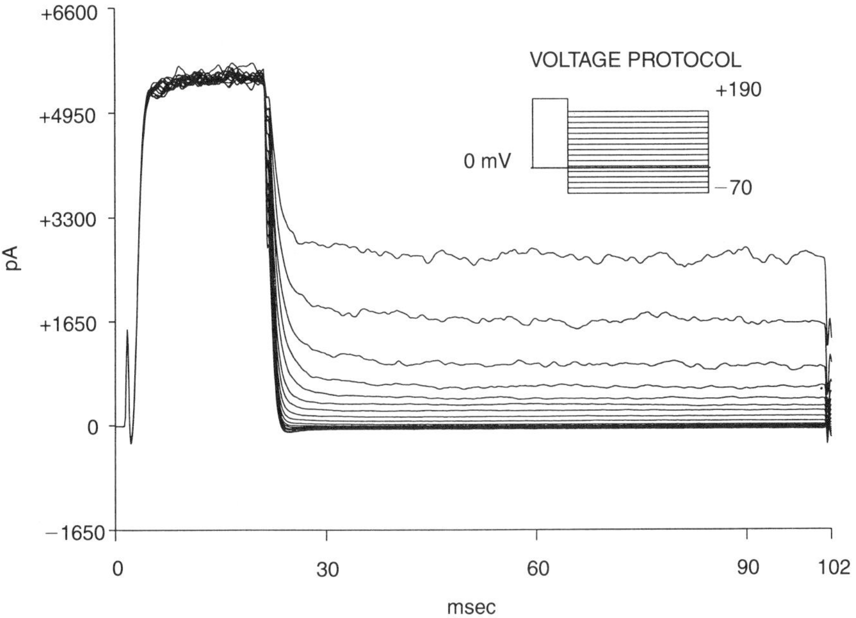

FIGURE 5 Continued

(Fig. 5b) the deactivation time constant is very fast in this channel and so the channels close at voltages negative to 0 mV so fast that inward currents are either not generated or result only in a spikelike tail current. These characteristics alone will usually justify classifying the current as BK, the large conductance calcium-activated K^+ channel. For completeness, one can verify that voltage-dependent blockade occurs at 1 m*M* external TEA (Villaroel *et al.*, 1988; Latorre *et al.*, 1989; Rae *et al.*, 1990b), a concentration too low to have much effect on delayed rectifiers. Charybdotoxin is usually an effective blocker as well. Finally, one can verify that a rise in intracellular calcium shifts the voltage dependence of gating to more negative voltages.

4. A-Type Currents

Currents that might be referred to as *A-currents* also occur occasionally in human lens epithelium (Fig. 6) (Rae, 1994). In general, the currents observed show rapid inactivation at voltages positive to 0 mV. In general, the inactivation time course is relatively slower in human lens epithelium when compared to classical A-currents from neurons. The inactivation time is quite variable from cell to cell (although always slow). Due to their rareness, these currents have not been well characterized electrophysiologically and have not yet been identified unequivocally by molecular biological techniques.

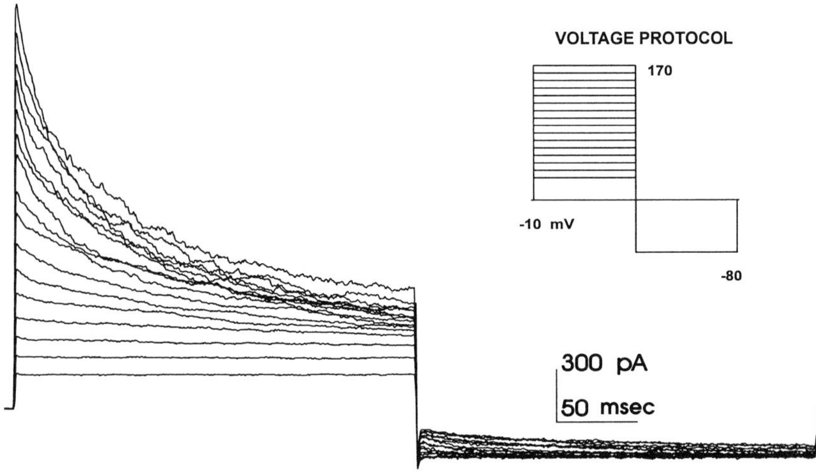

FIGURE 6 Whole-cell currents measured from a cultured human lens epithelial cell bathed in a Ringer's solution containing 2.5 m*M* $CaCl_2$, 5 m*M* HEPES, 4 m*M* KCl, and 150 m*M* NaCl (Rae and Shepard, unpublished). Measurements come from a perforated patch whole-cell recording configuration where the pipette contained amphotericin B, 15 m*M* NaCl, 5 m*M* KCl, 130 m*M* $KMeSO_4$, 5 m*M* HEPES, and 2 m*M* EGTA. The voltage protocol used is shown in the inset. The currents come from an A-type potassium channel.

III. MOLECULAR BIOLOGICAL CHARACTERIZATION

A. Introduction

Having identified channel currents by patch-clamp recording, how might one determine the protein sequence of the contributing channel? It is much easier to sequence the cDNA coding for the protein than to sequence the protein itself. Given the cDNA sequence, the amino acid sequence can be determined unambiguously. The crystallography of channel proteins to determine folding, etc., has proven extremely difficult or impossible as it has for most membrane proteins attempted to date.

The cDNA sequence can be learned by two different approaches: expression cloning or homology cloning. *Expression cloning* involves injecting messenger RNA into an expression vector (often *Xenopus* oocytes) and determining with an appropriate functional assay if the proper protein is expressed. This usually involves splitting the mRNA into pools by some appropriate protocol and then resplitting the pools, which produce functional products until a final pool is left containing mRNA coding for the protein of interest (Soreq and Seidman, 1992). This involves countless injections and functional tests but, more important, consumes large amounts of mRNA. Given the small tissue mass of ocular epithelia and the low copy number anticipated for the mRNA for ion channels, this is not a desirable approach.

The second approach, *homology cloning,* requires that one know *a priori* at least a small portion of the nucleotide or amino acid sequence. Because K channels fall into functionally distinctive subfamilies and these subfamilies have similar pore region sequences, homology screening is a feasible approach. Given the tiny amount of channel protein expected in these epithelia, it is unlikely that sequencing protein fragments of the purified channel protein would be possible. Therefore, homology cloning using nucleotide sequence is the obvious way to start.

B. Constructing cDNA Libraries

Much of the work of isolating potassium channel cDNA sequences can be done using RNA (total or message) and reverse transcriptase–polymerase chain reaction (RT–PCR). However, some of the potentially more powerful techniques require the construction of cDNA libraries. Here, each of the cDNA sequences of interest is cloned into a plasmid that can be used to transform bacteria (Fig. 7). The plasmid gets replicated in the bacterium and so by growing bacteria, one can obtain a substantial stock of all of the sequences in the cell. Each DNA resides in a circularized form so that techniques such as inverse PCR or GSP-vector PCR can be used (see later discussions). Making libraries from ocular tissue is a considerable problem. Most companies that make commercial libraries require the mRNA from roughly 10^8 cells to construct a representative library, whereas lens epithelium or corneal endothelium yield only 10^5–10^6 cells per preparation. To make quality libraries from our preparations, we had to customize highly efficient procedures that would allow representative library construction from the small amount of mRNA we could get from our cells. We were able to do that (Shepard and Rae, 1995) but those procedures are not covered here.

C. Multiple Primer Pair Reverse Transcriptase–Polymerase Chain Reaction

Even if one knows a nucleotide sequence from the Genbank, it may still be difficult to find the protein or message in a specific cell if it is from a different species than that from which the original sequence was obtained. This is, of course, because even though two cells may contain the same

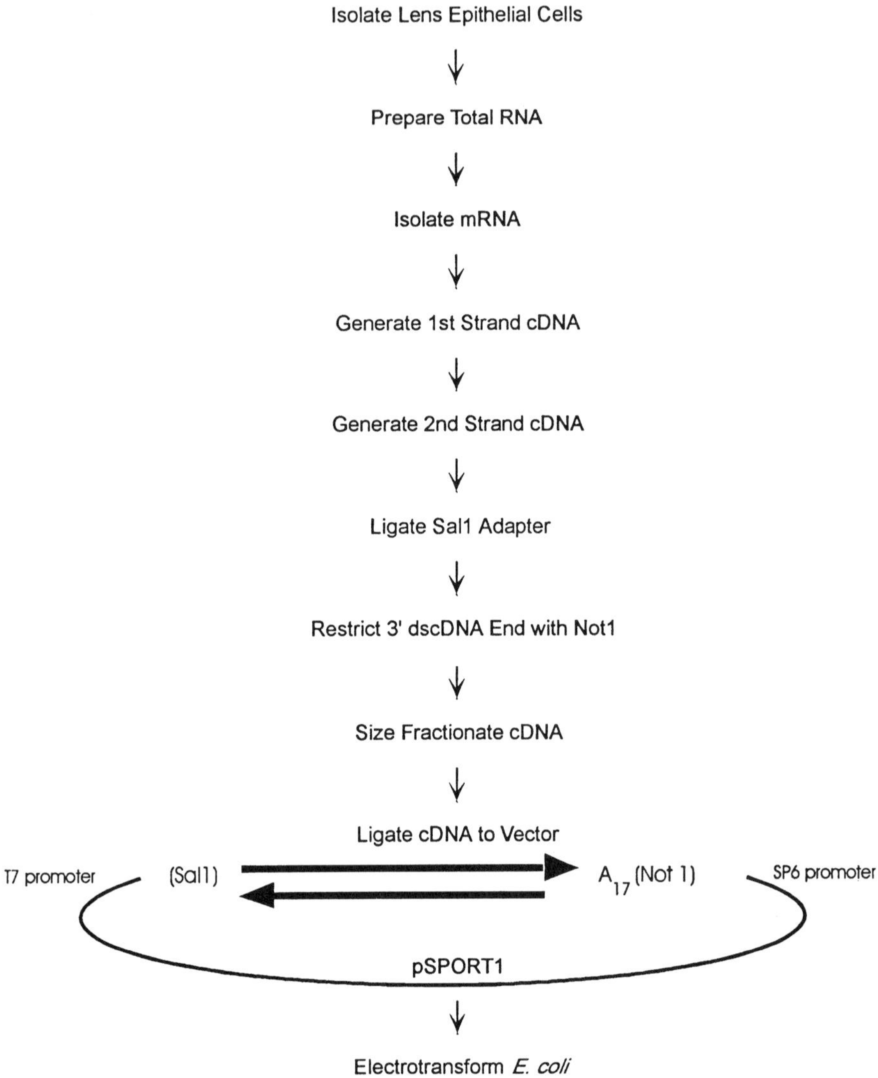

FIGURE 7 A simple flowchart to delineate the sequential steps in the generation of cDNA libraries from human lens epithelium.

protein with the same amino acid sequence, the nucleotide sequence may be quite different due to the redundancy of the genetic code and codon usage differences from one species to another. Therefore, PCR primers made to a published nucleotide sequence may or may not work in RT–PCR, cDNA library PCR, or genomic DNA PCR from another species. However, it is well known that PCR primers do not have to be a perfect match to the sequence one is trying to amplify to work. In general, the first six 3′ bases are the most important and in fact become less important as one moves 5′ to the very first 3′ base. After several hundred RT–PCR reactions in which we used species mismatched primers, it became obvious that an 18 mer primer could have six or seven mismatches with the actual sequence and still work with reasonable efficiency in PCR reactions as long as most of the mismatches were 5′ to the first six 3′ bases. This is not a perfect rule, however, because many times we found that PCR primers did not work even if they were perfect matches to the sequence and were designed using computer software that presumably selected chemically optimized primers. So, how well primers work depends a great deal on the particular template being studied. Still, this understanding that perfectly matched primer sequences were not required allowed us to use an approach that we refer to as *multiple primer pair PCR* (Fig. 8). In this approach, we use an oligo design program to make 6–10 sense strand primers and a like number of antisense strand primers, all of which are reasonably temperature

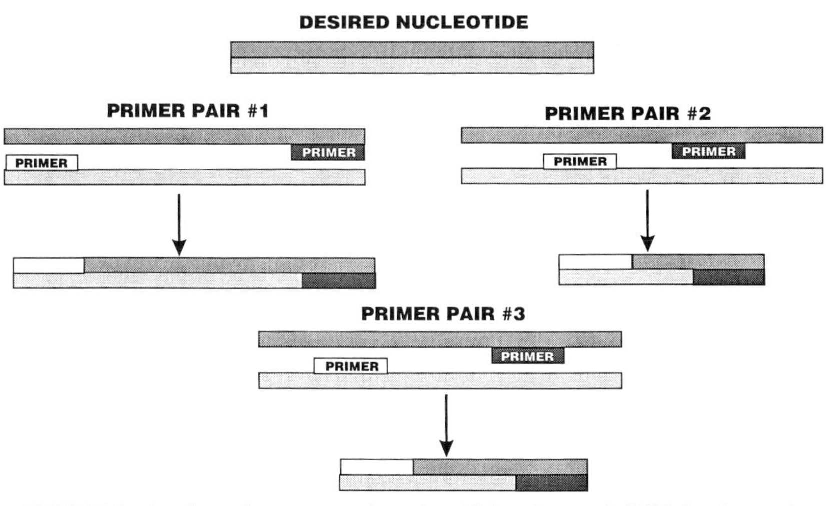

FIGURE 8 A schematic representation of multiple primer pair PCR for three primer pairs producing sequence from three different parts of the DNA molecule. Generally, six sense strand and six antisense strand primers are used giving a potential of 36 different combinations of primer pairs. Following each PCR reaction, the products are separated on agarose gels, purified, and sequenced. Typically, some of the primer pairs will work even if there are sequence mismatches due to species differences or difficult regions in the cDNA.

matched to each other. The sense and antisense primers can then be paired to make products of many different lengths. PCR reactions are done with all of the chosen primer pairs. Some do not work and thus make no product. However, some are sufficiently good sequence matches, even between species, such that discernible PCR products of proper length can be seen on agarose gels and can be subsequently purified and sequenced. If a portion of the molecule is not covered by any of the primer pairs, one can use the sequences determined by multiple primer pair PCR to redesign primer pairs that are perfect matches to the determined sequence. These can be made to cover the portion of the sequence that is unknown after the original PCR amplifications. This approach was sufficient to allow us to determine almost the entire sequence of inwardly rectifying potassium channels from mouse, rat, rabbit, chick, pig, bovine, monkey, human, and guinea-pig lens epithelia. This approach has a minor disadvantage in that messenger RNA has both 5′ and 3′ untranslated nucleotide sequences, which are not necessarily conserved from species to species. Primers made to these untranslated regions will usually not work in any species other than the one for which the primers were designed. Primers made to the extreme ends of the translated cDNA region unduly constrain the optimization. By this approach, all species will be constrained to have the same nucleotide sequence within the first 20 bases or so from the start codon and stop codon. The amino acids coded will be those from the species to which the primers were designed and may or may not be different than the species under consideration. In our experience, this did not produce difficulty in expressing clones containing these sequences but the failure to know or make the actual sequence the species used in nature could be a decided drawback to this approach.

D. Redundant Polymerase Chain Reaction

When expression cloning cannot be done, it can be difficult to isolate a particular DNA. For homology cloning, it is necessary for one to know at least a small part of the nucleotide or amino acid sequence. For potassium channels, the sequence of the pore region is expected to be somewhat conserved since it is believed that the selectivity of the channel for potassium largely resides in the pore sequence. Because all potassium channels are highly selective for potassium over sodium, it is believed that there must be substantial sequence homology in potassium channel pores. How then might this homology to isolate, sequence, and clone a potassium channel be exploited?

It was initially believed that it would easily be possible using 3′ and 5′ RACE techniques (Frohman, 1994). In these approaches mRNA is reverse transcribed and in the process of doing second strand cDNA production, defined oligonucleotide groups are placed either on the 5′ or the 3′ end of the molecule. These groups are such that a specific PCR primer should be able to be made for the added sequence. Then using a sense or antisense primer for the pore sequence and an opposite strand primer for the added sequence, PCR amplification can be done. In principle, this would allow sequencing the unknown molecule from the pore to the 3′ end or from the pore to the 5′ end. However, being unable to get either 3′ or 5′ RACE to work, the approach was abandoned.

For delayed rectifier-type potassium channels, amino acid sequence homology was found not only in the pore region but also in a region 60 bases or so 3′ to the pore. We made fully redundant primers to these two regions (Fig. 9a). Redundant primers were necessary because we did not know the nucleotide sequence of human lens epithelial channels and because we planned to use the primers for different cell types from different animal species. In either case, because of codon usage differences, we could not *a priori* know the actual nucleotide sequences. Redundant primers consisted of all possible codons for the amino acid sequences. Following PCR amplification with these primers (Fig. 9b), a substantial number of sequence tags were identified. Each of these was submitted to the BLAST server, which returned all of the sequences in the Genbank that were similar to ours. In some cases, the potassium channel was identified. In others, the sequence was close enough to a channel previously reported that it was reasonable to make a full set of primer pairs to the published sequence followed by start-stop or multiple primer pair PCR to determine the remaining 5′ and 3′ sequence in our cells.

E. Start-Stop Polymerase Chain Reaction

Having identified possible candidates for the channel proteins from the previously described redundant pore primer PCR or, as from the inward rectifier, having matched the properties of the recorded current in our cells with the properties of a known clone (Kubo *et al.,* 1993), cloning an expressible version of the channel can be straightforward and rapid. A simple download of the nucleotide sequence of the suspected channel from the Genbank and the design of a sense strand primer that begins at the start codon and an antisense strand primer that ends at the stop codon allows start-stop PCR to be performed. For cloning purposes, one can simply add to each primer a 5′ end sequence that codes for a suitable

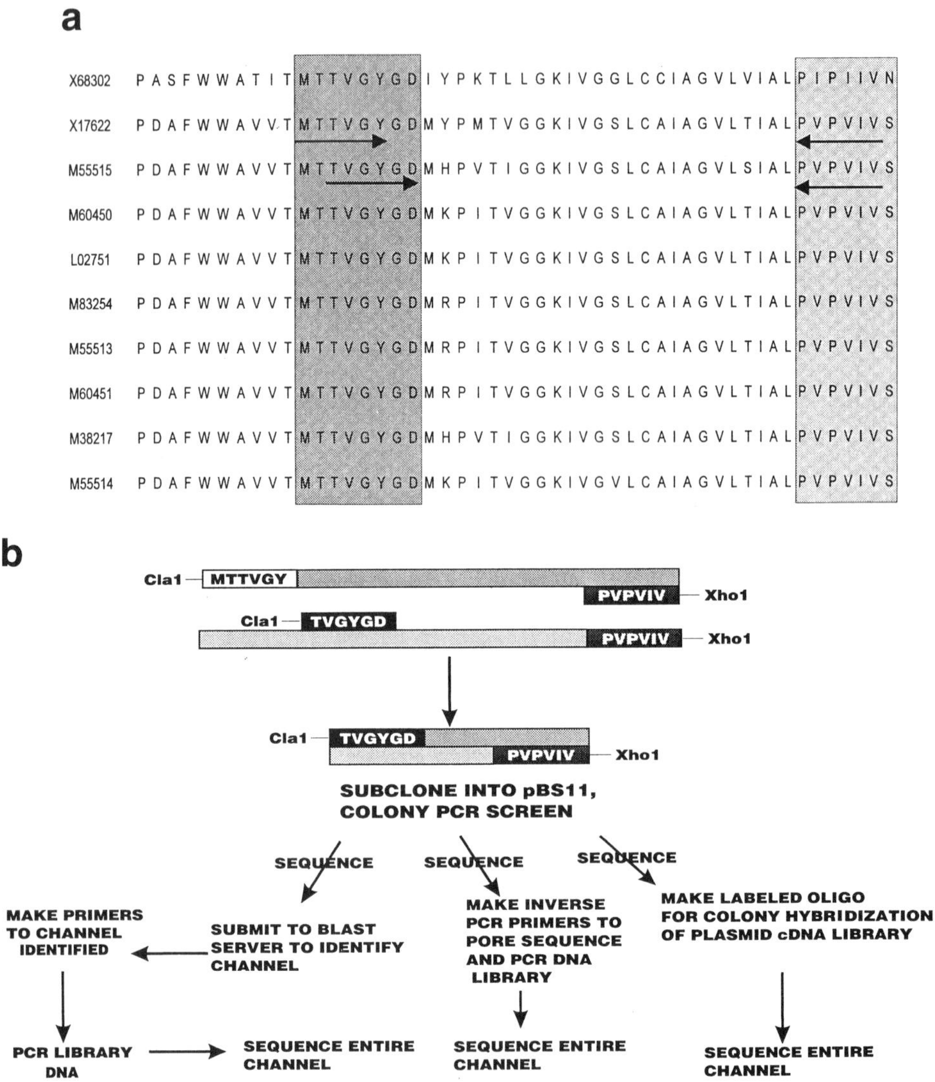

FIGURE 9 (a) Conserved amino acid sequences in the region of the pore (left-hand shaded region) from several published potassium sequences (accession numbers in first column) or a conserved region 3′ to the pore (right-hand shaded sequence). Redundant primers were made for the sequences over each arrow. These were subsequently used in PCR reactions (either RT–PCR or cDNA PCR on library DNA). (b) Flowchart for the subsequent PCR studies.

restriction enzyme site. We use Sal1 with the start codon primer and Not1 with the antisense stop codon primer since these are rarely occurring sequences. In addition, the sense strand primer also contains a consensus Kozak sequence (GCCACC) between the start codon and the Sal1 site to facilitate translation of the cloned product in a mammalian cell expression system (Fig. 10). Using either RNA extracted from the cells of interest and RT–PCR or using a cDNA library constructed from the cell's messenger RNA, a simple 30-cycle or so PCR can generate the entire channel sequence with a Sal1 site and Kozak sequence on the 5′ end and a Not1 site on the 3′ end. For long sequences (>3 kb), this cannot be done easily with Taq polymerase but we have found that "long PCR" with rTth done under conditions of high fidelity will often generate error-free sequences of 2700 bases or so which can be cloned into a common expression vector such as pcDNA1/Amp. For sequences longer than this, it has proven impossible to produce error-free clones due to the relatively high error rates of Taq or rTth polymerases. We have been able to produce clones of 3600 or so bases with one error, which could be corrected by inverse PCR as is discussed later.

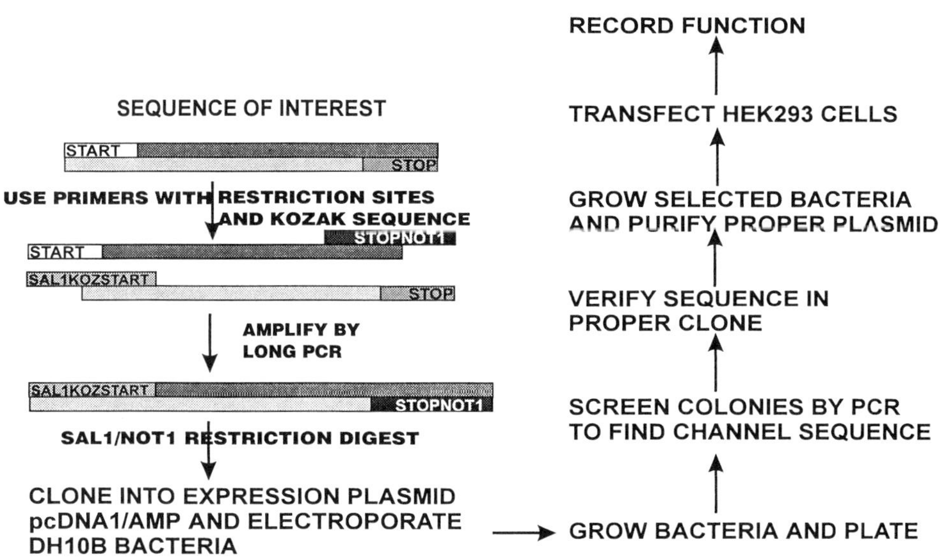

FIGURE 10 Flowchart for the start-stop PCR strategy. This produces products that can be sequenced or can be directly "sticky-end" subcloned into the pcDNA1/AMP expression vector. The Kozac consensus sequence added by PCR promotes binding to ribosomal RNA at the time of protein synthesis and so enhances the expression of the desired protein in an expression system.

F. Inverse Polymerase Chain Reaction

Inverse PCR can be a very valuable technique for determining largely unknown DNA sequences (Rosenthal *et al.*, 1994). It was originally used for determining genomic DNA sequences but is useful whenever the DNA to be searched can be circularized. As originally applied (Fig. 11), genomic DNA is digested with a restriction enzyme such as *Eco*R1 or *Bam*H1 and the resulting sticky-end DNA fragments are then ligated under conditions that promote intramolecular circularization of the molecules. A sense strand primer (30 nucleotides or so) was synthesized with its 5′ end directly abutting the 5′ end of an antisense strand primer. Polymerase chain reaction amplification with these back-to-back primers then results in the entire sequence of the circularized DNA being amplified. With this approach, it is only necessary that one know ≈60 bases of sequence. The sense and antisense strand primers are then made for this known sequence and from the resulting amplification, the rest of the sequence can be determined. This approach also works well with plasmid DNA and can be applied directly to cDNA libraries (Fig. 11). The 5′ end of the primers are phosphorylated for subsequent recircularization and bacterial transformation/amplification.

Inverse PCR can also be used to correct the sequence in a clone. As stated previously, our start-stop PCR approach results in, on average, one

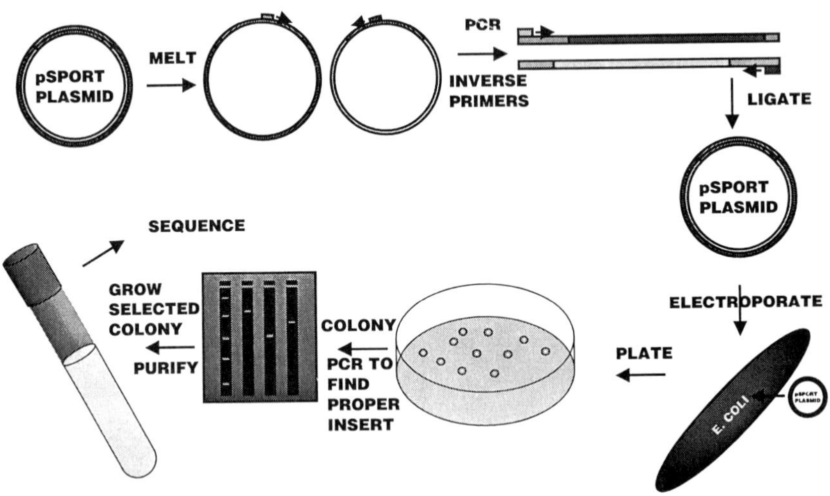

FIGURE 11 Flowchart for inverse PCR showing not only the PCR steps but the subsequent steps to amplify the plasmid and select it from background.

error in a sequence as long as 3600 bases. To correct the error, the correct sequence is included in the 5′ end of one of the PCR primers and PCR performed as described earlier. PCR amplification should be done with an error-correcting polymerase such as *Pfu* (Stratagene). This will result in the entire cDNA insert being copied except that the mistaken base is corrected. Of course, the entire plasmid is also copied. This works well for insert plus plasmid sequences of 10 kb or less and thus is sufficient for most channel sequences that have been reported.

G. Transient Expression

Once a channel sequence has been determined and the cDNA cloned, it is necessary to express the cDNA to verify that it makes a protein with the correct properties. While this is often done in *Xenopus* oocytes, we chose to use mammalian cells from which we could record by whole-cell

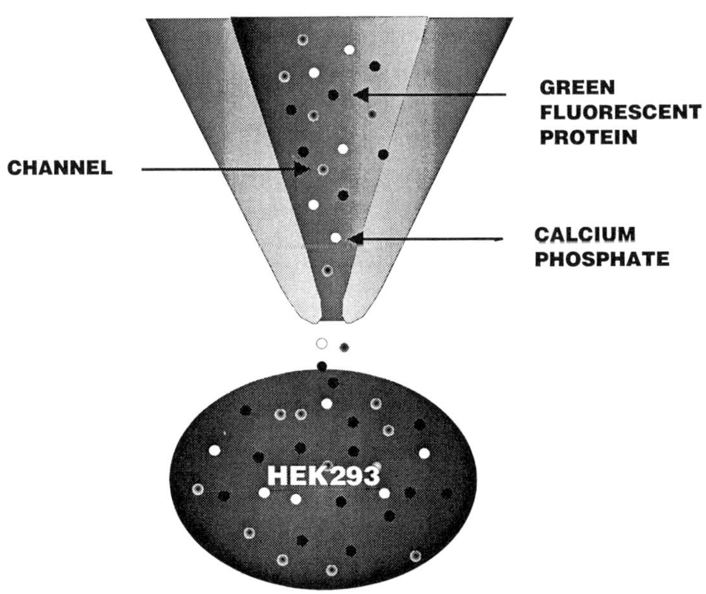

FIGURE 12 Schematic representation of the procedure to cotransfect HEK 293 cells with the expression plasmid for GFP protein and the expression plasmid for the appropriate channel protein by calcium phosphate precipitation. This was actually done in a petri dish of HEK cells covered with tissue culture medium where the plasmids and calcium phosphate were "dripped" over the surface of the culture medium by use of a Pasteur pipette.

patch clamping or by single-channel recording techniques. HEK 293, tsA-201, CHO, or αTN4 cells all worked well for transient expression of the channel of interest although CHO cells showed lower expression levels than the others. The cDNA of interest was cloned into the pcDNA1/Amp vector and then transfected into the cells either by lipofection techniques or by calcium phosphate coprecipitation (Fig. 12). This resulted in transfection rates of 5–38%. Therefore, more cells failed to receive the construct than those that received it. This created a problem in choosing the cells to patch clamp. This problem was solved by cotransfecting the cells at the same time with an expression plasmid containing the green fluorescent protein (GFP). At the time of patch-clamp recording, it was necessary only to shine blue light on the cells to identify the cells that emit green fluorescence due to their having made GFP (Fig. 13). To date, every cell that made GFP has also been transfected with our channel sequence and so also expressed the channel. In fact, cotransfected cells nearly quantitatively take up both types of plasmid DNA.

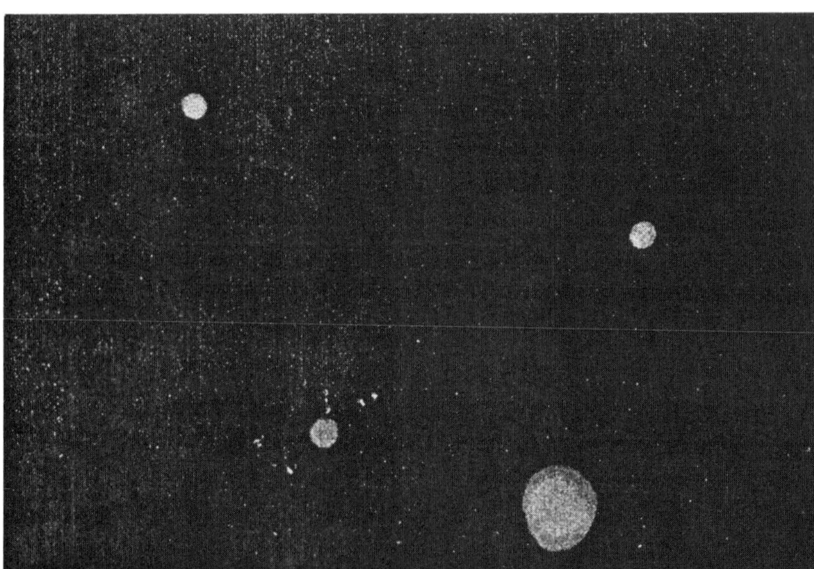

FIGURE 13 A fluorescence micrograph of transfected HEK 293 cells that contain GFP to identify the transfected cells (Rae and Shepard, unpublished). In every case where the cells fluoresced, they also contained the channel protein of interest. The transfection rate with calcium phosphate precipitation varied between 3–10% and provided more than enough cells to patch clamp. The GFP had no discernible effect on the currents.

TABLE I

Human Inward Rectifier Amino Acid Sequence

MGSVRTNRYS	IVSSEEDGMK	LATMAVANGF	GNGKSKVHTR	QQCRSRFVKK	50
DGHCNVQFIN	VGEKGQRYLA	DIFTTCVDIR	WRWMLVIFCL	AFVLSWLFFG	100
CVFWLIALLH	GDLDASKEGK	ACVSEVNSFT	AAFLFSIETQ	TTIGYGFRCV	150
TDECPIAVFM	WFQSIVGCI	IDAFIIGAVM	AKMAKPKKRN	ETLVFSHNAV	200
IAMRDGKLCL	MWRVGNLRKS	HLVEAHVRAQ	LLKSRITSEG	EYIPLDQIDI	250
NVGFDSGIDR	IFLVSPITIV	HEIDEDSPLY	DLSKQDIDNA	DFEIVVILEG	300
MVEATAMTTQ	CRSSYLANEI	LWGHRYEPVL	FEEKHYYKVD	YSRFHKTYEV	350
PNTPLCSARD	LAEKKYILSN	ANSFCYENEV	ALTSKEEDDS	ENGVPESTST	400
DTPPDIDLHN	QASVPLEPRP	ERRESEIX			428

Shaded M and X are start and stop of translated sequence. Light shaded region is pore sequence, whereas darkly shaded regions are transmembrane spanning loops.

1. Inward Rectifiers

The sequence we determined for the human lens inward rectifier (Table I) is >95% identical to Genback accession number M73052 (Kubo *et al.*, 1993). We inserted the coding sequence into our pcDNA1/Amp expression vector and expressed it, along with the GFP expression plasmid, in HEK 293 cells. The resulting current is shown in Fig. 14. The protein expressed is a fully functional inward rectifier based on the criteria described earlier. Even in the presence of 4 m*M* K^+ in the bathing medium, inward and transient outward currents are easily resolved. Again, this is possible because of the enormous number of channels expressed per cell. In fact, the expressed channel density was so high that it was impossible to record single-channel currents. On-cell patches had so many channels in the patch membrane that the recorded currents were essentially macroscopic-like. The currents were, as expected, blocked by submillimolar concentrations of either Cs^{2+} or Ba^{2+} (Cooper *et al.*, 1991).

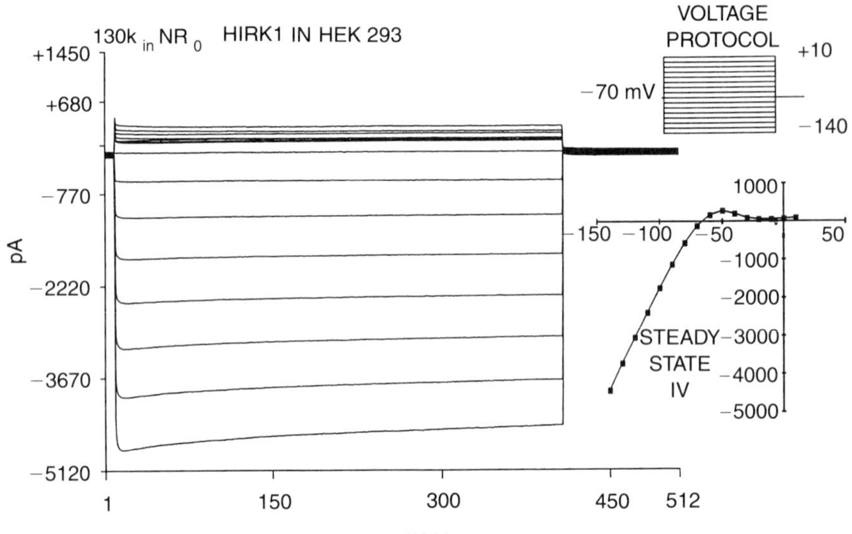

FIGURE 14 14Whole-cell currents measured from an HEK 293 cell transfected with an expression plasmid containing the human IRK1 sequence (Rae and Shepard, unpublished). The cell is bathed in a Ringer's solution containing 2.5 m*M* $CaCl_2$, 5 m*M* HEPES, 4 m*M* KCl, and 150 m*M* NaCl. Measurements come from a standard whole-cell recording configuration where the pipette contained 15 m*M* NaCl, 5 m*M* KCl, 130 m*M* $KMeSO_4$, 5 m*M* HEPES, and 2 m*M* EGTA. The voltage protocol used is shown in the upper inset. The steady-state current–voltage relationship is shown in the lower inset. These currents definitely come from an inwardly rectifying-type potassium channel. Note that because of the enormous channel density expressed in these cells, the small steady-state outward current is easily identified.

TABLE II

Human DRK1 Amino Acid Sequence

MPAGMTKHGS	RSTSSLPPEP	MEIVRSKACS	RRVRLNVGGL	AHEVLWRTLD	50
RLPRTRLGKL	RDCNTHDSLL	EVCDDYSLDD	NEYFFDRHPG	AFTSILNFYR	100
TGRLHMMEEM	CALSFSQELD	YWGIDEIYLE	SCCQARYHQK	KEQMNEELKR	150
EAETLREREG	EEFDNTCCAE	KRKKLWDLLE	KPNSSVAAKI	LAIISIMFIV	200
LSTIALSLNT	LPELQSLDEF	GQSTDNPQLA	HVEAVCIAWF	TMEYLLRFLS	250
SPKKWKFFKG	PLNAIDLLAI	LPYYVTIFLT	ESNKSVLQFQ	NVRRVVQIFR	300
IMRILRILKL	ARHSTGLQSL	GFTLRRSYNE	LGLLILFLAM	GIMIFSSLVF	350
FAEKDEDDTK	FKSIPASFWW	ATITMTTVGY	GDIYPKTLLG	KIVGGLCCIA	400
GVLVIALPIP	IIVNNFSEFY	KEQKRQEKAI	KRREALERAK	RNGSIVSMNM	450
KDAFARSIEM	MDIVVEKNGE	NMGKKDKVQD	NHLSPNKWKW	TKRTLSETSS	500
SKSFETKEQG	SPEKARSSSS	PQHLNVQQLE	DMYNKMAKTQ	SQPILNTKES	550
AAQSKPKEEL	EMESIPSPVA	PLPTRTEGVI	DMRSMSSIDS	FISCATDFPE	600
ATRFSHSPLT	SLPSKTGGST	APEVGWRGAL	GASGGRFVEA	NPSPDASQHS	650
SFFIESPKSS	MKTNNPLKLR	ALKVNFMEGD	PSPLLPVLGM	YHDPLRNRGS	700
AAAAVAGLEC	ATLLDKAVLS	PESSIYTTAS	AKTPPRSPEK	HTAIAFNFEA	750
GVHQYIDADT	DDEGQLLYSV	DSSPPKSLPG	STSPKFSTGT	RSEKNHFESS	800
PLPTSPKFLR	QNCIYSTEAL	TGKGPSGQEK	CKLENHISPD	VRVLPGGGAH	850
GSTRDQSIX					859

Shaded M and X are start and stop of translated sequence. Light shaded region is pore sequence, whereas darkly shaded regions are transmembrane spanning loops.

2. Delayed Recifiers

One of the human lens epithelial cell delayed rectifier sequences (Table II) proved identical to that published in the Genbank under accession number

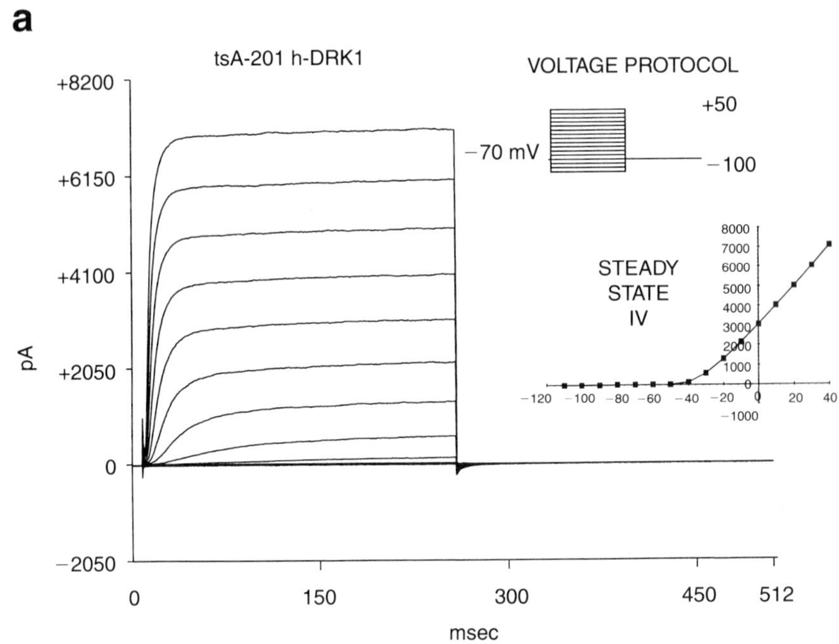

FIGURE 15 15(a) Whole-cell currents measured from a tSA-201 cell transfected with an expression plasmid containing the human DRK1 sequence (Rae and Shepard, unpublished). The cell is bathed in a Ringer's solution containing 2.5 m*M* $CaCl_2$, 5 m*M* HEPES, 4 m*M* KCl, and 150 m*M* NaCl. Measurements come from a standard whole-cell recording configuration where the pipette contained 15 m*M* NaCl, 5 m*M* KCl, 130 m*M* $KMeSO_4$, 5 m*M* HEPES, and 2 m*M* EGTA. The voltage protocol used is shown in the upper inset. The steady-state current–voltage relationship is shown in the lower inset. For these records, no series resistance compensation is used and, in fact, currents this large cannot be seriously measured by whole-cell patch clamping due to the large voltage and dynamic errors associated with so much current flowing through a single pipette electrode. However, the currents clearly have the form of DRK1-like currents. (b) Essentially macroscopic currents measured from a perforated outside-out vesicle from an αTN4 cell transfected with an expression plasmid containing the human DRK1 sequence. The outer surface of the patch is bathed in a Ringer's solution containing 2.5 m*M* $CaCl_2$, 5 m*M* HEPES, 30 m*M* KCl, and 120 m*M* NaCl. The pipette contained 15 m*M* NaCl, 5 m*M* KCl, 130 m*M* $KMeSO_4$, 5 m*M* HEPES, and 2 m*M* EGTA. The voltage protocol used is shown in the upper inset. The steady-state current–voltage relationship is shown in the lower inset. The channel density is so large that many channels reside even in a tiny membrane patch and allow macroscopic currents to be measured from patches. Note the inward tail currents possible here because of the availability of 30 m*M* K^+ in the batch. The tail currents are also HDRK1-like.

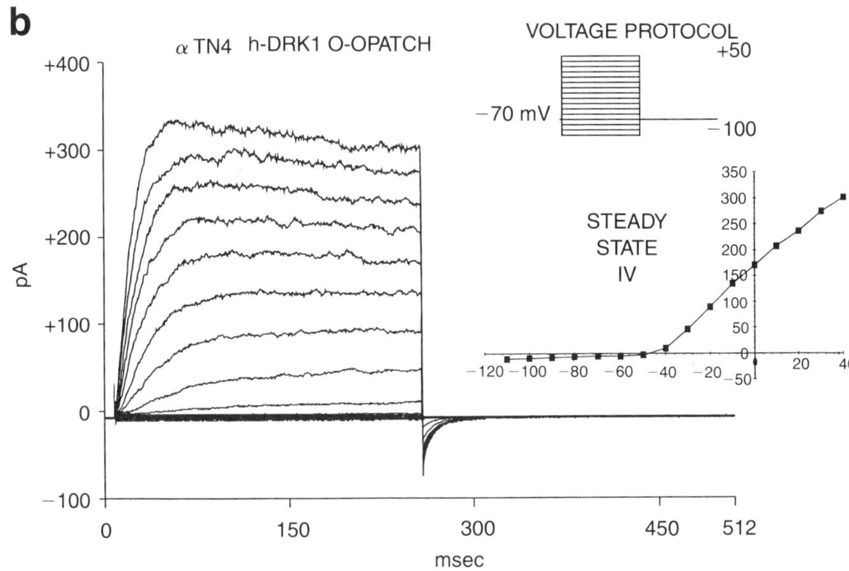

FIGURE 15 Continued

X68302. The cDNA was inserted into pcDNA1/Amp and expressed in either αTN4 cells or tsA-201 cells. In both cases, a current like that shown in Fig. 15a resulted. This is a delayed rectifier that under these recording conditions, looks nearly identical to naturally occurring delayed rectifiers. No extensive comparison of properties of any of our expressed channels to naturally occurring channels in lens epithelium has yet been done, although studies are ongoing. Again, the per-cell channel density expressed is large so the whole-cell currents are really too large to be recorded faithfully by a patch-clamp technique. Figure 15b shows the currents in a perforated outside-out vesicle from αTN4 cells when the bath contains about 30 m*M* K^+. Again, these currents are essentially macroscopic due to the large channel density even in the tiny membrane patch. Again, the currents look decidedly like delayed rectifiers (Albrecht *et al.*, 1993; Bendorf *et al.*, 1994; Frech *et al.*, 1989; Swanson *et al.*, 1990).

3. Calcium-Activated Potassium Channels

One of the α-subunits of BK from human lens epithelium (Table III) is the same as that published in the Genbank under accession number HSU11058. The β-subunit (Table IV) is the same as accession number HSU25138 (Knaus *et al.*, 1994; McManus *et al.*, 1995). When the plasmid containing the α-subunit of BK is transfected into tsA-201 cells, currents

TABLE III

Human Bkα Amino Acid Sequence

MDALIIPVTM	EVPCDSRGQR	MWWAFLASSM	VTFFGGLFII	LLWRTLKYLW	50
TVCCHCGGKT	KEAQKINNGS	SQADGTLKPV	DEKEEAVAAE	VGWMTSVKDW	100
AGVMISAQTL	TGRVLVVLVF	ALSIGALVIY	FIDSSNPIES	CQNFYKDFTL	150
QIDMAFNVFF	LLYFGLRFIA	ANDKLWFWLE	VNSWDFFTV	PPVFVSVYLN	200
RSWLGLRFLR	ALRLIQFSEI	LQFLNILKTS	NSIKLVNLLS	IFISTWLTAA	250
GFIHLVENSG	DPWENFQNNQ	ALTYWECVYL	LMVTMSTVGY	GDVYAKTTLG	300
RLFMVFFILG	GLAMFASYVP	EIIELIGNRK	KYGGSYSAVS	GRKHIVVCGH	350
ITLESVSNFL	KDFLHKDRDD	VNVEIVFLHN	ISPNLELEAL	FKRHFTQVEF	400
YQGSVLNPHD	LARVKIESAD	ACLILANKYC	ADPDAEDASN	IMRVISIKNY	450
HPKIRIITQM	LQYHNKAHLL	NIPSWNWKEG	DDAICLAELK	LGFIAQSCLA	450
QGLSTMLANL	FSMRSFIKIE	EDTWQKYYLE	GVSNEMYTEY	LSSAFVGLSF	550
PTVCELCFVK	LKLLMIAIEY	KSANRESRIL	INPGNHLKIQ	EGTLGFFIAS	600
DAKEVKRAFF	YCKACHDDIT	DPKRIKKCGC	KRLEDEQPST	LSPKKKQRNG	650
GMRTNSPNTSP	KLMRHDPLLI	PGNDQIDNMD	SNVKKYDSTG	MFHWCAPKEI	700
EKVILTRSEA	AMTVLSGHVV	VCIFGDVSSA	LIGLRNLVMP	LRASNFHYHE	750
LKHIVFVGSI	EYLKREWETL	HNFPKVSILP	GTPLSRADLR	AVNINLCDMC	800
VILSANQNNI	DDTSLQDKEC	ILASLNIKSM	QFDDSIGVLQ	ANSQGFTPPG	850
MDRSSPDNSP	VHGMLRQPSI	TTGVNIPIIT	ELVNDTNVQF	LDQDDDDDPD	900
TELYLTQPFA	CGTAFAVSVL	DSLMSATYFN	DNILTLIRTL	VTGGATPELE	950
ALIAEENALR	GGYSTPQTLA	NRDRCRVAQL	ALLDGPFADL	GDGGCYGDLF	1000
CKALKTYNML	CFGIYRLRDA	HLSTPSQCTK	RYVITNPPYE	FELVPTDLIF	1050
CLMQFDHNAG	QSRASLSHSS	HSSQSSSKKS	SSVHSIPSTA	NRQNRPKSRE	1100
SRDKQKYVQE	ERLX				1114

Shaded M and X are start and stop of translated sequence. Light shaded region is pore sequence, whereas darkly shaded regions are transmembrane spanning loops.

TABLE IV

Human Bkβ Amino Acid Sequence

MVKKLVMAQK	RGETRALCLG	VTMVVCAVIT	YYILVTTVLP	LYQKSVWTQE	50
SKCHLIETNI	RDQEELKGKK	VPQYPCLWVN	VSAAGRWAVL	YHTEDTRDQN	100
QQCSYIPGSV	DNYQTARADV	EKVRAKFQEQ	QVFYCFSAPR	GNETSVLFQR	150
LYGPQALLFS	LFWPTFLLTG	GLLIIAMVKS	NQYLSILAAQ	KX	192

Shaded M and X are start and stop of translated sequence.

of the form shown in Fig. 16a result. The currents are noisy, suggesting they come from a channel with a large single-channel conductance. As expected, when the intracellular calcium is buffered at a level of less than 10 n*M*, voltages of greater than +50 mV are required to activate the channel. The whole-cell currents take <10 msec to reach steady state (somewhat faster than expected) and there is no evidence of inactivation. In addition, the current at voltages near 0 mV is almost completely blocked

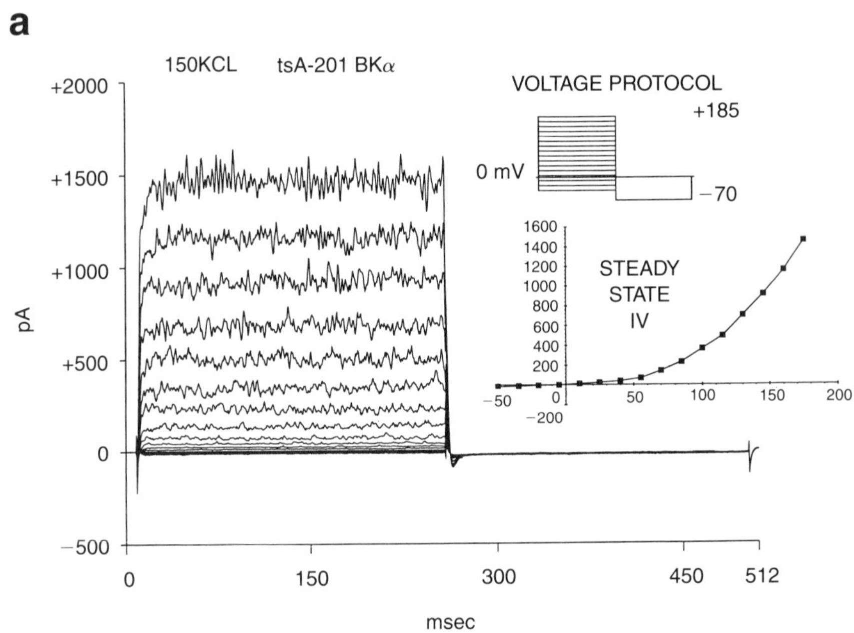

FIGURE 16 (a) Whole-cell currents measured from a tSA-201 cell transfected with an expression plasmid containing the human BK α-subunit sequence (Rae and Shepard, unpublished). The cell is bathed in a Ringer's solution containing 2.5 m*M* $CaCl_2$, 5 m*M* HEPES, and 150 m*M* KCl. Measurements come from a standard whole-cell recording configuration where the pipette contained 15 m*M* NaCl, 5 m*M* KCl, 130 m*M* $KMeSO_4$, 5 m*M* HEPES, and 2 m*M* EGTA. The voltage protocol used is shown in the upper inset. The steady-state current–voltage relationship is shown in the lower inset. Note the noise currents and the failure to generate a significant tail current with steps negative to E_k. (b) A tail current protocol (upper inset) applied to the same cell shown in Part (a). As with the naturally occurring full BK sequence, the tail currents deactivate rapidly, increasing such that the tail current generating voltage is made more negative. No inward currents are generated, either instantaneously or steady state. (c) The single-channel current–voltage relationship generated from a representative BK α-subunit channel in a perforated outside-out vesicle with 150 m*M* K^+ on the outside of the channel and 135 m*M* on the inside (Rae and Shepard, unpublished). The single-channel conductance is 218 pS. The data from parts (a), (b), and (c) identify this current as BK like.

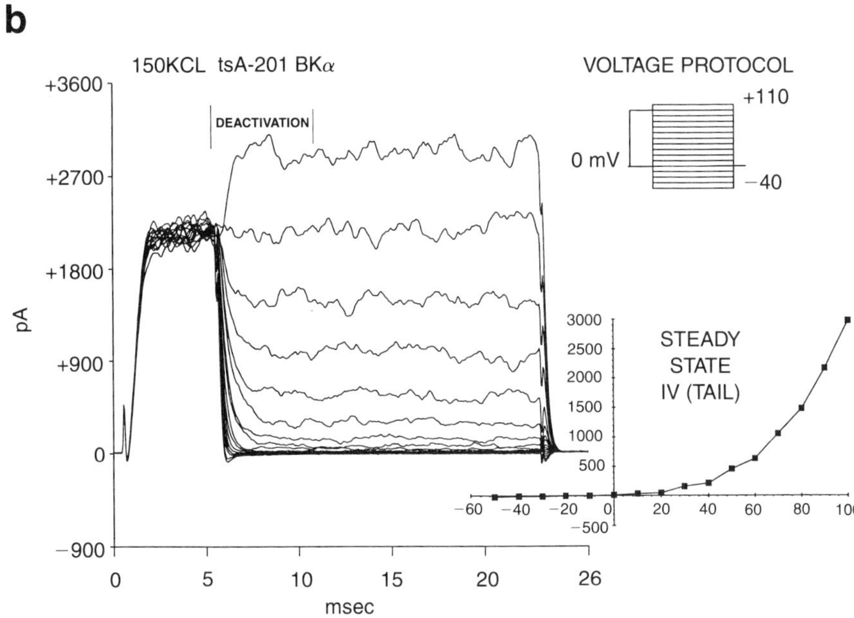

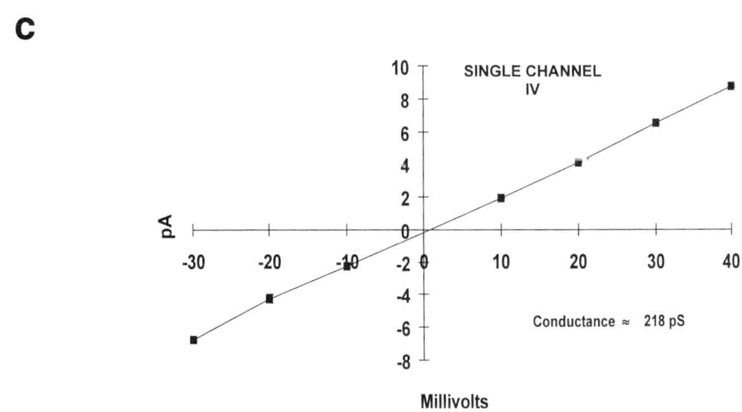

FIGURE 16 Continued

by external 1 mM TEA, a result quite specific for BK. Using a tail current protocol, it can be demonstrated that the deactivation time course is very voltage dependent, getting faster and faster as the voltage is made more negative. At voltages negative to 0 mV, the channels close so fast that no inward tail current can be recorded, again a result in keeping with BK (Fig. 16b). Every outside-out patch studied had several large-conductance single channels that were highly selective for potassium and the single-channel conductance was some 220 pS in symmetrical 150 mM K^+ (Fig. 16c), again compatible with BK. The single channels were sensitive to internal Ca^{2+} but this was not quantified since a full study will require the simultaneous expression of the β-subunit. From the composite data, the expressed current can be identified reasonably as BK (Adelman *et al.,* 1992; Atkinson *et al.,* 1991; Butler *et al.,* 1993; Dworetzky *et al.,* 1994; Haylett and Jenkinson, 1990; Lagrutta *et al.,* 1994; McManus and Magleby, 1988, 1991; Pallanck and Ganetzky, 1994; Shen *et al.,* 1994; Tseng-Crank *et al.,* 1994; Vogalis *et al.,* 1996; Wei *et al.,* 1994; Yellen, 1984).

4. A-Type Channels

To date, we have not determined the sequence for or expressed any A-type channels from human lens epithelium.

IV. SUMMARY

It is clear that the three channels that we have cloned and sequenced from human lens epithelium are similar to the inward rectifiers, delayed rectifiers, and calcium-activated potassium channels expected from the previous patch-clamp recordings from these cells (Fig. 17). Some of the contrib-

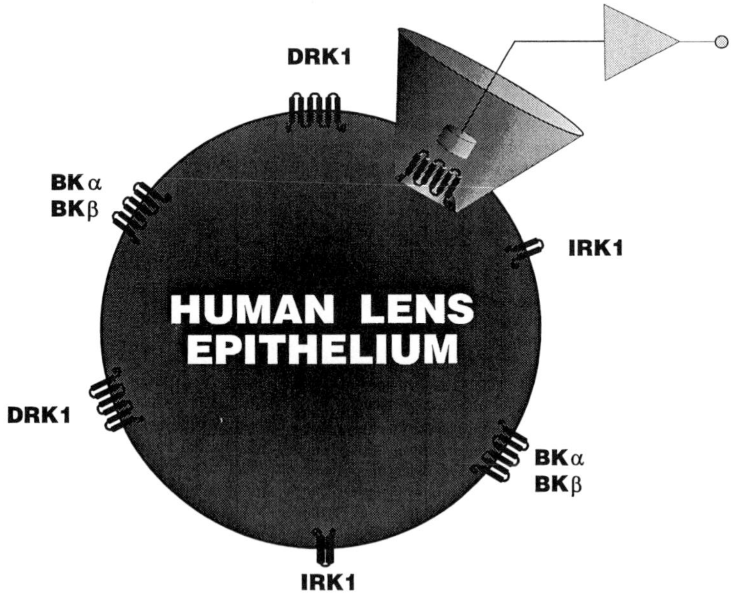

FIGURE 17 A schematic summary of the channels identified in human lens epithelium both by electrophysiology and molecular biology.

uting channels can now be considered to be identified at a molecular level. This does not mean that either the normal function or regulation of the channels is known. Because most ion channels have many parts of their sequence that allow them to be specifically regulated, exactly how they work in any particular cell type will be determined not only by the properties of the native channel but also by how those properties are altered by regulatory biochemistry the cell possesses. In addition, it is reasonable to believe from both electrophysiology and molecular biology that these are not the only relevant channels in lens epithelium. However, we clearly have a good start to an ultimate understanding of the detailed functioning of potassium channels in human lens epithelium.

Acknowledgments

This work was supported by National Institutes of Health grants EY03282 and EY06005 and an unrestricted award from Research to Prevent Blindness. We thank Jerry Dewey and Helen Hendrickson for technical assistance, Joan Rae for sequence alignment, primer design and figure production, and Kristy Zodrow for secretarial help.

References

Adelman, J. P., Shen K.-Z., Kavanaugh, M. P., Warren, R. A., Wu, Y.-N., Lagrutta, A., Bond, C. T., and North, R. A. (1992). Calcium-activated potassium channels expressed from cloned complementary DNAs. *Neuron* **9,** 209–216.

Albrecht, B., Lorra, C., Stocker, M., and Pongs, O. (1993). Cloning and characterization of a human delayed rectifier potassium channels gene. *Receptors and Channels* **1,** 99–110.

Atkinson, N. S., Robertson, G. A., and Ganetzky, B. (1991). A component of calcium-activated potassium channels encoded by *Drosophilia Slo* locus. *Science* **253,** 551–555.

Bendorf, K., Koopman, R., Lorra, C., and Pongs, O. (1993). Electrophysiological properties of the human delayed rectifier channel h-DRK1 expressed in frog oocytes. *Pflugers Arch* **422,** R26.

Bendorf, K., Koopmann, R., Lorra, C., and Pongs, O. (1994). Gating and conductance properties of a human delayed rectifier K^+ channel expressed in frog oocytes. *J. Physiol.* **477,** 1–14.

Butler, A., Tsunoda, S., McCobb, D. P., Wei, A., and Salkoff, L. (1993). *mSlo,* a complex mouse gene encoding "maxi" calcium-activated potassium channels. *Science* **261,** 221–224.

Cooper, K., Gates, P., Rae, J. L., and Dewey, J. (1990). Electrophysiology of cultured human lens epithelial cells. *J. Membr. Biol.* **117,** 285–298.

Cooper, K., Rae, J. L., and Dewey, J. (1991). Inwardly rectifying potassium current in mammalian lens epithelial cells. *Am. J. Physiol.* **261,** C115–C123.

Cooper, K., Rae, J. L., and Dewey, J. (1991). Inwardly rectifying potassium current in mammalian lens epithelial cells. *Am. J. Physiol.* **261,** C115–C123.

Dworetzky, S. I., Trojnacki, J. T., and Gribkoff, V. K. (1994). Cloning and expression of a human large-conductance calcium-activated potassium channel. *Mol. Brain Res.* **27,** 189–193.

Fakler, B., Brändle, U., Bond, C., Glowatzki, E., König, C., Adelman, J. P., Zenner, H.-P., and Ruppersberg, J. P. (1994). A structural determinant of differential sensitivity of cloned inward rectifier K^+ channels to intracellular spermine. *FEBS Lett.* **356,** 199–203.

Fakler, B., Brändle, U., Glowatzki, E., Weidemann, S., Zenner, H.-P., and Ruppersberg, J. P. (1995). Strong voltage-dependent inward rectification of inward rectifier K^+ channels is caused by intracellular spermine. *Cell* **80,** 149–154.

Ficker, E., Taglialatela, M., Wible, B. A., Henley, C. M., and Brown, A. M. (1994). Spermine and spermidine as gating molecules for inward rectifier K^+ channels. *Science* **266,** 1068–1071.

Frech, G. C., VanDongen, J. M. J., Schuster, G., Brown, A. M., and Joho, R. H. (1989). A novel potassium channel with delayed rectifier properties isolated from rat brain by expression cloning. *Nature* **340,** 642–645.

Frohman, M. A. (1994). Cloning PCR products. *In* "The Polymerase Chain Reaction" (K. B. Mullis, F. Ferré, and R. A. Gibbs, Eds.), pp. 14–37. Birkhauser, Boston.

Haylett, D. G., and Jenkinson, D. H. (1990). Calcium-activated potassium channels. *In* "Potassium Channels: Structure, Classification, Function and Therapeutic Potential" (N. S. Cook, Ed.), pp. 70–95. Ellis Horwood, Chichester.

Heginbotham, L., and Mackinnon, R. (1992). The aromatic binding site for tetraethylammonium ions on potassium channels. *Neuron* **8,** 483–491.

Horn, R., and Marty, A. (1988). Muscarinic action of ionic currents measured by a new whole-cell clamp method. *J. Gen. Physiol.* **92,** 145–159.

Kirsch, G. E., Taglialetela, M., and Brown, A. M. (1991). Internal and external TEA block in single cloned K^+ channels. *Am. J. Physiol.* **261,** C583–C590.

Knaus, H. G., Folander, K., Garcia-Calvo, M., Garcia, M. L., Kaczorowski, G. J., Smith M., and Swanson, R. (1994). Primary sequence and immunological characterization of β-subunit of high conductance Ca(2+)-activated K^+ channel from smooth muscle. *J. Biol. Chem.* **269,** 17274–17278.

Kubo, Y., Baldwin, T. J., Jan, Y. N., and Jan, L. Y. (1993). Primary structure and functional expression of a mouse inward rectifier potassium channel. *Nature* **362,** 127–133.

Lagrutta, A., Shen, K.-Z., North, R. A., and Adelman, J. P. (1994). Functional differences among alternatively spliced variants of *Slowpoke,* a *Drosophila* calcium-activated potassium channel. *J. Biol. Chem.* **269,** 20347–20351.

Latorre, R., Oberhauser, A., Labarca, P., and Alvarez, O. (1989). Varieties of calcium-activated potassium channels. *Annu. Rev. Physiol.* **51,** 385–399.

Levitan, E. S., and Kramer, R. H. (1990). Neuropeptide modulation of a single calcium and potassium channels detected with a new patch clamp configuration. *Nature* (*London*) **348,** 545–547.

Lopatin, A. N., Makhina, E. N., and Nichols, C. G. (1994). Potassium channel block by cytoplasmic polyamines as the mechanism of intrinsic rectification. *Nature* **372,** 366–369.

Lu, Z., and MacKinnon, R. (1994). Electrostatic tuning of Mg^{2+} affinity in an inward-rectifier K^+ channel. *Nature* **371,** 243–246.

MacKinnon, R., and Yellen, G. (1990). Mutations affecting TEA blockade and ion permeation in voltage-activated K^+ channels. *Science* **250,** 276–279.

Matsuda, H. (1991). Effects of external and internal K+ ions on magnesium block of inwardly rectifying K^+ channels in guinea-pig heart cells. *J. Physiol.* **435,** 83–99.

McManus, O. B., and Magleby, K. L. (1988). Kinetic states and modes of single large-conductance calcium-activated potassium channels in cultured rat skeletal muscle. *J. Physiol.* (*London*) **402,** 79–120.

McManus, O. B., and Magleby, K. L. (1991). Accounting for the Ca(2+)-dependent kinetics of single large-conductance Ca(2+)-activated K^+ channels in rat skeletal muscle. *J. Physiol.* (*London*) **443,** 739–777.

McManus, O. B., Helms, L. M. H., Pallanck, L., Ganetzky, B., Swanson, R., and Leonard, R. J. (1995). Functional role of the β-subunit of high conductance calcium-activated potassium channels. *Neuron* **14,** 645–650.

Pallanck, L., and Ganetzky, B. (1994). Cloning and characterization of human and mouse homologs of the *Drosophila* calcium-activated potassium channel gene, *slowpoke. Hum. Mol. Genet.* **3,** 1239–1243.

Peterson, O. H., and Maruyama, Y. (1984). Calcium-activated potassium channels and their role in secretion. *Nature* **307,** 693–696.

Pongs, O. (1992). Molecular biology of voltage-dependent potassium channels. *Physiol. Rev.* **72,** 69–88.

Rae, J. L. (1994). Outwardly rectifying potassium currents in lens epithelial cell membranes. *Curr. Eye Res.* **13,** 679–686.

Rae, J. L., and Fernandez, J. (1991). Perforated patch recordings in physiology. *NIPS* **6,** 273–277.

Rae, J. L., and Rae, J. S. (1992). Whole-cell currents from noncultured human lens epithelium. *Invest. Ophthalmol. Vis. Sci.* **33,** 2262–2268.

Rae, J., Cooper, K., Gates, G., and Watsky, M. (1990a). Low access resistance perforated patch recordings using amphotericin B. *J. Neurosci. Methods* **37,** 15–26.

Rae, J. L., Dewey, J., Rae, J. S., and Cooper, K. (1990b). A maxi calcium-activated potassium channel from chick lens epithelium. *Curr. Eye Res.* **9,** 847–861.

Rae, J. L., Shepard, A. R., and Rich, A. (1996). The molecular basis of potassium conductance in human lens epithelium [Abstract]. ARVO annual meeting, April 21–26, 1996, Fort Lauderdale, FL, p. S7.

Rosenthal, A., Platzer, M., and Charnock-Jones, D. S. (1994). Capture PCR. An efficient method of walking along chromosomal DNA and cDNA. *In* "The Polymerase Chain Reaction" (K. B. Mullis, F. Ferré, and R. A. Gibbs, Eds.), pp. 222–229. Birkhauser, Boston.

Saigusa, A., and Matsuda, H. (1988). Outward currents through the inwardly rectifying potassium channel of guinea-pig ventricular cells. *Jpn. J. Physiol.* **38,** 77–91.

Shen, K.-Z., Lagrutta, A., Davies, N. W., Standen, N. B., Adelman, J. P., and North, R. A. (1994). Tetraethylammonium block of *Slowpoke* calcium-activated potassium channels expressed in *Xenopus* oocytes: Evidence for tetrameric channel formation. *Pflugers Arch.* **426,** 440–445.

Shepard, A. R., and Rae, J. L. (1995). Identification of K^+ channels from corneal and lens tissues by PCR screening of plasmid cDNA libraries [Abstract]. ARVO annual meeting, May 14–19, 1995, Fort Lauderdale, FL, p. S264.

Soreq, H., and Seidman, S. (1992). *Xenopus* oocyte microinjection: From gene to protein. *In* "Methods of Enzymology" (B. Rudy and L. E. Iverson, Eds.), Vol. 207, pp. 225–265. Academic Press, San Diego.

Stanfield, P. R., Davies, N. W., Shelton, P. A., Khan, I. A., Brammar, W. J., Standen, N. B., and Conley, E. C. (1994). The intrinsic gating of inward rectifier K^+ channels expressed from murine IRK1 gene depends on voltage K^+ and Mg^+. *J. Physiol.* **475,** 1–7.

Stühmer, W., Ruppersberg, J. P., Schöter, K. H., Sakmann, B., Stocker, M., Giese, K. P., Perschke, A., Baumann, A., and Pongs, O. (1989). Molecular basis of functional diversity of voltage gated potassium channels in mammalian brain. *EMBO J.* **8,** 3235–3244.

Swanson, R., Marshall, J., Smith, J. S., Williams, J. B., Boyle, M. B., Folander, K., Luneau, C. J., Antanavage, J., Oliva, C., Buhrow, S. A., Bennett, C., Stein, R. B., and Kaczmarek, L. K. (1990). Cloning and expression of cDNA and genomic clones encoding three delayed rectifier potassium channels in rat brain. *Neuron* **4,** 929–939.

Tagliatatela, M., Vandongen, M. J., Drewe, J. A., Joho, R. H., Brown, A. M., and Kirsch, G. E. (1991). Patterns of internal and external tetraethylammonium block in four homologous K^+ channels. *Mol. Pharmacol.* **40,** 299–307.

Tagliatatela, M., Wible, B. A., Caporaso, R., Brown, A. M. (1994). Specification of pore properties by the carboxyl terminus of inwardly rectifying K^+ channels. *Science* **264,** 844–847.

Tajima, Y., Ono, K., and Akaike, N. (1996). Perforated patch-clamp recording in cardiac myocytes using cation-selective ionophore gramicidin. *Am. J. Physiol.* **271,** C524–C532.

Tseng-Crank, J., Foster, C. D., Krause, J. D., Mertz, R., Godinot, N., DiChiara, T. J., and Reinhart, P. H. (1994). Cloning, expression, and distribution of functionally distinct Ca^{2+}-activated K^+ channel isoforms from human brain. *Neuron* **13,** 1315–1330.

VandenBerg, C. A. (1987). Inward rectification of a potassium channel in cardiac ventricular cells depends on internal magnesium ions. *Proc. Natl. Acad. Sci. U.S.A.* **84,** 2560–2564.

VanDongen, A. M. J., French, G. C., Drewe, J. A., Joho, R. H., and Brown, A. M. (1990). Alterations and restoration of K^+ channel function by deletions at the N- and C-termini. *Neuron* **5,** 433–443.

Villaroel, A., Alvarez, O., Oberhauser, A., and LaTorre, R. (1988). Probing a Ca^{2+}-activated K^+ channel with quaternary ammonium ions. *Pflugers Arch.* **413,** 118–126.

Vogalis, F., Vincent, T., Qureshi, I., Schmalz, F., Ward, M. W., Sanders, K. M., and Horowitz, B. (1996). Cloning and expression of the large-conductance Ca^{2+}-activated K^+ channel from clonic smooth muscle. *Am. J. Physiol.* **271,** G629–G639.

Wei, A., Solaro, C., Lingle, C., and Salkoff, L. (1994). Calcium sensitivity of BK-type K_{Ca} channels determined by a separable domain. *Neuron* **13,** 671–681.

Wible, B. A., Taglialatela, M., Ficker, E., and Brown, A. M. (1994). Gating of inwardly rectifying K^+ channels localized to a single negatively charged residue. *Nature* **371,** 246–249.

Yang, J., Jan, Y. N., and Jan, L. Y. (1995). Control of rectification and permeation by residues in two distinct domains in an inward rectifier K^+ channel. *Neuron* **14,** 1047–1054.

Yellen, G. (1984). Ion permeation and blockade in Ca^{2+}-activated K^+ channels of bovine chromaffin cells. *J. Gen. Physiol.* **84,** 157–186.

Yellen, G., Jurman, M. E., Abramson, T., and MacKinnon, R. (1991). Mutations affecting internal TEA blockade identify the probably pore-forming region of a K^+ channel. *Science* **251,** 939–942.

CHAPTER 5

Aquaporin Water Channels in Eye and Other Tissues*

M. Douglas Lee,* Landon S. King,* and Peter Agre†

*Division of Pulmonary and Critical Care Medicine, †Department of Biological Chemistry, Department of Medicine, Johns Hopkins University, School of Medicine, Baltimore, Maryland 21205

I. INTRODUCTION

Passage of water across cell membranes is accomplished by two distinct mechanisms (reviewed in depth by Finkelstein, 1987): (1) simple diffusion of water through the phospholipid bilayer and (2) rapid transit of water through specialized membrane-spanning water transport proteins known as *aquaporins.* Diffusional water permeability (P_d) is measured in the absence of a salt gradient and is relatively low (~10 μm/sec). Osmotic water

* Adapted with permission from Lee, M. D., King, L. S., and Agre, P. (1997). The Aquaporin family of water channels proteins in clinical medicine. *Medicine* **76,** 141–156.

1063-5823/98 $25.00

permeability (P_f) represents the rate by which water is driven across membranes in response to an osmotic gradient. Biophysical studies performed 40 years ago on red blood cells predicted the existence of membrane water channels, since P_f was much larger than P_d, and renal proximal tubules were found to have high water permeability ($P_f > 1000$ μm/sec) (reviewed by Solomon *et al.*, 1983). The ability of mercurial agents to inhibit reversibly water transport in red cells and renal proximal tubules provided additional evidence for the existence of specialized water-transport proteins (Macey and Farmer, 1970).

Despite significant interest in the water channel hypothesis, the molecular basis of membrane water transport remained unknown until recently. A membrane protein of approximately 30 kDa in size was predicted to be the water channel based on radiation inactivation studies (van Hoek *et al.*, 1991, 1992). Nevertheless, the ubiquity of water, the lack of highly specific inhibitors, and the presence of high-background diffusional water permeability precluded molecular identification of the water channel protein. The past decade has seen multiple membrane transport proteins proposed as water transporters, including the band 3 anion exchanger of red cells (Solomon *et al.*, 1983), cystic fibrosis transmembrane regulator (CFTR) (Hasegawa *et al.*, 1992), and the sodium-independent glucose transporter GLUT1 (Fischbarg *et al.*, 1989), but subsequent investigation failed to confirm water channel activity by these proteins (reviewed by van Os *et al.*, 1994). The recent discovery of CHIP28 and its identification as a molecular water channel has led to its redesignation as *Aquaporin-1* (abbreviated AQP1), the first characterized member of the family of water channels expressed throughout the plant and animal kingdoms (reviewed by Agre *et al.*, 1993; Knepper, 1994; Chrispeels and Agre, 1994).

II. DISCOVERY OF THE AQUAPORINS

Recognition of AQP1 occurred in 1988 as a result of biochemical purifications of the 32-kDa human red cell Rh protein (Saboori *et al.*, 1988). The Rh preparations also contained a 28-kDa integral membrane protein that was found to be abundant in red cells and renal proximal tubules (Denker *et al.*, 1988; Smith and Agre, 1991). Although the 28-kDa protein was originally believed to be a breakdown product of the 32-kDa Rh protein, it became apparent that the protein (initially referred to as "CHIP28" for channel-forming integral protein of 28 kDa) has an altogether different function (hence, the designation "AQP1"). Isolation of the CHIP28 complementary DNA (Preston and Agre, 1991) permitted injection of the cRNA into *Xenopus* oocytes, which responded with a 20-fold increase in osmotic water permeability (P_f), which was partially inhibited by mercurial agents (Preston *et al.,* 1992). Reconstitution of highly purified CHIP28 into proteoliposomes confirmed that this molecule is a functional water channel (Zeidel *et al.,* 1992), and the rapid changes in volume indicate that the water permeability is a constitutive feature of AQP1. Although some investigators have proposed that AQP1 also transports glycerol (Abrami *et al.,* 1996) or cations (Yool *et al.,* 1996), these studies have not been confirmed (Agre *et al.,* 1997a), and the results of numerous other studies have established that AQP1 is freely permeated by water and does not permit transport of small uncharged molecules, ions, or even protons.

Analysis of the nucleotide sequence of *AQP1* revealed 20–40% homology with a family of proteins related to major intrinsic protein (MIP) of bovine lens (Park and Saier, 1996). Although functionally undefined, MIP was believed to form membrane channels (Ehring *et al.,* 1990). MIP family members all contain conserved DNA sequences; these allowed the design of degenerate oligonucleotide primers for polymerase chain amplifications (Preston, 1993). Using this homology cloning technique and cDNA libraries from other tissues, multiple laboratories identified related mammalian water channel cDNAs (Fushimi *et al.,* 1993; Ishibashi *et al.,* 1994; Ma *et al.,* 1994; Echevarria *et al.,* 1994; Hasegawa *et al.,* 1994; Jung *et al.,* 1994b; Raina *et al.,* 1995). Aquaporins have subsequently been identified in plants (Maurel *et al.,* 1993), bacteria (Calamita *et al.,* 1995), and yeast, but the focus of this review is on the characterized mammalian aquaporins (Fig. 1).

In addition to AQP1, which is expressed in red cells, kidney, eye, and numerous tissues, five other mammalian aquaporins have subsequently

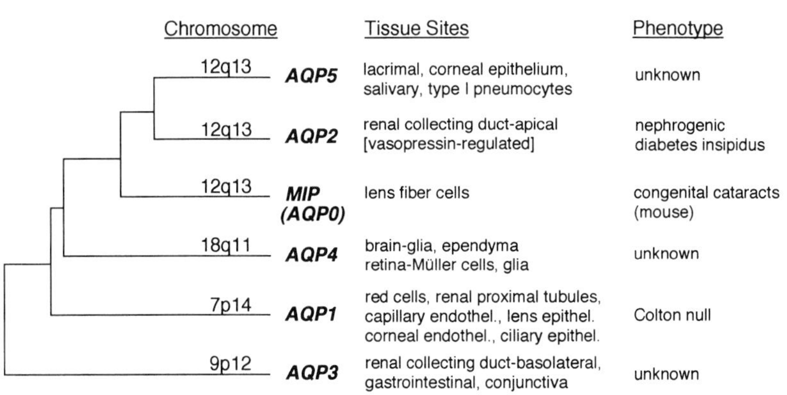

FIGURE 1 Phylogenetic tree of the mammalian aquaporins. Indicated are the sites on human chromosomes, the predominant tissue sites of expression, and the known mutant phenotypes.

been defined. Although the lens protein MIP (AQP0) was recognized in the mid-1970s, a role in water transport has been shown only recently (Mulders *et al.,* 1995; Kushmerick *et al.,* 1995; Chandy *et al.,* 1995). The renal collecting duct is known to be the site where the antidiuretic hormone vasopressin regulates water permeability of the apical membrane, a physiological event now known to involve AQP2 (Fushimi *et al.,* 1993; Nielsen *et al.,* 1993c). The outflow of water through the basolateral domains of renal collecting duct is the result of AQP3, a channel permeated by water and glycerol (Ishibashi *et al.,* 1994; Ma *et al.,* 1994; Echevarria *et al.,* 1994). The cDNA encoding AQP4 was isolated from brain (Jung *et al.,* 1994b); however, AQP4 is also present at low concentrations in other tissues, including the retina (Hasegawa *et al.,* 1994). The cDNA encoding AQP5 was isolated from submandibular salivary gland and is present also in lacrimal tissue, corneal epithelium, and lung (Raina *et al.,* 1995).

III. MOLECULAR STRUCTURE

Multiple studies have been undertaken to establish the molecular structure of AQP1. Hydropathy analysis of the deduced amino acid sequence predicted a protein with six transmembrane-spanning domains and intracellular amino and carboxy termini (Preston and Agre, 1991). The linear sequence is formed by two tandem repeats corresponding approximately to the amino-terminal and the carboxy-terminal halves of the polypeptide, each containing the three amino acid motif asparagine-proline-alanine (NPA) flanked by other conserved residues within two hydrophobic loops B and E (Fig. 2). The tandem repeats are predicted to have an obverse symmetry oriented at 180° to one another (Preston *et al.,* 1994a). Site-directed mutagenesis identified the mercury-sensitive residue at cysteine-189 in loop E (Preston *et al.,* 1993), and further mutagenesis studies revealed that mercury sensitivity resulted when cysteine was substituted for the native alanine at residue 73 (the loop B position corresponding to residue 189 in loop E) (Jung *et al.,* 1994a). Loop B is predicted to fold into the bilayer from the cytosolic face of the membrane and loop E from the extracellular face. The overlap of loop B and loop E domains between the leaflets of the lipid bilayer creates a single aqueous pathway for passage of water molecules in a single file. This structure is referred to as the "hourglass model" (Jung *et al.,* 1994a) and is still subject to further experimental verification. The association of AQP1 into a noncovalently linked tetrameric assembly with glycosylation of one of the four subunits was established by lectin chromatography, sedimentation, and filtration studies

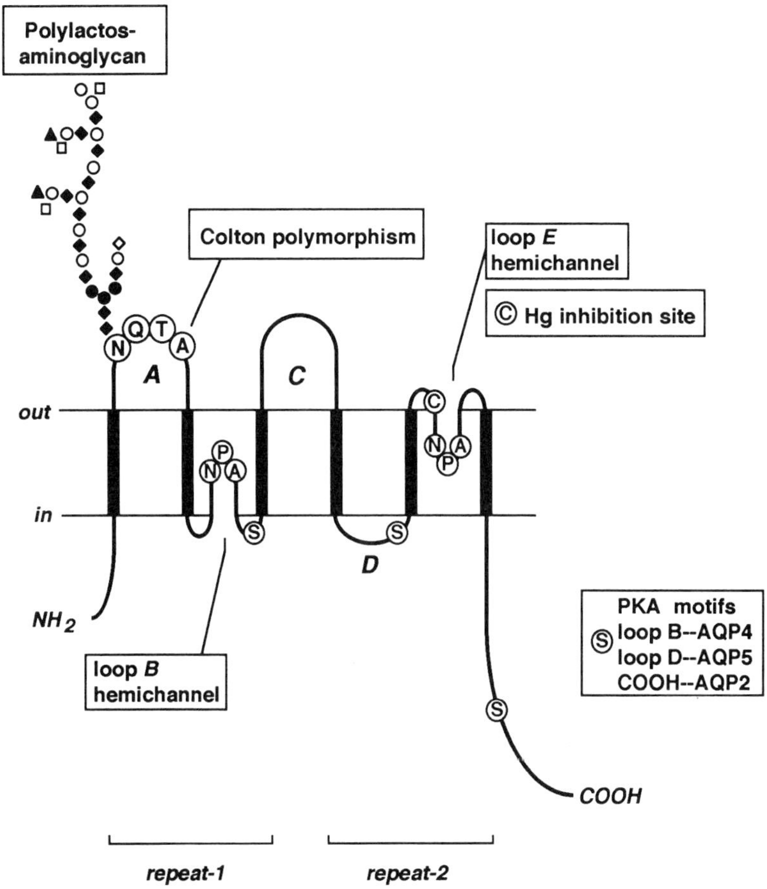

FIGURE 2 Model representing the membrane topology of an individual 28-kDa aquaporin subunit. The aqueous pore is believed to result when loop B and loop E are folded together forming an "hourglass." (Note that the Hg inhibition site is not present in MIP or AQP4.) The sites of glycosylation and the Colton polymorphism are specific for AQP1. (Note that *N*-glycosylation consensus sites of AQP2, AQP3, AQP4, and AQP5 are on loop C.)

(Smith and Agre, 1991). Evidence supporting the tetrameric assembly has now been reported by freeze-fracture of AQP1 in lipid bilayers (Verbavatz *et al.*, 1993; Zeidel *et al.*, 1994) and electron diffraction of two-dimensional membrane crystals (Walz *et al.*, 1994a; Mitra *et al.*, 1995; Jap and Li, 1995). The membrane crystallography studies promise high resolution of the AQP1 structure, and tilt analyses are revealing the three-dimensional shape of the molecule (Walz *et al.*, 1994b, 1995, 1996).

IV. GENETIC ORIGINS OF THE AQUAPORINS

Cloning and characterization of the genes for the mammalian aquaporins have promoted speculation about the ancestral origin of this family of proteins (Pisano and Chepelinsky, 1991; Moon *et al.*, 1993; Uchida *et al.*, 1994; Inase *et al.*, 1995; Lee *et al.*, 1996; Lu *et al.*, 1996). The tandem repeats comprising the amino-terminal and carboxy-terminal halves of AQP1 are believed to have resulted from an ancient gene duplication (Park and Saier, 1996), an event also believed to have occurred during the evolution of genes encoding other structural proteins (Maeda and Smithies, 1986; Henderson, 1991). The tandem repeat is present in all known members of the MIP family of proteins and also present in the bacterial homologs including AqpZ from *Escherichia coli* (Calamita *et al.*, 1995). So far, determination of microbial genomes has revealed aquaporin sequences in all species except the archaebacterium *Methanococcus jannaschii* (Bult *et al.*, 1996). Previous studies of bacterial transport have established the presence of GlpF, the glycerol facilitator (Heller *et al.*, 1980), which has a nucleotide and deduced amino acid sequence related to the aquaporins. Analysis of bacterial genomes is revealing that each contains an aquaporin gene and a glycerol facilitator gene; functional studies have confirmed the specificity of each (Maurel *et al.*, 1994; Calamita *et al.*, 1995).

A. Gene Organization

Five of the mammalian aquaporin genes (*MIP, AQP1, AQP2, AQP4,* and *AQP5*) are organized similarly. Each has a large first exon encoding the amino-terminal half of the molecule; exons 2–4 encode smaller segments of the carboxy-terminal half (Lee *et al.*, 1996). Intron-exon boundaries are located at identical sites within the respective coding regions of these aquaporins despite considerable variability in deduced amino acid sequences. In contrast, the *AQP3* gene consists of six exons and shares greater homology with the glycerol facilitator of *E. coli* than the other aquaporins (Fig. 1), suggesting a different evolutionary origin for this gene. In addition to water, AQP3 transports urea and glycerol, but at much lower rates (Echevarria *et al.*, 1996), suggesting that the protein may serve additional roles in cellular homeostasis. Another exception to classical aquaporin gene structure is found in the gene encoding the amino-terminus of AQP4, which contains an additional exon (exon 0) without homology to other aquaporin genes (Lu *et al.*, 1996). Alternative expression of transcripts produces two overlapping polypeptides differing in length by 23 amino acids at the amino-terminus. This alternative splicing of *AQP4* mRNA may provide a means for generating diversity at the amino-terminus of the protein, a phenomenon described for other types of transport proteins (Brosius *et al.*, 1989; Linn *et al.*, 1992; Wang *et al.*, 1996).

Gene duplication may have provided additional means of generating diversity among the aquaporins. Amino-terminal halves of the aquaporins share much more homology with each other than the carboxy-terminal halves, suggesting that the first half of the molecule serves characteristics basic to all the family members while the carboxy-terminal half of the molecule codes for specific functional roles of the protein. The recent identification of the chromosomal loci for aquaporin genes within the human genome supports this hypothesis (Moon *et al.*, 1993; Saito *et al.*, 1995; Lee *et al.*, 1996; Lu *et al.*, 1996; Mulders *et al.*, 1996a). The *AQP1, AQP3,* and *AQP4* genes are all located on different chromosomes while the recent identification of *MIP, AQP2,* and *AQP5* at chromosome 12q13 identifies the site of an aquaporin gene cluster. Interestingly, all three of the latter aquaporins are more closely aligned with each other at both the gene and amino acid levels than to other aquaporins (Fig. 1). AQP2, MIP, and AQP5 have protein kinase A consensus sequences, and preliminary evidence indicates phosphorylation is involved in the regulation of MIP and AQP2 (Ehring *et al.*, 1991; Kuwahara *et al.*, 1995; Katsura *et al.*, 1996; Nishimoto *et al.*, 1996). Whether cyclic nucleotides have a role in the physiology of AQP5 remains unclear. Another member of the MIP family has been identified in kidney and assigned to 12q13 (WCH3 and hKID); however, the negligible water permeability and incompatible nucleotide sequences preclude interpretation of their signficance (Ma *et al.*, 1993, 1996). It remains possible that additional aquaporins may be identified within the 12q13 locus, similar to the keratin gene cluster also located at 12q13 (Yoon *et al.*, 1994).

B. *AQP1 Mutations*

Although hundreds of blood antigens have been defined, they belong to only 24 blood groups, each corresponding to a single genetic locus. The *AQP1* gene was localized to chromosome 7p14 (Moon *et al.*, 1993). The Colton blood group had previously been linked to the short arm of human chromosome 7 (Zelinski *et al.*, 1990), suggesting a possible linkage to *AQP1*. Immunoprecipitations performed with anti-Co^a and anti-Co^b antibodies on Co^a and Co^b specific blood indicated a physical association of Co antigens with an epitope on AQP1 (Smith *et al.*, 1994). Sequencing of multiple DNA samples from individuals with defined Colton phenotypes confirmed that the Colton antigen results from an alanine/valine polymorphism at residue 45 on the first extracellular loop of AQP1 (Fig. 2).

Worldwide blood group referencing has led to the identification of exceedingly few individuals who are AQP1 deficient (Mollison *et al.*, 1987). A youngster with an unusual variant of congenital dyserythropoietic anemia was found to be partially deficient in Co antigens and totally deficient in CD44, but the etiological significance remains unknown (Agre *et al.*, 1994). In addition, members of five different kindreds have been identified with total lack of Co antigens; the probands were women who were identified because of the presence of high titers of circulating anti-Co antibodies that apparently developed during pregnancy. Blood and urine specimens were obtained from three probands from three different kindreds, and DNA analysis confirmed that each was homozygous for a different mutation in *AQP1* (Preston *et al.*, 1994b). Two Colton null individuals had no detectable AQP1 in red cells or renal sediment: One was homozygous for deletion of the entire exon 1; the second was homozygous for a frameshift mutation after glycine-104. A third Co null individual was homozygous for the missense mutation proline-38-leucine at the top of the first bilayer-spanning domain. This mutation resulted in an unstable AQP1 protein when expressed in oocytes and corresponded to a 99% reduction in AQP1 expression in red cells. It is unlikely that another water channel may compensate for the lack of red cell AQP1, since the water permeability of their red cells was markedly depressed (Mathai *et al.*, 1996). Surprisingly, none of these individuals suffered severe effects from the absence of the AQP1 protein, but definitive clinical studies of these Colton null individuals may uncover subtle functional abnormalities such as a renal concentrating defect or subclinical deficiency in the clearance of ocular or pulmonary fluid.

Although these studies suggest that AQP1 is not essential, the extreme rarity of the null phenotype may suggest that their status could be more complex. The identification of only five homozygote Colton null individuals among the hundreds of millions of blood donors and transfusion recipients is surprisingly rare. The incidence of partial Co deficiency is unknown, but a high ratio of heterozygote to homozygote frequency would suggest that individuals with the Colton null phenotype survive because of some compensatory mechanism such as up-regulation of other aquaporins in kidney or other tissues. If this is the case, Colton null red cells do not appear to express another aquaporin, since their membrane water permeability is markedly reduced (Mathai *et al.*, 1996). Targeted disruption of the mouse *Aqp1* gene should reveal the direct consequences of *AQP1* deletion.

V. DISTRIBUTION AND PHYSIOLOGY

Presently six mammalian aquaporins have been identified. Definition of the tissue sites of expression, ontogeny, mechanisms of regulation, and gene structure have revealed the participation of these molecules in multiple normal physiological processes, and aquaporins have been implicated in several disease states (reviewed by King and Agre, 1996). Each aquaporin has a unique pattern of tissue expression with minimal overlap in distribution (Fig. 1), and the site of each aquaporin expression is believed to represent a site where rapid transit of water occurs in response to changes in osmotic gradients across epithelia. Although our understanding of aquaporins is becoming advanced, some controversies still exist regarding protein structure and function, and the subcellular sites of expression are not fully known (reviewed by Agre *et al.,* 1995). The distribution of aquaporins throughout the body suggests possible roles in several disease states (reviewed by King and Agre, 1996). Humans have been identified with mutations in *AQP1* and *AQP2,* and two mouse mutation shave been identified that cause a deficiency of Mip protein (Preston *et al.,* 1994b; Deen *et al.,* 1994; Shiels and Bassnett, 1996). Severe clinical phenotypes were observed only in the *AQP2* and *Mip* mutants raising questions as to the relevance of the AQP1 protein. In this section we review current knowledge about the distribution of the aquaporins and discuss the clinical implications of altered aquaporin expression with special emphasis on eye.

A. Aquaporins in Eye

The eye contains multiple different aquaporins in defined, nonoverlapping sites. At present five mammalian aquaporins (MIP, AQP1, AQP3, AQP4, and AQP5) are known to be expressed in the eye, explaining several physiological and pathological events in eye (Fig. 3).

1. Lens

The story of the aquaporins could have begun with eye, since the first recognized member of the gene family is MIP, major intrinsic protein of lens. The lens is enclosed by an epithelium, the basement membrane forms a capsule, and the interior is filled with fiber cells that are formed from the lateral margins of the lens. During this process, the fiber cells lose their nuclei, become filled with the cytosolic proteins known as *crystallins,* and express MIP at the plasma membrane. It has been estimated that more than 50% of the membrane protein of the lens fiber cells is comprised of MIP (Gorin *et al.,* 1984). The result of this process is the compression of

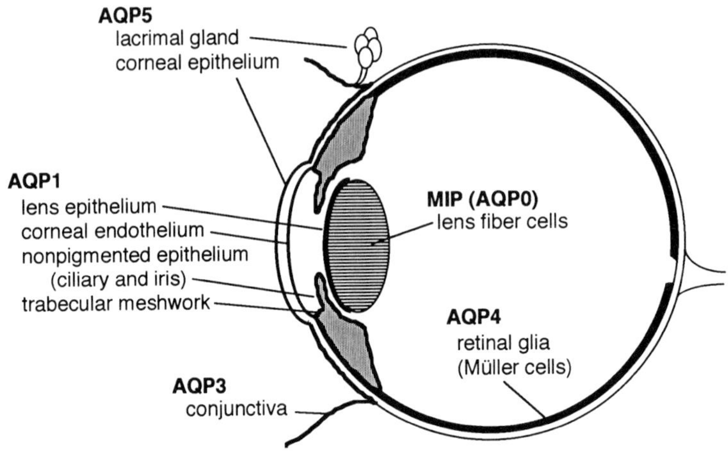

FIGURE 3 Model representing distribution of aquaporins in eye. Shown is a model depicting the sites of aquaporin expression of the four identified water channels in the eye.

old fibers (forming the central nucleus) as new fibers cells are formed. Eventually, only about 5% of the lens volume remains as interstitial space (reviewed by Hart, 1992; Paterson and Delamere, 1992), and the MIP protein is most heavily expressed in membranes surrounding pockets of interstitial fluid (Zampighi *et al.*, 1989). Initially believed to be a gap junction protein or a voltage-gated ion channel (Ehring *et al.*, 1990), MIP was recently shown to be a functional aquaporin, although its water channel activity is several-fold less than other aquaporins (Mulders *et al.*, 1995; Kushmerick *et al.*, 1995; Chandy *et al.*, 1995).

Lens is known to have a lower water content than most other tissues, and MIP may play a significant role in maintenance of lens transparency by enhancing the removal of water, thereby reducing the scatter of light. Although definitive physiological studies are lacking, water is presumed to move from the interstitial space into the fiber cells and out through anterior epithelium where AQP1 is present. Lens water content increases with age (Siebinga *et al.*, 1991), and this is believed to reflect changes in MIP content or structure. In calf lens, the MIP protein is the native 26-kDa form, but in adult cattle, the protein has degraded into a smaller 22-kDa polypeptide (Zampighi *et al.*, 1989), providing additional evidence for age-related changes in MIP structure. It is still unclear whether this phenomenon leads to pathological changes in lens water content, presbyopia, or development of cataracts. The band 3 protein of red cell is known to have an important structural role as the membrane attachment site for ankyrin, as well as a transport function as the chloride–bicarbonate exchanger. It has been speculated that MIP may also serve as a membrane structural protein, thus explaining the need for a huge number of copies (Michea *et al.*, 1994).

The gene encoding Mip has been established as the site of two naturally occurring mouse mutations that were recognized because they caused congenital cataracts (Shiels and Bassnett, 1996). Mice homozygous for either the cataract Fraser mutation, *Cat*^{Fr}, or the lens opacity mutation, *Lop* (Lyon *et al.*, 1981; Muggleton-Harris *et al.*, 1987), developed bilateral cataracts and degeneration of lens fiber cells. Both of these mutations were mapped near the end of mouse chromosome 10, a region syntenic with human chromosome 12q13-q14, the site of *MIP*. Linkage analysis in mice suggested the *Mip* gene colocalized with these two mutations (Griffin and Shiels, 1992). Analysis of the Mip protein from these two mutations revealed two distinct products. The *Cat*^{Fr} mutation leads to a splicing error between the third and fourth exons, resulting in a truncated Mip protein fused at the carboxy-terminal end to a 55 amino acid segment of the long terminal repeat (LTR) of the embryo transposon sequences of the mouse (Shiels and Bassnett, 1996). The Mip-LTR fusion protein was not identified by immunofluorescence, since the missing carboxy-terminal domain of Mip is the epitope to which the antibodies are known to react. The *Lop* cDNA sequence demonstrated an unconserved amino acid substitution (proline for alanine) at residue 51 in the second bilayer-spanning domain of the MIP protein. The resulting protein was localized to the endoplasmic reticulum of the lens fiber cells, a process that could lead to abnormal trafficking of other membrane and cellular proteins. Similar pathology was noted from the two different mutations, suggesting that defects in protein trafficking or stability may lead to the resulting phenotypes. Although the heterozygotes were less severely affected, both the *Lop* and *Cat*^{Fr} mice developed cataracts, thus defining the traits as semidominant. This provides further support for the hypothesis that MIP may also serve a structural role, since it is known that mutations in most genes encoding transporter proteins, such as AQP2, are expressed as recessive traits, whereas mutations in genes encoding structural proteins, such as crystallins, are usually expressed as dominant traits.

2. Anterior Eye

AQP1, AQP3, and AQP5 are expressed in nonoverlapping domains that correspond to known water-permeable tissues in the anterior chamber, conjunctiva, and lacrimal glands. Although aquaporins are considered important mediators of water movements in eye, individuals with Colton null phenotype suggest that aquaporins may not all be rate-limiting water transporters (Preston *et al.*, 1994b). Nevertheless, since none of these individuals has undergone extensive visual examinations, the lack of an obvious clinical phenotype does not rule out the possibility of latent or impending ocular dysfunction.

The cornea has evolved to provide the eye with an anterior margin that is both tough and transparent. Like lens, the cornea is known to have a water content that is lower than that of most other tissues. The corneal endothelium facilitates the removal of water from the overlying stroma, and since AQP1 is abundant in this tissue, it is believed to contribute to maintenance of transparency (Nielsen *et al.,* 1993b; Echevarria *et al.,* 1993). In corneal endothelium, AQP1 mRNA is dramatically reduced after birth; this may explain the development of edema known to occur after trauma to the cornea in adults (Bondy *et al.,* 1993). Demonstration that the Colton blood group is a protein polymorphism on AQP1 (Smith *et al.,* 1994) raises the possibility that it may be a source of immune-mediated graft rejection after corneal transplant. Approximately 90% of the population have Co^{a} antigen, 9.7% have $Co^{a,b}$, and only 0.3% have Co^{b} (Mollison *et al.,* 1987), so anti-Colton antibodies are rarely encountered in clinical practice. Nevertheless, anti-Colton antibodies have been implicated in transfusion reactions and hemolytic disease of the newborn (Simpson, 1973; Kurtz *et al.,* 1982). Although still unproven, the small subset of patients who experience graft rejection after corneal transplant may represent a result of Colton incompatibility, since other studies have suggested that ABH blood group antigens play a role in this process (CCTS, 1992; Maguire *et al.,* 1994). Only 11% of transplants will be potentially Co incompatible, and almost all of these will be Co^{a} individuals receiving a cornea from a $Co^{a,b}$ donor. Only rare transplants will involve Co^{b} individuals receiving a cornea from Co^{a} donors. A retrospective study of patients enrolled in the high-risk corneal transplantation study is under way to establish if these Colton frequencies are found with increased frequency among patients with histories of repeated corneal transplant rejections (Harris *et al.,* 1996).

Open-angle glaucoma is a common but serious disease characterized by elevated intraocular pressure and is believed to be multifactorial (reviewed by Hart, 1992). Although it is not present in the ciliary or retinal pigmented epithelium, AQP1 is abundant in the nonpigmented epithelium of the anterior ciliary body where aqueous humor is secreted (Nielsen *et al.,* 1993b). AQP1 is also present in trabecular meshwork epithelium and the canals of Schlemm (Nielsen *et al.,* 1993b; Stamer *et al.,* 1994), the outflow tracts where reabsorption of aqueous humor occurs (reviewed by Brubaker, 1991; Caprioli, 1992). Since the pathogenesis of glaucoma may represent a subtle, long-standing elevation of anterior chamber pressure, it is easily imagined that subtle increases in aqueous humor secretion or perturbations of aqueous humor reabsorption may contribute to the pathogenesis of some forms of glaucoma. AQP1 is also abundant in the nonpigmented epithelium of iris, where its high water permeability may assist in the rapid volume changes necessary for pupillary constriction.

Aquaporins are present at the surface of the eye where they are presumed to participate in the maintenance of tear film. Although less well understood than the corneal endothelium, the corneal epithelium site has been shown to exhibit apical chloride and water permeability in response to cAMP mediators or catecholamines (Candia and Zamudo, 1995), and AQP5 is present in corneal epithelium (Raina *et al.,* 1995). The conjunctiva may also participate in the maintenance of tear film, and AQP3 is abundant in the basolateral domains of the conjunctival epithelium (Frigeri *et al.,* 1995a).

Originally cloned from a salivary gland cDNA library (Raina *et al.,* 1995), AQP5 is also abundantly expressed on the apical membranes of lacrimal gland acini, where the protein may contribute to the aqueous component of tear formation, which is regulated by neurohormonal factors (King *et al.,* 1997). The similarity at the amino acid level between AQP5 and AQP2, coupled with their similar consensus sequences for cAMP protein kinase A, suggests that AQP5 may be regulated in a similar manner. Rapid shifts in fluid across salivary and lacrimal glands in response to neurohormonal stimuli make this an appealing hypothesis. The distribution of AQP5 in lung, lacrimal, and salivary tissues may provide an explanation for the pathogenesis of Sjögren's syndrome, a disease characterized by autoimmune destruction of salivary and lacrimal glands and respiratory tissue (Fox, 1995). Although the pathogenesis of Sjögren's is not known, the existence of a tissue-specific target antigen has been suggested. The presence of Colton antigens on the surface of AQP1 suggests that other aquaporins may have their own surface epitopes, and existence of an epitope on AQP5 would make it suspect. Dry eye is a much more common problem, often affecting postmenopausal women, and no definitive treatments are available. Although the studies are preliminary, recent advances have been made in genetic reconstitution with adenoviral mediated transfer of aquaporins to damaged salivary glands (Delporte *et al.,* 1996, 1997), raising the hope that similar treatments can be developed for lacrimal glands.

3. Retina

Aquaporins have not yet been identified in the pigmented epithelium, although AQP4 was identified in retina by *in situ* hybridizations (Hasegawa *et al.,* 1994). Immunohistochemical and immunoelectron microscopic studies have defined AQP4 in retinal glial cells (Agre *et al.,* 1997b). This distribution is similar to that of AQP4 in brain, since the perivascular glial cells in retina are extensively labeled with anti-AQP4. Moreover, AQP4 is very abundant in retinal Müller cells, which are known to surround and support the photoreceptor cells. Thus AQP4 may contribute to visual activity by regulating the light-dependent hydration of space surrounding photoreceptors (Li *et al.*, 1994).

B. Brain

Alterations in water distribution within the brain and central nervous system are unfortunately common for patients with stroke, head trauma, brain tumors, normal pressure hydrocephalus, or pseudotumor cerebri (reviewed by Adams and Victor, 1989; Fishman, 1995; Klatzo, 1994; Milhorat, 1992; Prockop, 1995). Encasement of the brain within the rigid structure of the bony cranium makes brain edema a potentially catastrophic event, and dictates the need for better understanding of the mechanisms underlying water flux and cell volume changes within the central nervous system. Although widely distributed among other tissues, AQP1 is only present in choroid plexus epithelium within the central nervous system (Nielsen *et al.*, 1993b), and AQP4 is the major water channel for the remainder of brain (Jung *et al.*, 1994b). Initial *in situ* hybridizations and immunolocalizations demonstrated abundant AQP4 expression in the cerebellum, hypothalamus, spinal cord, and ependymal cells lining the ventricles (Hasegawa *et al.*, 1994; Jung *et al.*, 1994b; Frigeri *et al.*, 1995a,b). The cellular and subcellular locations of AQP4 were determined in certain ependymal cells and glial cells by high-resolution immunocytochemistry and immunogold electron microscopy (Nielsen *et al.*, 1997). These distributional studies may provide insight into physiological and pathological roles for AQP4.

The mechanisms controlling body water balance are known to reside in brain. Increased plasma osmolality is sensed by yet-undefined molecules termed *osmoreceptors* that signal for the secretion of vasopressin by the neurohypophysis. Analysis of the magnocellular neurons in the supraoptic nucleus revealed a mechanosensitive ion channel that is triggered by increases in osmolality and may be propagated onto the axon terminals in pituitary (Oliet and Bourque, 1993). Although AQP4 was not found in magnocellular neurons (nor any other neurons), the protein is heavily expressed in both membranes of the adjacent glial lamellae of osmosensory areas, which are known to be involved in the regulation of water balance including the supraoptic nucleus and subfornical region (Nielsen *et al.*, 1997). AQP4 at these sites may be involved in the osmoregulatory response by enhancing rapid changes of cell volume in response to local changes in osmolality. Since these glial lamellae are highly redundant, surrounding the magnocellular neurons like a boa constrictor around its prey, they may amplify an otherwise subtle mechanical response to minor changes in osmolality. While it is still unclear, changes in AQP4 distribution may account for some disorders of vasopressin release within the brain, such as central diabetes insipidus or syndrome of inappropriate antidiuretic hormone.

AQP4 is also heavily expressed in astroglial foot processes surrounding capillaries throughout the brain parenchyma, where the protein exhibits a highly polarized distribution localized to the perivasular membrane opposite from the membrane facing the neuropil (Nielsen *et al.,* 1997). Although AQP1 is present in endothelial cells in several organs (Nielsen *et al.,* 1993b), it has not been identified in brain capillary endothelia where another aquaporin may possibly reside. This distribution strongly suggests a role for AQP4 in moving water into or out of the brain. Thus, AQP4 may play a role in either the pathogenesis or amelioration of cerebral edema. Disruptions of AQP4 metabolism, such as dephosphorylation or disordered distribution, may result in altered permeability of the blood–brain barrier that occurs in association with tumors, stroke, ischemia, or infection (reviewed by Klatzo, 1994; Fishman, 1995). AQP4 may provide the entrance port for water resulting in brain edema caused by renal failure, acute plasma hypoosmolality, or diabetic hyperosmolar states, which may be directly linked to AQP4 expression. Alternatively, AQP4 in the perivascular foot processes may provide the mechanism for ridding the brain of excess fluid with the driving force being the increased hydrostatic pressure of the edematous brain.

The presence of AQP4 in cerebellum and ependymal cells suggests that the protein may be involved in the pathophysiology of ataxia or disorders of cerebrospinal fluid reabsorption. The presence of AQP4 in the Purkinje layer of the cerebellum suggested involvement of *Aqp4* and the mouse mutation *ataxia.* Although gene linkage studies revealed proximity of the mouse *Aqp4* gene to *ataxia,* no differences in DNA coding sequence or protein levels of *ataxia* mice were noted when compared to unaffected wild-type mice (Turtzo *et al.,* 1997). Production of cerebrospinal fluid occurs in the choroid plexus of the lateral ventricles (reviewed by Lyons and Meyer, 1990; Segal, 1993), where AQP1 colocalizes with the sodium pump on apical membrane microvilli (Ernst *et al.,* 1986; Nielsen *et al.,* 1993b). In contrast, AQP4 resides in the basolateral membranes of ependymal cells lining the ventricles, implicating AQP4 in the reabsorption of cerebral spinal fluid (normally 140 mL in volume), which turns over four to five times a day (reviewed by Lyons and Meyer, 1990). Studies designed to evaluate whether AQP4 is involved in the pathogenesis or dissipation of normal pressure hydrocephalus or pseudotumor cerebri are currently under way. Of particular interest in all disorders related to *AQP4* is the significance of the alternative transcripts in the distribution and regulation of the protein (Lu *et al.,* 1996). Development of a mouse *Aqp4* gene knockout and use of anti-peptide antibodies may be helpful in identifying whether developmental abnormalities of water balance or brain edema in adults are associated with alterations in AQP4 mRNA or protein.

C. Other Tissues and Mutations

Like the eye and brain, tissues such as kidney and lung are known to express multiple aquaporins.

1. Kidney

Much of our current understanding of the pathophysiology of aquaporins derives from biophysical and immunohistochemical studies performed on the mammalian kidney, a site of abundant aquaporin expression (reviewed by Nielsen and Agre, 1995). The complex concentrating mechanisms of the adult kidney underscore the importance of the roles that aquaporins play in maintaining the body salt and water balance. Of the 200 liters of glomerular filtrate produced daily by an average adult human, approximately 80% is reabsorbed by the proximal tubule and thin limb of Henle's loop (Fig. 4), sites where AQP1 is abundant on both the apical and basolateral membranes (Nielsen *et al.*, 1993a, 1995b; Sabolic *et al.*, 1992). AQP1 in the apical membrane of proximal tubule epithelia provides a molecular explanation for the high osmotic water permeability ($P_f > 1200$ μm/sec), which cannot be accounted for by simple diffusion through the lipid bilayer (Maeda *et al.*, 1995). The tight junctions between cells in this region suggest that paracellular routes of fluid absorption are unlikely to play a significant role in this process. Absence of known aquaporins in the relatively water impermeable ascending thin and thick limbs, and the distal convoluted and connecting tubules explains the low P_f measured at these sites in the distal nephron.

The ~20% of glomerular filtrate that is not reabsorbed by the proximal kidney may be reabsorbed in the renal collecting duct when stimulated by vasopressin (Fig. 4). In the basal state, water permeability at the apical membrane of collecting duct principal cells is relatively low due to the absence of AQP2 at the apical cell membrane. In response to vasopressin, AQP2 is shuttled from subapical vesicles located within the cytoplasm of collecting duct principal cells to the apical cell membrane (DiGiovanni *et al.*, 1994; Nielsen *et al.*, 1995a). The molecular cascade responsible for this translocation is currently being delineated in several laboratories: (1) interaction of vasopressin with V2 receptors, (2) activation of adenylyl cyclase through a G-protein; (3) kinase A phosphorylation of a site on the carboxy-terminus of AQP2, (4) transit to the apical membrane in associa-

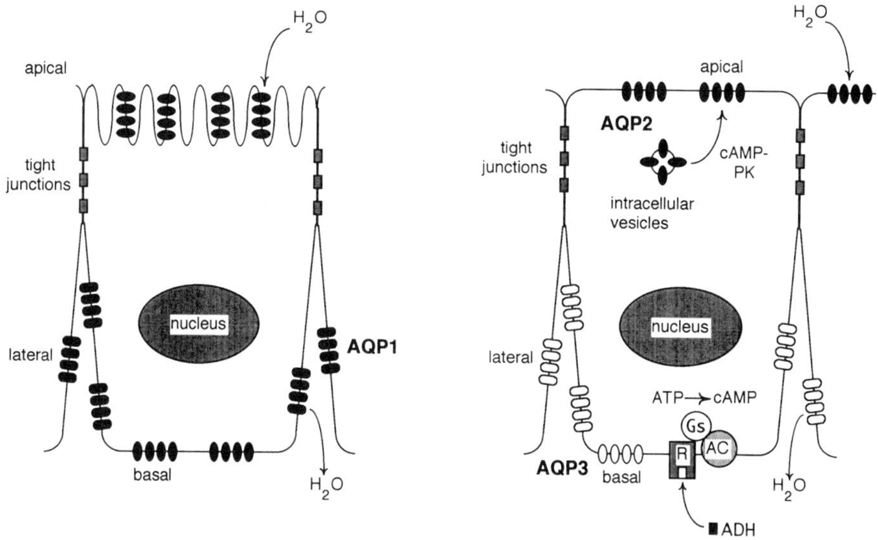

FIGURE 4 Models representing function of renal aquaporins. Shown is a proximal nephron epithelial cell (left) and collecting duct principal cell (right). In both examples, water moves in the direction of an osmotic gradient (from apical to basal) created by the vectorial arrangement of salt and sugar transporters (not shown). Thus AQP1 functions as a constitutively active water transport pathway, permitting the entrance and exit of water (left). In contrast, AQP2 resides in internal vesicles that are targeted to the cell surface in response to the vasopressin-regulated cascade. [Modified from Nielsen and Agre (1995).]

tion with proteins previously defined for their roles in synaptic vesicle membrane targeting (reviewed by Nielsen and Agre, 1995). Together these observations provide a molecular explanation for the "shuttle hypothesis" proposed more than a decade and a half ago to explain vasopressin regulation of water reabsorption in the renal collecting duct (Wade *et al.*, 1981). Water exits from the basolateral membrane of principal cells through AQP3 in most levels of collecting duct, or AQP4 in the inner medulla (Ecelbarger *et al.*, 1995; Terris *et al.*, 1995) (Fig. 4). Throughout the kidney, water is absorbed back into the vascular system through AQP1 present in endothelial cells of the vasa recta (Nielsen *et al.*, 1995b). Thus, multiple distinct mechanisms of aquaporin function are present within the kidney: (1) constitutive presence of AQP1 on both the apical and basolateral surfaces of renal proximal tubule epithelia, (2) hormonally regulated shuttling of AQP2 from subapical vesicles to the apical membrane of collecting duct principal cells, (3) outflow of water through AQP3 (or AQP4) in the basolateral membranes of collecting duct principal cells, and (4) uptake of water from the interstitium into the vasa recta.

The lack of a clinical phenotype associated with human AQP1 deficient individuals raises questions as to the importance of water channels in the mammalian kidney (Preston *et al.,* 1994b). In contrast to AQP1, individuals lacking AQP2 protein suffer from a severe form of nephrogenic diabetes insipidus confirming the suspicion that this protein is the vasopressin water channel (Deen *et al.,* 1994; van Lieburg *et al.,* 1994). Previously most cases of nephrogenic diabetes insipidus were found to result from an X-linked trait resulting from mutations in the V_2*-receptor* gene (Pan *et al.,* 1992; Rosenthal *et al.,* 1992). Several patients with nephrogenic diabetes insipidus that were found to have normal V_2*-receptor* genes are now known to have mutations in the *AQP2* gene resulting in defective AQP2 proteins that do not traffic normally or form functional water channels (Deen *et al.,* 1995). Additional evidence supporting the requirement for the AQP2 protein in concentrating the urine was provided by studies of kidney from the Brattleboro rat, a model for central diabetes insipidus (DiGiovanni *et al.,* 1994), and by examinations of urine from patients with nephrogenic diabetes insipidus or central diabetes insipidus (Kanno *et al.,* 1995). When compared to normal urine, AQP2 protein is markedly reduced in urine from either nephrogenic or central diabetes insipidus patients. As expected, administration of vasopressin increased urinary AQP2 concentrations only in the central diabetes insipidus patients but not in those with nephrogenic diabetes insipidus. Thus, AQP2 expression at the apical membrane of collecting duct principal cells is essential to the normal concentrating ability of the human kidney. Recently, several individuals were identified with nephrogenic diabetes insipidus resulting from mutations in the *AQP2* gene, but these cRNAs functioned normally when expressed in *Xenopus* oocytes (Mulders *et al.,* 1996b). These findings have raised additional questions about the precise targeting events necessary for the normal expression of AQP2.

Congenital nephrogenic diabetes insipidus is a rare disorder, but acquired forms of nephrogenic diabetes insipidus are frequently encountered in clinical practice. Approximately 1% of the U.S. public suffers from bipolar disorder (manic depressive illness), and lithium carbonate is the mainstay of treatment. Approximately half of the patients taking lithium have defective urine concentration, and inhibition of cAMP generation by collecting duct principal cells has been proposed as the mechanism behind this defect (Christensen *et al.,* 1985; Boton *et al.,* 1987). Chronic exposure of rats to lithium was found to dramatically reduce expression of AQP2, an effect partially corrected by vasopressin administration (Marples *et al.,* 1995). After cessation of lithium dosing, AQP2 levels increased and the renal concentrating mechanisms improved in a manner similar to that observed in humans (Howard *et al.,* 1992). Hypokalemia and bilateral ureteral obstruction have both been shown to induce water-losing nephropathy, and very recent studies have demonstrated a marked down-regulation of AQP2 protein expression in rat models of both disorders (Marples *et al.,* 1996; Frøkiær *et al.,* 1996). Future studies may uncover additional causes of acquired nephrogenic diabetes insipidus associated with down-regulation of AQP2 expression. Better understanding of the mechanisms involved in the regulation of *AQP2* gene transcription and AQP2 protein trafficking may assist investigators with the design of future strategies to treat nephrogenic diabetes insipidus.

Congestive heart failure, cirrhosis, and the syndrome of inappropriate antidiuretic hormone are among the most common causes of impaired water excretion (reviewed by Bichet *et al.,* 1992). In animal models of cirrhosis and congestive heart failure, AQP2 protein levels were found to be up-regulated (Asahina *et al.,* 1995; Xu *et al.,* 1996; Teitelbaum *et al.,* 1996; Ma and Lin, 1996). In contrast, increased vasopressin expression occurs in the syndrome of inappropriate antidiuretic hormone. Other more recent reports have implicated AQP2 in the pathophysiology of the altered water metabolism of pregnancy (Ohara *et al.,* 1996) and acclimation to high altitudes (Ramirez *et al.,* 1996), predicting that AQP2 expression may be commonly involved in disorders of water balance. Although most cases of impaired water excretion probably involve AQP2 as a secondary phenomenon, modulation of AQP2 expression may confer therapeutic benefits in some of these clinical settings.

Despite the separate distributions of AQP2 and AQP1 in kidney tubule epithelia, recent studies indicate that a complete understanding of the pathophysiology of some kidney diseases may involve a consideration of both AQP1 and AQP2 expression in concert. Autosomal dominant polycystic kidney disease is a genetic disease leading to the collection of fluid-filled cysts surrounded by tubular epithelium (Dalgard, 1957). Recent reports localized both AQP1 and AQP2 to autosomal dominant polycystic kidney disease cysts, although both proteins were never found within the same cyst (Bachinsky *et al.,* 1995; Devuyst *et al.,* 1996). Identification of AQP2 in tubule epithelium from renal cell tumors implicates the collecting duct epithelia in the origin of renal cell carcinoma (Kageyama *et al.,* 1996). Thus, AQP1 and AQP2 are useful markers that may enable investigators to better understand the etiology of abnormal cell differentiation as it relates to salt and water transport at sites other than the kidney.

2. Lung

The complex fluid movements occurring in the respiratory tract are believed to involve several different aquaporins acting in concert. The pathophysiology of several pulmonary disorders, including lung prematurity (De-

Lemos *et al.*, 1970; Liggins and Howie, 1972), cystic fibrosis (Boucher, 1994; Smith *et al.*, 1996), asthma (Anderson and Togias, 1994; Widdicombe, 1995), pulmonary edema (Staub *et al.*, 1967; Schoene *et al.*, 1994), and acute respiratory distress syndrome (reviewed by Matthay, 1996), are known to involve movements of water into or out of the upper and lower respiratory tracts. Four aquaporins have been identified in the respiratory tract of rat (AQP1, AQP3, AQP4, and AQP5), and the driving forces behind water movements are believed to include osmotic pressure and vapor pressure (reviewed by Lee *et al.*, 1997a,b). Osmotic gradients are generated in lung by ion channels, such as CFTR and sodium channels (reviewed by Boucher, 1994). Identification of multiple different aquaporins at several different cellular and subcellular locations throughout the pulmonary tree has suggested specific pathways by which water movement may contribute to the pathophysiology of pulmonary disease.

3. Other Tissues

In addition to the above-mentioned sites of expression, aquaporins have also been identified in other organ systems. After the original cloning of AQP1 from human red cells, speculation has focused on the role of AQP1 in hematological disease. Red cells in sickle cell disease are dehydrated compared to normal cells and exhibit decreased diffusional water permeability (Fung *et al.*, 1989; Joiner, 1993). Preliminary studies indicate decreased water permeability and decreased AQP1 content in red cells from a subset of sickle cell patients (Agre and Mathai, unpublished). Study of the mouse erythroleukemia cell line indicates that biosynthesis of AQP1 is stimulated by hydroxyurea, an agent that ameliorates the course of the disease by unknown mechanisms (Moon *et al.*, 1997). AQP1 is also present in the biliary epithelium (Nielsen *et al.*, 1993b; Roberts *et al.*, 1994) and in cultured cholangiocytes where secretin appears to increase the cell surface distribution of the protein (Marinelli *et al.*, 1997). AQP3, and AQP4 have been identified in colon (Frigeri *et al.*, 1995a), and a human AQP4 cDNA was recently cloned from a human stomach cDNA library (Misaka *et al.*, 1996). AQP1 has been identified in several other tissue sites, including the nonfenestrated capillary endothelium in skeletal, smooth, and cardiac muscle, as well in the submucosal space in gut and elsewhere (Nielsen *et al.*, 1993b). AQP1 was recently identified in cells surrounding male epididymis and inner ear (Brown *et al.*, 1993; Stankovic *et al.*, 95). These sites of expression are currently under investigation to determine the role of AQP1 expression in health and disease. Discovery of additional sites of aquaporin expression in the future will undoubtedly expand our knowledge of the aquaporins.

VI. SUMMARY

To date no mutations have been identified in *AQP3, AQP4,* or *AQP5.* The development of targeted gene disruptions in mice may provide clues to the importance of AQP4 in brain and AQP5 in diseases of glandular tissue, although it is not certain that instructive phenotypes will result from these studies. Because both AQP4 and AQP5 are first expressed postnatally, it is unlikely that the null phenotypes will be embryonic lethals (Jung *et al.,* 1994b; King *et al.,* 1997).

Identification of the aquaporin gene family has provided new insight into how water moves across the plasma membranes of many tissues and cell types. Definition of the gene structures for some of these water channels has established both primary and secondary roles for these proteins in health and disease. At present our understanding of the physiology of the aquaporins in the eye and other tissues remains incomplete. Additional aquaporin family members will likely be discovered in specific sites within the eye and other tissues, which will provide a better understanding of the complex pathways through which water is distributed in response to osmotic gradients. Thus, the aquaporins may be molecular answers to the etiology of important clinical problems. As outlined in this review, it is our challenge to identify in which of the many disease states aquaporins are involved, to define their involvement mechanistically, and to search for ways in which they may be exploited therapeutically.

References

Abrami, L., Berthonaud, V. Deen, P. M., Rousselet, G., Tacnet, F., and Ripoche, P. (1996). Glycerol permeability of mutant aquaporin 1 and other AQP-MIP proteins: Inhibition studies. *Pflügers Archiv. Eur. J. Phys.* **431,** 408–414.

Adams, R. D., and Victor, M. (1989). Disturbances of cerebrospinal fluid circulation, including hydrocephalus and meningeal reactions. *In* "Principles of Neurology," pp. 501–515. McGraw-Hill, New York.

Agre, P., Preston, G. M., Smith, B. L., Jung, J. S., Raina, S., Moon, C., Guggino, W. B., and Nielsen, S. (1993). Aquaporin CHIP, the archetypal molecular water channel. *Am. J. Physiol.* **265,** F463–F476.

Agre, P., Smith, B. L., Baumgarten, R., Preston, G. M., Pressman, E., Wilson P., Ilum, N., Anstee, D. N., Lande, M. B., and Zeidel, M. L. (1994). Human red cell aquaporin CHIP: II. Expression during normal fetal development and in a novel form of congenital dyserythropoietic anemia. *J. Clin. Invest.* **94,** 1050–1058.

Agre, P., Brown, D., and Nielsen, S. (1995). Aquaporin water channels: Unanswered questions and unresolved controversies. *Curr. Opin. Cell. Biol.* **7,** 472–483.

Agre, P., Lee, M. D., Devidas, S., and Guggino, W. B. (1997a). Aquaporins and ion conductance. *Science* **275,** 1490.

Agre, P., Lee, M. D., and Nielsen, S. (1997b). Aquaporin water channels in eye. *Keystone Symp. Ocular Cell Mol. Biol.* (abstract).

Anderson, S. D., and Togias, A. G. (1994). Dry air and hyperosmolar challenge in asthma and rhinitis. *In* "Asthma and Rhinitis" (W. W. Busse and S. T. Holgate eds.), pp. 1178–1195, Blackwell, Boston.

Asahina, Y., Izumi, N., Enomoto, N., Sasaki, S., Fushimi, K., Marumo, F., and Sato, C. (1995). Increased gene expression of water channel in cirrhotic rat kidneys. *Hepatology* **21,** 169–173.

Bachinsky, D. R., Sabolic, I., Emmanouel, D. S., Jefferson, D. M., Carone, F. A., Brown, D., and Perrone, R. D. (1995). Water channel expression in human ADPKD kidneys. *Am. J. Physiol.* **268,** F398–F403.

Bichet, D. G., Kluge, R., Howard, R. L., and Schrier, R.W. (1992). Hyponatremic states. *In* "The Kidney: Physiology and Pathophysiology" (D. W. Seldin and G. Giebisch, eds.), pp. 1727–1751. Raven Press, New York.

Bondy, C., Chin, E., Smith, B. L., Preston, G. M., and Agre, P. (1993). Developmental gene expression and tissue distribution of the CHIP28 water channel protein. *Proc. Natl. Acad. Sci. USA.* **90,** 4500–4504.

Boton, R., Gaviria, M., and Battle, D. C. (1987). Prevalence, pathogenesis, and treatment of renal dysfunction associated with chronic lithium therapy. *Am. J. Kidney Dis.* **10,** 329–345.

Boucher, R. C. (1994). Human airway ion transport: Parts I and II. *Am. J. Respir. Crit. Care Med.* **150,** 271–281, 581–593.

Brosius, F. C., Alper, S. L., Garcia, A. M., and Lodish, H. F. (1989). The major kidney band 3 gene transcript predicts an amino-terminal truncated band 3 polypeptide. *J. Biol. Chem.* **264,** 7784–7787.

Brown, D., Verbavatz, J. M., Valenti, B., Lui, B., and Sabolic, I. (1993). Localization of the CHIP28 water channel in reabsorptive segments of the rat male reproductive tract. *Eur. J. Cell Biol.* **61,** 264–273.

Brubaker, R. F. (1991). Flow of aqueous humor in humans. *Invest. Ophthalmol. Vis. Sci.* **32,** 3145–3166.

Bult, C. J., White, O., Olsen, G. J., Zhou, L., Fleischmann, R. D., Sutton, G. G., Blake, J. A., FitzGerald, L. M., Clayton, R. A., Gocayne, J. D., Kerlavage, A. R., Dougherty, B. A., Tomb, J. F., Adams, M. D., Reich, C. I., Overbeek, R., Kirkness, E. F., Weinstock, K. G., Merrick, J. M., Glodek, A., Scott, J. L., Geoghagen, S. M., Weidman, J. F., Fuhrmann, J. L., Nguyen, D., Utterback, R. R., Kelley, J. M., Peterson, J. D., Sadow, P. W., Hanna, M. C., Cotton, M. D., Roberts, K. M., Jurst, M. A., Kaine, B. P., Borodovsky, M., Klenk, H.-P., Fraser, C. M., Smith, H. O., Woese, C. R., and Venter, J. C. (1996). Complete genome sequence of the methanogenic archaeon, *Methanococcus jannaschii. Science* **273,** 1058–1073.

Calamita, G., Bishai, W. R., Preston, G. M., Guggino, W. B., and Agre, P. (1995). Molecular cloning and characterization of AqpZ, a water channel from *Escherichia coli. J. Biol. Chem.* **270,** 29063–29066.

Candia, O. A., and Zamudo, A. C. (1995). Chloride-activated water permeability in the frog corneal epithelium. *J. Membr. Biol.* **143,** 259–266.

Caprioli, J. (1992). The ciliary epithelium and aqueous humor. *In* "Adler's Physiology of the Eye" (W. M. Hart, Jr., ed.), pp. 228–247. Mosby, St. Louis.

Chandy, G., Kreman, M., Laidlaw, D. L., Zampighi, G. A., and Hall, J. E. (1995). The water permeability per molecule of MIP is less than that of CHIP. *Biophys. J.* **68,** 353 (abstract).

Chrispeels, M. J., and Agre, P. (1994). Aquaporins: Water channel proteins of plant and animal cells. *Trends Biochem. Sci.* **19,** 421–425.

Christensen, S., Kusano, E., Yusufi, A.N., Murayama, N., and Dousa, T.P. (1985). Pathogenesis of nephrogenic diabetes insipidus due to chronic administration of lithium in rats. *J. Clin. Invest.* **75,**1869–1879.

Collaborative Corneal Transplantation Studies Research Group (CCTS). (1992). Effectiveness of histocompatibility matching in high-risk corneal transplantation. *Arch. Ophthalmol.* **110,** 1392–1403.

Dalgard, O. Z. (1957). Bilateral polycystic disease of the kidneys: A follow-up study of 284 patients and their families. *Acta Med. Scand.* **S323,** 326–329.

Deen, P. M. T., Verdijk, M. A. J., Knoers, N. V. A. M., Wieringa, B., Monnens, L. A. H., van Os, C. H., and van Oost, B. A. (1994). Requirement of human renal water channel aquaporin-2 for vasopressin-dependent concentration of urine. *Science* **264,** 92–95.

Deen, P. M. T., Croes, H., van Aubel, R. A. M. H., Ginsel, L. A., and van Os, C. H. (1995). Water channels encoded by mutant aquaporin-2 genes in nephrogenic diabetes insipidus are impaired in their cellular routing. *J. Clin. Invest.* **95,** 2291–2296.

DeLemos, R. A., Shermata, D. W., Knelson, J. H., Kotas, R., and Avery. M. E. (1970). Acceleration of appearance of pulmonary surfactant in the fetal lamb by administration of corticosteroids. *Am. Rev. Resp. Dis.* **102,** 459–461.

Delporte, C., O'Connell, B. C., He, X., Ambudkar, I. S., Agre, P., and Baum, B. J. (1996). Adnovirus-mediated expression of aquaporin-5 in epithelial cells. *J. Biol. Chem.* **271,** 22070–22075.

Delporte, C., O'Connell, B. C., He, X., Lancaster, H. E., O'Connell, A. C., Agre, P., and Baum, B. J. (1997). Increased fluid secretion following adenoviral-mediated transfer of the aquaporin-1 cDNA to irradiated rat salivary glands. *Proc. Natl. Acad. Sci. U.S.A.* **94,** 3268–3273.

Denker, B. M., Smith, B. L., Kuhajda, F. P., and Agre, P. (1988) Identification, purification, and characterization of a novel M_r 28,000 integral membrane protein from erythrocytes and renal tubules. *J. Biol. Chem.* **263,** 15634–15642.

Devuyst, O., Burrow, C. R., Smith, B. L., Agre, P., Knepper, M. A., and Wilsonk, P. I. D. (1996). Expression of aquaporins 1 and 2 during human nephrogenesis and in autosomal dominant polycystic kidney disease. *Am. J. Physiol.* **271,** F169–F183.

DiGiovanni, S. R., Nielsen, S., Christensen, E. I., and Knepper, M. A. (1994). Regulation of collecting duct water channel expression by vasopressin in Brattleboro rat. *Proc. Natl. Acad. Sci. U.S.A.* **91,** 8984–8988.

Ecelbarger, C. A., Terris, J., Frindt, G., Echevarria, M., Marples, D., Nielsen, S., and Knepper, M. A. (1995). Aquaporin-3 water channel localization and regulation in rat kidney. *Am. J. Phys.* **269,** F663–F672.

Echevarria, M., Kuang, K., Preston, G. M., Agre, P., and Fischbarg, J. (1993). Cultured bovine corneal endothelial cells express CHIP28 water channels. *Am. J. Physiol. Cell* **265,** C1349–C1355.

Echevarria, M., Windhager, E. E., Tate, S. S., and Frindt, G. (1994). Cloning and expression of AQP3, a water channel from the medullary collecting duct of rat kidney. *Proc. Natl. Acad. Sci. U.S.A.* **91,** 10997–11001.

Echevarria, M., Windhager, E. E., and Frindt, G. (1996). Selectivity of the renal collecting duct water channel aquaporin-3. *J. Biol. Chem.* **271,** 25079–25082.

Ehring, G. R., Zampighi, G., Horwitz, J., Bok, D., and Hall, J. (1990). Properties of channels reconstituted from the major intrinsic protein of lens fiber membranes. *J. Gen. Physiol.* **96,** 631–664.

Ehring, G. R., Lagos, N., Zampighi, G. A., and Hall, J. E. (1991). Phosphorylation modulates the voltage dependence of channels reconstituted from the major intrinsic protein of lens fiber membranes. *J. Memb. Biol.* **126,** 75–88.

Ernst, S. A., Palacios, J. R., and Siegel, G. J. (1986). Immunocytochemical localisation of Na^+, K^+ ATPase catalytic polypeptide in mouse choroid plexus. *J. Histochem. Cytochem.* **34,** 189–195.

Finkelstein, A. (1987). "Water Movement Through Lipid Bilayers, Pores, and Plasma Membranes." Wiley, New York.

Fischbarg, J., Kuang, K., Hirsch, J., Lecuona, S., Rogozinski, L., Silverstein, S. C., and Loike, J. (1989). Evidence that the glucose transporter serves as a water channel in J774 macrophages. *Proc. Natl. Acad. Sci. U.S.A.* **86,** 8397–8401.

Fishman, R. A. (1995). Brain edema and disorders of intracranial pressure. *In* "Merrit's Textbook of Neurology" (L. P. Rowland, ed.), 9th ed. pp., 302–310, Williams & Wilkins, Baltimore.

Fox, R. I. (1995). Sjögren's syndrome. *Curr. Opin. Rheumatol.* **7,** 409–416.

Frigeri, A., Gropper, M. A., Turck, C. W., and Verkman, A. S. (1995a). Immunolocalization of the mercurial-insensitive water channel and glycerol intrinsic protein in epithelial plasma membranes. *Proc. Natl. Acad. Sci. U.S.A.* **92,** 4328–4331.

Frigeri, A., Gropper, M. A., Umenishi, F., Kawashima, M., Brown, D., and Verkman, A. S. (1995b). Localization of MIWC and GLIP water channel homologs in neuromuscular, epithelial and glandular tissues. *J. Cell. Sci.* **108,** 2993–3002.

Frøkiær, J., Marples, D., Knepper, M. A., and Nielsen, S. (1996). Bilateral ureteral obstruction downregulates expression of the vasopressin sensitive aquaporin-2 water channel in rat kidney. *Am. J. Physiol.* **270,** F657–F668.

Fung, L. W. M., Narasimhan, C., Lu, H. Z., and Westerman, M. P. (1989). Reduced water exchange in sickle cell anemia red cells: A membrane abnormality. *Biochim. Biophys. Acta* **982,** 167–172.

Fushimi, K., Shinichi, U., Hara, Y., Hiratya, Y., Marumo, F., and Sasaki, S. (1993). Cloning and expression of apical membrane water channel of rat kidney collecting tubule. *Nature* **361,** 549–552.

Gorin, N. B., Yancey, S. B., Cline, J., Revel, J. P., and Horwitz, J. (1984). The major intrinsic protein (MIP) of the bovine lens fiber membrane. *Cell* **39,** 49–59.

Griffin, C. S., and Shiels, A. (1992). Localisation of the gene for the major intrinsic protein of eye lens-fibre cell membranes to mouse chromosome 10 by *in situ* hybridisation. *Cytogenet. Cell Genet.* **59,** 300–302.

Harris, L., Agre, P., and Stark, W. (1996). Colton antigens and aquaporin-1 proteins: Implications in cornea allograft rejection. *55th Annu. Wilmer Residents Meeting* (abstract).

Hart, W. M. (1992). Intraocular pressure. *In* "Adler's Physiology of the Eye" (W. M. Hart, Jr., ed.), pp. 246–267. Mosby, St. Louis.

Hasegawa, H., Skach, W., Baker, O., Calayag, M. C., Lingappa, V., and Verkman, A. S. (1992). A multifunctional aqueous channel formed by CFTR. *Science* **258,** 1477–1479.

Hasegawa, H., Ma, T., Skach, W., Matthay, M. A., and Verkman, A. S. (1994). Molecular cloning of a mercurial-insensitive water channel expressed in selected water-transporting tissues. *J. Biol. Chem.* **269,** 5497–5500.

Heller, K. B., Lin, E. C. C., and Wilson, T. H. (1980). Substrate specificity and transport properties of the glycerol facilitator of *Escherichia coli. J. Bacteriol.* **144,** 274–278.

Henderson, P. J. F. (1991). Sugar transport proteins. *Curr. Opin. Struct. Biol.* **1,** 590–601.

Howard, R. L., Bichet, D. G., and Schrier, R. W. (1992). Hypernatremic and polyuric states. *In* "The Kidney: Physiology and Pathophysiology" (D. W. Seldin and G. Giebisch, eds.), pp. 1753–77, Raven Press, New York.

Inase, N., Fushimi, K., Ishibashi, K., Uchida, S., Ichioka, M., Sasaki, S., and Marumo, F. (1995). Isolation of human *aquaporin 3* gene. *J. Biol. Chem.* **270,** 17913–17916.

Ishibashi, K., Sasaki, S., Fushimi, K., Uchida, S., Kuwahara, M., Saito, H., Furukawa, T., Nakajima, K., Yamaguchi, Y., Gojobori, T., and Marumo, F. (1994). Molecular cloning and expression of a member of the aquaporin family with permeability to glycerol and urea in addition to water expressed at the basolateral membrane of kidney collecting duct cells. *Proc. Natl. Acad. Sci. U.S.A.* **91,** 6269–6273.

Jap, B. K., and Li, H. (1995). Structure of the osmo-regulated H_2O-channel, AQP-CHIP, in projection at 3.5 A resolution. *J. Mol. Biol.* **251,** 413–420.

Joiner, C. H. (1993). Cation transport and volume regulation in sickle red blood cells. *Am. J. Physiol.* **264,** C251–C270.

Jung, J. S., Preston, G. M., Smith, B. L., Guggino, W. B., and Agre, P. (1994a). Molecular structure of the water channel through aquaporin CHIP: The hourglass model. *J. Biol. Chem.* **269,** 14648–14654.

Jung, J. S., Bhat, R. V., Preston, G. M., Baraban, J. M., and Agre, P. (1994b). Molecular characterization of an aquaporin cDNA from brain: a candidate osmoreceptor and regulator of water balance. *Proc. Natl. Acad. Sci. U.S.A.* **91,** 13052–13056.

Kageyama, Y., Sasaki, S., Yamamura, Y., Oshima, H., and Ikawa, Y. (1996). Water channel protein subtype suggests the origin of renal cell carcinoma. *J. Urol.* **156,** 291–295.

Kanno, K., Sasaki, S., Hirata, Y., Ishikawa, S., Fushimi, K., Nakanishi, S., Bichet, D.G., and Marumo, F. (1995). Urinary excretion of aquaporin-2 in patients with diabetes insipidus. *N. Engl. J. Med.* **332,** 1575–1576.

Katsura, T., Ausiello, D. A., and Brown, D. (1996). The ser256 protein kinase A (PKA) phosphorylation site is required for regulated exocytosis of aquaporin-2 (AQP2) in transfected LLC-PK cells. *J. Am. Soc. Nephrol.* **7,** A0100.

King, L. S., and Agre, P. (1996). Pathophysiology of the aquaporin water channels. *Annu. Rev. Physiol.* **58,** 619–648.

King, L. S., Nielsen, S., and Agre, P. (1996). Aquaporin-1 water channel protein in lung: Ontogeny, steroid-induced expression, and distribution in rat. *J. Clin. Invest.* **97,** 2183–2191.

King, L. S., Nielsen, S., and Agre, P. (1997). Aquaporins in complex tissues: I. Developmental patterns in respiratory tract and glandular tissue of rat. *Am. J. Physiol.* in press.

Klatzo, I. (1994). Evolution of brain edema concepts. *Acta Neurochir.* **60,** Suppl., 3–6.

Knepper, M. A. (1994). The aquaporin family of molecular water channels. *Proc. Natl. Acad. Sci. U.S.A.* **91,** 6255–6258.

Kurtz, S. R., Kuszaj, T., Ouellet, R., and Valeri, C. R. (1982). Survival of homozygous Co^a (Colton) in a patient with anti-Co^{a1}. *Vox Sang.* **43,** 28–30.

Kushmerick, C., Rice, S. J., Baldo, G. J., Haspel, H. C., and Mathias, R. T. (1995). Ion, water and neutral solute transport in *Xenopus* oocytes expressing frog lens MIP. *Exp. Eye Res.* **61,** 351–362.

Kuwahara, M., Fushimi, K., Terada, Y., Bai, L., Marumo, F., and Sasaki, S. (1995). cAMP-dependent phosphorylation stimulates water permeability of aquaporin-collecting duct water channel protein expressed in *Xenopus* oocytes. *J. Biol. Chem.* **270,** 10384–10387.

Lee, M. D., Bhakta, K. Y., Raina, S., Yonescu, R., Griffin, C. A., Copeland, N. G., Gilbert, D. J., Jenkins, N. A., Preston, G. M., and Agre, P. (1996). The human *aquaporin-5* gene: Molecular characterization and chromosomal localization. *J. Biol. Chem.* **271,** 8599–8604.

Lee, M. D., King, L. S., and Agre, P. (1997a). The aquaporin family of water channel proteins in clinical medicine. *Medicine* **76,** 141–156.

Lee, M. D., King, L. S., Nielsen, S., and Agre, P. (1997b). Genomic organization and developmental expression of Aquaporin-5 in lung. *Chest* 1115–1135.

Li, J.-D., Govardovskii, B. K., and Steinberg, R. H. (1994). Light-dependent hydration of the space surrounding photopreceptors in the cat retina. *Vis. Neurosci.* **11,** 743–752.

Liggins, G. C., and Howie, R. N. (1972). A controlled trial of antepartum glucocorticoid treatment for prevention of the respiratory distress syndrome in premature infants. *Pediatrics* **50,** 515–525.

Linn, S. C., Kudrycki, K. E., and Shull, G. E. (1992). The predicted translation product of a cardiac AE3 mRNA contains an N terminus distinct from that of the brain AE3 Cl-/HCO3- exchanger. Cloning of a cardiac AE3 cDNA, organization of the AE3 gene, and identification of an alternative transcription initiation site. *J. Biol. Chem.* **267,** 7927–7935.

Lu, M., Lee, M. D., Smith, B. L., Jung, J. S., Agre, P., Verdijk, M. A. J., Merkx, G., Rijss, J. P. L., and Deen, P. M. T. (1996). The human *AQP4* gene: Definition of the locus encoding two water channel polypeptides in brain. *Proc. Natl. Acad. Sci. U.S.A.* **93,** 10908–10912.

Lyon, M. F., Jarvis, S. E., Sayers, I., and Holmes, R. S. (1981). Lens opacity: A new gene for congenital cataract on chromosome 10 of the mouse. *Genet. Res.* **38,** 337–341.

Lyons, M. K., and Meyer, F. B. (1990). Cerebrospinal fluid physiology and the management of increased intracranial pressure. *Mayo. Clin. Proc.* **65,** 684–707.

Ma, J., and Lin, S. (1996). Studies on AQP2 distribution and changes of renal AQP2 mRNA expression in nephrotic and heart failure rats. *J. Am. Soc. Nephrol.* **7,** A0105.

Ma, T., Frigeri, A., Skach, W., and Verkman, A. S. (1993). Cloning of a cDNA from rat kidney with homology to CHIP28 and WCH-CD water channels. *Biochim. Biophys. Res. Commun.* **197,** 654–659.

Ma, T., Frigeri, A., Hasegawa, H., and Verkman, A. S. (1994). Cloning of a water channel homolog expressed in brain meningeal cells and kidney collecting duct that functions as a stilbene sensitive glycerol transporter. *J. Biol. Chem.* **269,** 21845–21849.

Ma, T., Yang, B., Kuo, W.-L., and Verkman, A. S. (1996). cDNA cloning and gene structure of a novel water channel expessed exclusively in human kidney: Evidence for a gene cluster of aquaporins at chromosome locus 12q13. *Genomics* **35,** 543–550.

Macey, R. I., and Farmer, R. E. L. (1970). Inhibition of water and solute permeability in human red cells. *Biochim. Biophys. Acta* **211,** 104–106.

Maeda, N., and Smithies, O. (1986). The evolution of multigene families: Human haptoglobin genes. *Annu. Rev. Genet.* **20,** 81–108.

Maeda, Y., Smith, B. L., Agre, P., and Knepper, M. A. (1995). Quantification of aquaporin-CHIP water channel protein in microdissected renal tubules by fluorescence-based ELISA. *J. Clin. Invest.* **95,** 422–428.

Maguire, M. G., Stark, W. J., Gottsch, J. D., Stulting, R. D., Sugar, A., Fink, N. E., and Schwartz, A. (1994). Risk factors for corneal graft failure and rejection in the collaborative corneal transplantation studies. Collaborative Corneal Transplantation Studies Research Group. *Ophthalmology* **101,** 1536–1547.

Marinelli, R. A., Pham, L., Agre, P., and LaRusso, N. F. (1997). Secretin promotes osmotic water transport in rat cholangiocytes by increasing aquaporin-1 water channels in plasma membrane: Evidence for a secretin-induced vesicular translocation of aquaporin-1. *J. Biol. Chem.* **272,** 12984–12988.

Marples, D., Christensen, S., Christensen, E. I., Ottosen, P. D., and Nielsen, S. (1995). Lithium-induced downregulation of aquaporin-2 water channel expression in rat kidney medulla. *J. Clin. Invest.* **95,** 1838–1845.

Marples, D., Dørup, J., Knepper, M. A., and Nielsen, S. (1996). Hypokalemia-induced down-regulation of aquaporin-2 water channel expression in rat kidney medulla and cortex. *J. Clin. Invest.* **97,** 1960–1968.

Mathai, J. C., Mori, S., Smith, B. L., Preston, G. M., Mohandas, N., Collins, M., van Zijl, P. C. M., Zeidel, M. I., and Agre, P. (1996). Functional analysis of aquaporin-1 deficient red cells: The Colton-null phenotype. *J. Biol. Chem.* **271,** 1309–1313.

Matthay, M. A. (1996). The acute respiratory distress syndrome. *N. Engl. J. Med.* **334,** 1469–1470.

Maurel, C., Reizer, J., Schroeder, J. I., and Chrispeels, M. J. (1993) .The vacuolar membrane protein gamma-TIP creates water specific channels in *Xenopus* oocytes. *EMBO J.* **12,** 2241–2247.

Maurel, C., Reizer, J., Schroeder, J. I., Chrispeels, M. J., and Saier, M. H., Jr. (1994). Functional characterization of the *Escherichia coli* glycerol facilitator, GlpF, in *Xenopus* oocytes. *J. Biol. Chem.* **269,** 11869–11872.

Michea, L. F., de la Fuente, M., and Lagos, N. (1994). Lens major intrinsic protein promotes adhesion when reconstituted into large unilamellar liposomes. *Biochemistry* **33,** 7663–7669.

Milhorat, T. H. (1992). Classification of the cerebral edemas with reference to hydrocephalus and pseudotumor cerebri. *Child's Nerv. Syst.* **8,** 301–306.

Misaka, T., Abe, K., Iwabuchi, K., Yuko, K., Ichinose, M., Miki, K., Emori, Y., and Arai, S. (1996). A water channel closely related to rat brain aquaporin 4 is expressed in acid- and pesinigen-secretory cells of human stomach. *FEBS Lett.* **381,** 208–212.

Mitra, A. K., van Hoek, A. N., Wiener, M. C., Verkman, A. S., and Yeager, M. (1995). The CHIP28 water channel visualized in ice by electron crystallography. *Nature Struct. Biol.* **2,** 726–729.

Mollison, P. L., Engelfriet, C. P., and Contreras, M. (1987). Other red cell antigens. *In* "Blood Transfusion in Clinical Medicine," 8th ed., p. 399. Blackwell, Oxford.

Moon, C., Preston, G. M., Griffin, C. A., Jabs, E. W., and Agre, P. (1993). The aquaporin-CHIP28 gene: Structure, organization and chromosomal localization. *J. Biol. Chem.* **268,** 15772–15778.

Moon, C., King, L. S., and Agre, P. (1997). Expression of aquaporin-1 water channel protein in mouse erythroleukemia cells: Genetic regulation of corticosteroid and chemical induction. *Am. J. Physiol.* in press.

Muggleton-Harris, A. L., Festing, M. F. W., and Hall, M. (1987). A gene location for the inheritance of the *cataract Fraser* (*Cat*^{Fr}) mouse congenital cataract. *Genet. Res.* **49,** 235–238.

Mulders, S. M., Preston, G. M., Deen, P. M. T., Guggino, W. B., van Os, C. H., and Agre, P. (1995). Water channel properties of major intrinsic protein of lens. *J. Biol. Chem.* **270,** 9010–9016.

Mulders, S. M., Olde Weghuis, D., van Boxtel, J. A. F., Guerst van Kessel, A., Echevarria, M., van Os, C. H., and Deen, P. M. T. (1996a). Localization of the human gene for aquaporin 3 (AQP3) to chromosome 9, region p21Æp12, using fluorescent *in situ* hybridization. *Cytogenet. Cell Genet.* **72,** 303–305.

Mulders, S., Van Lieburg, A. F., Rijss, J. P. L., Knoers, N. V. A. M., Monnens, L. A. H., van Os, C. H., and Deen, P. M. T. (1996b). New mutant AQP2 proteins in nephrogenic diabetes insipidus confer water permeability in oocytes. *J. Am. Soc. Nephrol.* **7,** A0111.

Nielsen, S., and Agre, P. (1995). The aquaporin family of water channels in kidney. *Kidney Int.* **48,** 1057–1068.

Nielsen, S., Smith, B. L., Christensen, E. I., Knepper, M. A., and Agre, P. (1993a). CHIP28 water channels are localized in constitutively water-permeable segments of nephron. *J. Cell. Biol.* **120,** 371–383.

Nielsen, S., Smith, B. L., Christensen, E. I., and Agre, P. (1993b). Distribution of the aquaporin CHIP in secretory and resorptive epithelia and capillary endothelia. *Proc. Natl. Acad. Sci. U.S.A.* **90,** 7275–7279.

Nielsen, S., DiGiovanni, S. R., Christensen, E. I., Knepper, M. A., and Harris, H. W. (1993c). Cellular and subcellular immunolocalization of vasopressin-regulated water channel in rat kidney. *Proc. Natl. Acad. Sci. U.S.A.* **90,** 11663–11667.

Nielsen, S., Chou, C. L., Marples, D., Christensen, E. I., Kishore, B. K., and Knepper, M. A. (1995a). Vasopressin increases water permeability of kidney collecting duct by inducing translocation of Aquaporin-CD water channels to plasma membrane. *Proc. Natl. Acad. Sci. U.S.A.* **92,** 1013–1017.

Nielsen, S., Pallone, T., Smith, B. L., Christensen, E. I., Agre, P., and Maunsbach, A. B. (1995b). Aquaporin-1 water channels in short and long loop descending thin limbs and in descending vasa recta in rat kidney. *Am. J. Physiol.* **268,** F1023–F1037.

Nielsen, S., Nagelhus, E. A., Amiry-Moghaddam, M., Bourque, C., Agre, P., and Ottersen, O. P. (1997). Specialized membrane domains for water transport in glial cells: High resolution immunogold cytochemistry of aquaporin-4 in rat brain. *J. Neurosci.* **17,** 171–180.

Nishimoto, G., Nielsen, S., Yasui, M., Li, D., and Aperia, A. (1996). Arginine vasopressin (AVP) phosphorylates aquaporin-2 (AQP2) in rat renal papilla. *J. Am. Soc. Nephrol.* **7,** A0115.

Ohara, M., Martin, P. Y., Xu, D. L., St. John, J., Schrier, R. W. (1996). Expression of aquaporin-2 in altered water metabolism of pregnancy. *J. Am. Soc. Nephrol.* **7,** A0116.

Oliet, S. H. R., and Bourque, C. W. (1993). Mechanosensitive channels transduce osmosensitivity in supraoptic neurons. *Nature* **364,** 341–343.

Pan, Y., Metzenberg, A., Das, S., Jing, B., and Gitschier, J. (1992). Mutations in the V2 vasopressin receptor gene are associated with X-linked nephrogenic diabetes insipidus. *Nat. Genet.* **2,** 103–106.

Park, J. H., and Saier, M. H., Jr. (1996). Phylogenetic characterization of the MIP family of transmembrane channel proteins. *J. Membr. Biol.* **153,** 171–180.

Paterson, C. A., and Delamere, N. A. (1992). The lens. *In* "Adler's Physiology of the Eye" (W. M. Hart, Jr., ed.), pp. 348–390, Mosby, St. Louis.

Pisano, M. M., and Chepelinsky, A. B. (1991). Genomic cloning, complete nucleotide sequence, and structure of the human gene encoding the major intrinsic protein (MIP) of lens. *Genomics* **11,** 981–990.

Preston, G. M. (1993). Cloning by degenerative PCR. *In* "Methods in Molecular Biology" (B. A. White, ed.), Vol. 15, pp. 317–337. Humana Press, Totowa, NJ.

Preston, G. M., and Agre, P. (1991). Molecular cloning of the red cell integral protein of Mr 28,000: A member of an ancient channel family. *Proc. Natl. Acad. Sci. U.S.A.* **88,** 11110–11114.

Preston, G. M., Carroll, T. P., Guggino, W. B., and Agre, P. (1992). Appearance of water channels in *Xenopus* oocytes expressing red cell CHIP28 protein. *Science* **256,** 385–387.

Preston, G. M., Jung, J. S., Guggino, W. B., and Agre, P. (1993). The mercury-sensitive residue at cysteine-189 in the CHIP28 water channel. *J. Biol. Chem.* **268,** 17–20.

Preston, G. M., Jung, J. S., Guggino, W. B., and Agre, P. (1994a). Membrane topology of aquaporin CHIP: Analysis of functional epitope scanning mutants by vectorial proteolysis. *J. Biol. Chem.* **269,** 1668–1673.

Preston, G. M., Smith, B. L., Zeidel, M. L., Moulds, J. J., and Agre, P. (1994b). Mutations in *aquaporin-1* in phenotypically normal humans without functional CHIP water channels. *Science* **265,** 1585–1587.

Prockop, L. D. (1995). Hydrocephalus. *In:* "Merrit's Textbook of Neurology" (L. P. Rowland, ed.), 9th ed., pp. 294–302. Williams & Wilkins, Baltimore.

Raina, S., Preston, G. M., Guggino, W. B., and Agre, P. (1995). Molecular cloning and characterization of an aquaporin cDNA from salivary, lacrimal, and respiratory tissues. *J. Biol. Chem.* **270,** 1908–1912.

Ramirez, G., Pineda, D., Bittle, P. A., Rabb, H., Rosen, R., Vesely, D., and Sasaki, S. (1996). Partial renal resistance to vasopressin as an adaptation to high altitude living. *J. Am. Soc. Nephrol.* **7,** A0120.

Roberts, S. K., Yano, M., Ueno, Y., Pham, L., Alpini, G., Agre, P., and LaRusso, N. F. (1994). Cholangiocytes express the aquaporin CHIP and transport water via a channel-mediated mechanism. *Proc. Natl. Acad. Sci. U.S.A.* **91,** 13009–13013.

Rosenthal, W., Seibold, A., Antaramian, A., Lonergan, M., Arthus, M. F., Hendy, G. N., Birnbaumer, M., and Bichet, D. G. (1992). Molecular identification of the gene responsible for congenital nephrogenic diabetes insipidus. *Nature* **359,** 233–235.

Sabolic, I., Valenti, G., Verbavatz, J.-M., van Hoek, A., Verkman, A. S., Ausiello, D. A., and Brown, D. (1992). Localization of the CHIP28 water channel in rat kidney. *Am. J. Physiol.* **263,** C1225–C1233.

Saboori, A. M., Smith, B. L., and Agre, P. (1988). Polymorphism in the M_r 32,000 Rh protein purified from Rh(D)- positive and -negative erythrocytes. *Proc. Natl. Acad. Sci. U.S.A.*. **85,** 4042–4045.

Saito, F., Sasaki, S., Chepelinsky, A. B., Fushimi, K., Marumo, F., and Ikeuchi, T. (1995). Human AQP2 and MIP genes, two members of the MIP family, map within chromosome band 12q13 on the basis of two-color FISH. *Cytogenet. Cell Genet.* **68,** 45–48.

Schoene, R. B., Hackett, P. H., and Hornbein, T. F. (1994). High altitude. *In* "Textbook of Respiratory Medicine" (J. F. Murray, and J. A. Nadel, eds.), 2nd ed., pp. 2062–2098. W. B. Saunders, Philadelphia.

Segal, M. B. (1993). Extracellular and cerebrospinal fluids. *J. Inher. Metab. Dis.* **16,** 617–638.

Shiels, A., and Bassnett, S. (1996). Mutations in the founder of the MIP gene family underlie cataract development in the mouse. *Nat. Genet.* **12,** 212–215.

Siebinga, I., Vrensen, G. F. J. M., de Mul, F. F. M., and Greve, J. (1991). Age-related changes in local water and protein content of human eye lenses measured by Raman microspectroscopy. *Exp. Eye. Res.* **53,** 233–239.

Simpson, W. K. H. (1973). Anti-Co^a and severe haemolytic disease of the newborn. *S. Afr. Med. J.* **47,** 1302–1304.

Smith, B. L., and Agre, P. (1991). Erythrocyte M_r 28,000 transmembrane protein exists as a multi-subunit oligomer similar to channel proteins. *J. Biol. Chem.* **266,** 6407–6415.

Smith, B. L., Preston, G. M., Spring, F. A., Anstee, D. J., and Agre, P. (1994). Human red cell aquaporin CHIP: I. Molecular characterization of ABH and Colton blood group antigens. *J. Clin. Invest.* **94,** 1043–1049.

Smith, J. J., Travis, S. M., Greenberg, E. P., and Welsh, M. J. (1996). Cystic fibrosis airway epithelia fail to kill bacteria because of abnormal airway surface fluid. *Cell* **85,** 229–236.

Solomon, A. K., Chasan, B., Dix, J. A., Lukacovic, M. F., Toon, M. R., and Verkman, A. S. (1983). The aqueous pore in the red cell membrane. *Ann. N.Y. Acad. Sci.* **414,** 97–124.

Stamer, W. D., Snyder, R. W., Smith, B. L., Agre, P., and Regan, J. W. (1994). Localization of aquaporin CHIP in the human eye: Implications in the pathogenesis of glaucoma and other disorders of ocular fluid balance. *Invest. Ophthalmol. Vis. Sci.* **35,** 3867–3872.

Stankovic, K. M., Adams, J. C., and Brown, D. (1995). Immunolocalization of aquaporin CHIP in the guinea pig inner ear. *Am. J. Physiol.* **269,** C1450–1456.

Staub, N. C., Nagano, H., and Pearce, M. L. (1967). Pulmonary edema in dogs, especially the sequence of fluid accumulation in lungs. *J. Appl. Physiol.* **22,** 227–240.

Teitelbaum, I., Strasheim, A., and McGuinness, S. (1996). Decreased aquaporin (AQP)-2 content in chronic renal failure (CRF). *J. Am. Soc. Nephrol.* **7,** A0128.

Terris, J., Ecelbarger, C. A., Marples, D., Knepper, M. A., and Nielsen, S. (1995). Distribution of aquaporin-4 water channel expression within rat kidney. *Am. J. Physiol.* **269,** F775–F785.

Turtzo, L. C., Lee, M. D., Lu, M., Smith, B. L., Copeland, N. G., Gilbert, D. J., Jenkins, N. A., and Agre, P. (1997). Cloning and chromosomal localization of mouse *Aquaporin-4:* Exclusion of a candidate mutant phenotype, *Ataxia. Genomics* **41,** 267–270.

Uchida, S., Sasaki, S., Fushimi, K., and Marumo, F. (1994). Isolation of the aquaporin-CD gene. *J. Biol. Chem.* **269,** 23451–23455.

van Hoek, A. N., Hom, M. L., Luthjens, L. H., de Jong, M. D., Dempster, J. A., and van Os, C. H. (1991). Functional unit of 30 kDa for proximal tubule water channels as revealed by radiation inactivation. *J. Biol. Chem.* **266,** 16633–16635.

van Hoek, A. N., Luthjens, L. H., Hom, M. L., van Os, C. H., and Dempster, J. A. (1992). A 30 kDa functional size for the erythrocyte water channel determined by *in situ* radiation inactivation. *Biochem. Biophys. Res. Comm.* **184,** 1331–1338.

van Lieburg, A. F., Verdijk, M. A. J., Knoers, V. V. A. M., van Essen, A. J., Proesmans, W., Mallmann, R., Monnens, L. A., van Oost, B. A., van Os, C. H., and Deen, P. M. (1994). Patients with autosomal nephrogenic diabetes insipidus homozygous for mutations in the aquaporin-2 water-channel gene. *Am. J. Hum. Genet* **55,** 648–652.

van Os, C. H., Deen, P. M. T., and Dempster, J. A. (1994). Aquaporins: water selective channels in biological membranes. *Biochim. Biophys. Acta* **1197,** 291–309.

Verbavatz, J. M., Brown, D., Sabolic, I., Valenti, G., Ausiello, D. A., Van Hoek, A. N., Ma, T., and Verkman, A. S. (1993). Tetrameric assembly of CHIP28 water channels in liposomes and cell membranes: A freeze-fracture study. *J. Cell. Biol.* **123,** 605–618.

Wade, J. B., Stetson, D. L., and Lewis, S. A. (1981). ADH action: Evidence for a membrane shuttle hypothesis. *Ann. N.Y. Acad. Sci.* **372,** 106–117.

Walz, T., Smith, B. L., Zeidel, M. L., Engel, A., and Agre, P. (1994a). Biologically active two-dimensional crystals of Aquaporin CHIP. *J. Biol. Chem.* **269,** 1583–1586.

Walz, T., Smith, B. L., Agre, P., and Engel A. (1994b). The 3–D structure of human erythrocyte Aquaporin CHIP. E*MBO J.* **13,** 2985–2993.

Walz, T., Typke, D., Smith, B. L., Agre, P., and Engel, A. (1995). Projection map of aquaporin-1 determined by electron crystallography. *Nat. Struct. Biol.* **2,** 730–732.

Walz, T., Tittmann, P., Fuchs, K. H., Müller, D. J., Smith, B. L., Agre, P., Gross, H., and Engel. A. (1996). Surface topographies at subnanometer resolution reveal asymmetry and sidedness of aquaporin-1. *J. Mol. Biol.* **264,** 907–918.

Wang, S., Schultheis, P. J., and Shull, G. E. (1996). Three N-terminal variants of the AE2 Cl-/HCO3 exchanger are encoded by mRNAs transcribed from alternative promoters. *J. Biol. Chem.* **271,** 7835–7843.

Widdicombe, J. H. (1995). Structure and function of epithelial cells in controlling airway-lining fluid. *In* "Asthma and Rhinitis" (W. W. Busse and S. T. Holgate, eds.), pp. 565–571. Blackwell Scientific, Boston.

Xu, D. L., Martin, P. Y., Ohara, M., St. John, J., Pattison, T., and Schrier, R. W. (1996). Upregulation of aquaporin-2 water channel expression in the conjestive heart failure rat. *J. Am. Soc. Nephrol.* **7,** A0134.

Yool, A. J., Stamer, D., and Regan, J. W. (1996). Forskolin stimulation of water and cation permeabilility in aquaporin-1 water channels. *Science* **273,** 1216–1218.

Yoon, S. J., LeBlanc-Straceski, J., Ward, D., Krauter, K., and Kucherlapti, R. (1994). Organization of the human keratin type II gene cluster at 12q13. *Genomics* **24,** 502–508.

Zampighi, G. A., Hall, J. E., Ehring, G. R., and Simon, S. A. (1989). The structural organization and protein composition of lens fiber junctions. *J. Cell. Biol.* **108,** 2255–2275.

Zeidel, M. L., Ambudkar, S. V., Smith, B. L., and Agre, P. (1992). Reconstitution of functional water channels in liposomes containing purified red cell CHIP28 protein. *Biochemistry* **31,** 7436–7440.

Zeidel, M. L., Nielsen, S., Smith, B. L., Ambudkar, S. V., Maunsbach, A. B., and Agre, P. (1994). Ultrastructure, pharmacologic inhibition, and transport-selectivity of Aquaporin CHIP in proteoliposomes. *Biochemistry* **33,** 1606–1615.

Zelinski, T., Kaita, H., Gilson, T., Coghlan, G., Philips, S., and Lewis, M. (1990). Linkage between the Colton blood group locus and ASSP11 on chromosome 7. *Genomics* **6,** 623–625.

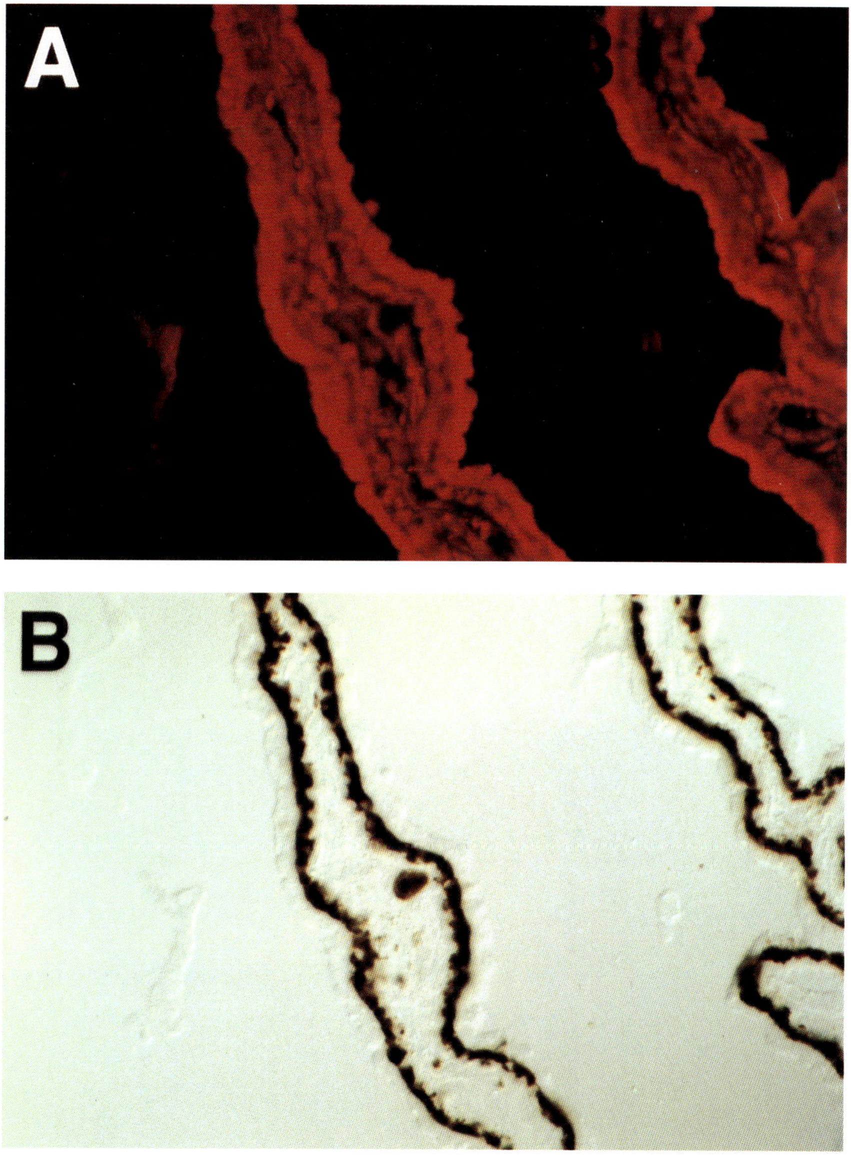

FIGURE 8 Indirect immunofluorescence of human ciliary epithelium (pars plicata) exposed to purified antiserum. (A) Strong fluorescence occurs in the nonpigmented inner layer of ciliary epithelium and fluorescence in the stroma can be attributed to reaction in the vascular endothelium. Pigment in the pigment cell layer may have partially quenched the reaction here. (B) Nomarski optics of specimen prepared for indirect immunofluorescence. [Reprinted from Wan *et al.* (1997) with permission from Academic Press.]

CHAPTER 6

Gap Junctions and Interlayer Communication in the Heterocellular Epithelium of the Ciliary Body

J. Mario Wolosin* and Michael Schütte†

*Departments of Ophthalmology, Physiology, and Biophysics, †Departments of Ophthalmology and Physiology, Mount Sinai School of Medicine, New York, New York 10029

I. INTRODUCTION

Aqueous humor is generated across the dual-layered epithelium that lines the surface of the ciliary body. The dominant anatomical feature of this epithelium is its unique arrangement in which the two distinct cell layers tightly oppose each other at their apical domains. The basolateral membrane of the pigmented epithelial (PE) layer faces the blood circula-

1063-5823/98 $25.00

tion; the basal membrane of the cells of the nonpigmented epithelial (NPE) layer abuts, contralaterally, on a prominent basement membrane that separates the ciliary body from the aqueous compartment. This unique arrangement originates from the folding of the optical vesicle during embryogenesis (Fig. 1).

An additional, critical aspect of the ciliary body anatomy is the exclusive location of an epithelial tight junctional seal at the apicolateral interface of the NPE cells (Bairati and Orzalesi, 1966); PE cells do not possess this membrane specialization. This feature establishes two functional domains within the context of a paradigm for simple secretory epithelia. The first domain, which can be equated with the basolateral membrane of a simple epithelial monolayer, comprises the entire PE cell membrane along with the apical membrane of the NPE cells (Fig. 2). The second domain, which within this formalism will correspond to the apical membrane of this simple epithelium, consists of the basolateral membrane of the NPE. The distinct transport functions of these two domains, parceled within the two cell layers, interact with each other to establish a net, energy-driven transport of salt, organic solutes, and water from the serosa to the aqueous chamber by mechanisms that have not yet been fully elucidated.

The relative contribution of the apical NPE membrane and of the PE cells to the system exchanges with the serosal surface is a function of two

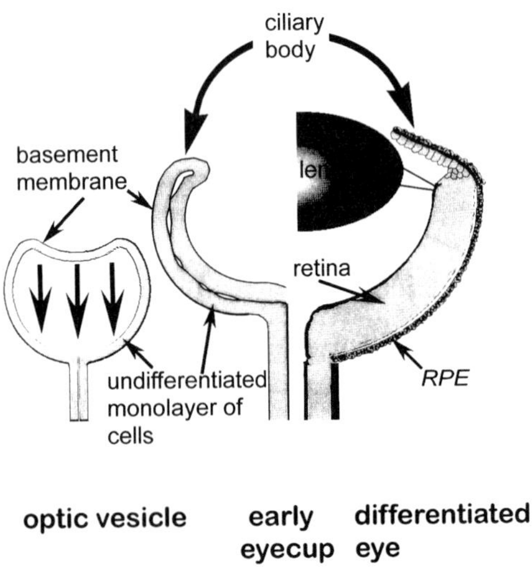

FIGURE 1 Illustration of the developmental origin of the CBE. Note that the basement membrane lines the outside of both layers, thus defining the polarity of both epithelial layers.

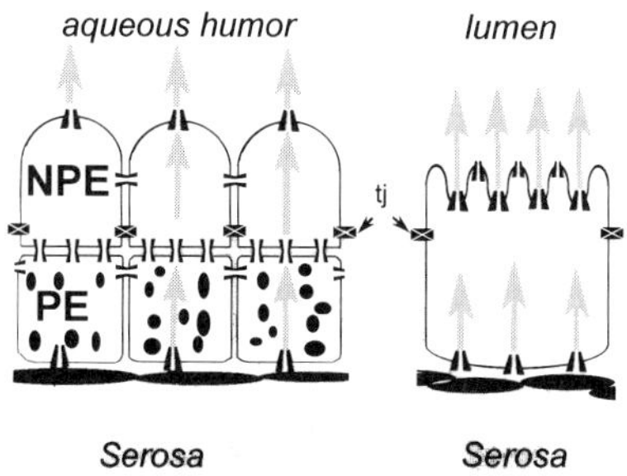

FIGURE 2 Domain relationships between the ciliary body epithelium and a generic, simple secretory epithelium. The location of tight junctions (tj) defines the separation between serosa and secretory compartments (aqueous humor or lumen, respectively).

parameters. The first relates to the intrinsic transport properties of each of these two subdomains, both in terms of individual transporting units present and their relative rate of turnover. The second parameter is the extent of transduction or translocation of activity between both cell types, that is, the extent and nature of the cell-to-cell communication between them. The anatomical correlate of this interlayer communication is the pronounced presence of gap junctions connecting the two distinct layers. These gap junctions are unique because of their apical rather than basolateral location within the cell membrane. Their molecular composition, physiological significance, and pharmacological modulation are the primary issue of this review.

II. THE GAP JUNCTION

A. Historical Overview

In regions of cell–cell contacts, cellular plasma membranes show ultrastructural specializations that distinguish these areas from the structure of the general plasma membrane. When the plasma membrane specialization of one cell meets a matching counterpart in an adjacent cell, "cell-to-cell" junctions are established. Until the general use of electron microscopy, no clear view of the ultrastructural properties of the majority of these cell junctions was possible. Owing to the limitations of light microscopy, the existence of cytoplasmic bridges in areas of cell-to-cell contact had been postulated (Zimmermann and von Palczewska, 1910). Only after the introduction of electron microscopy did it become clear that the plasma membrane of individual cells was continuous, dispelling the hypothesis of a supracellular cytosol (Fawcett, 1958).

Dewey and Barr (1962) were the first to describe a novel structure in smooth muscle cells of the jejunum which they called *nexus.* Farquhar and Palade (1963), in their landmark study, proposed a new terminology for cell junctions that was based on the width of the interspace between the adjacent plasma membranes. The term *occludens* was coined to describe junctions in which the membranes are in contact and occlude the extracellular space and the term *adherens* for junctions in which the membranes were merely adherent but not occluding the extracellular space. Simultaneously with Farquhar and Palade, Robertson (1963) described the presence of hexagonally packed subunits within electrical synapses in the goldfish brain. Later, Revel and Karnovsky (1967) showed that the category *macula occludens* actually consisted of two different junctional specializations, one the tight junction impermeable to colloidal lanthanum hydroxide, the other named the *gap junction* for the ability of the colloid to fill the intermembrane gap (Revel et al., 1967).

Unambiguous identification of gap junctions was achieved after the introduction of the freeze-fracture technique (McNutt and Weinstein, 1970, 1973; Wolburg and Rohlmann, 1995). Freeze-fracture experiments showed hexagonally arrayed cobblestone-like membrane particles in junctional plaques and also allowed a quantitative assessment of their distribution.

From a physiologic perspective, Furshpan and Potter (1959) were the first to report electrical coupling between two cells, the giant and motor neurons of the crayfish. Likewise, electrical coupling in vertebrate systems was demonstrated by Robertson (1963). However, the connection between the gap junction and the electrical synapse was not drawn until several years later by Revel and Sheridan (1968). More detailed comparison of electrophysiologic and ultrastructural data revealed a correlation between the particle (channel) density and the degree of electrical coupling (Pappas *et al.,* 1971; Revel *et al.,* 1971).

B. Physiology

The electrical properties of gap junctions have been investigated by several methods, including dual or multielectrode electrophysiologic recordings from different cells in intact tissue (Bennett, 1966) or pairs of isolated cells (Brightman and Reese, 1969). Another approach for the study of cell coupling is based on the capability of gap junctions to transfer solutes whose molecular weight is below 1.2 kDa from one cell to another (Loewenstein and Kanno, 1964; Lowenstein, 1981). Traceable solutes are introduced to a cellular point of origin by either intracellular injection of individual cells or by mechanical scraping (or cutting) of tissues to generate an array of source cells from which the tracer will diffuse onward. The rapidity or distance of spreading is then used to evaluate cell-to-cell communications. Several low molecular weight fluorescent dyes (e.g., Lucifer Yellow) or other tracers whose location could be established following tissue fixation (e.g., neurobiotin) have been shown to pass through gap junctions.

Factors regulating the permeability of gap junctions identified by the techniques mentioned previously comprise Ca^{2+} (Loewenstein *et al.,* 1967; Rose and Loewenstein, 1975), other divalent and trivalent cations (Oliveira-Castro and Loewenstein, 1971), pH (Peracchia and Peracchia, 1980) anesthetics such as n-alkanols (Johnston *et al.,* 1980; Bernardini *et al.,* 1984, Bastide *et al.,* 1995), and halothane (Burt and Spray, 1989), all of which inhibit permeability. Nitric oxide, however, has been alternatively postulated to act as a permeability enhancer (Loessberg-Stauffer *et al.,* 1993) or as an inhibitor (Murakami *et al.,* 1995). A direct inhibition effect of Ca^{2+} at the physiologically relevant range ($<1\ \mu M$) has been vigorously questioned (Spray *et al.,* 1982). Ca^{2+} effects may be mediated by calmodulin or calmodulin-like sequences (Peracchia, 1989). In addition to these effects related putatively to physical interactions, the permeability of individual gap junctions and/or of the overall junctional permeability of an electrical synapse may be controlled at the cellular level, through processes such as phosphorylation and/or membrane recycling initiated in response to neurotransmitters, hormones, and/or growth factors (Saez *et al.,* 1986; Piccolino *et al.,* 1984; Kurz-Isler and Wolburg, 1986; Murray and Gainer, 1989; Wolosin, 1991).

In spite of their high molecular weight cutoff, gap junctions exhibit many properties associated with membrane ion channels such as voltage gating (Giaume and Korn, 1984), rectification (Giaume and Korn, 1983; Margiotta and Walcott, 1983) and, with respect to organic solutes, charge and/or shape selectivity. For example, Lucifer Yellow and cadaverin, two substances with almost identical molecular weights, display very different intercellular transfer properties in amacrine cells (Mills and Massey, 1995). Relatively little is known about the signals mediating spreading of second messengers. In a recent series of elegant studies, Sáez *et al.* (1989) and later Sanderson and coworkers (Boitano *et al.,* 1992; Hansen *et al.,* 1995) found evidence for inositol trisphosphate (IP_3) as a critical intercellular messenger for the propagation of Ca^{2+} mobilization signals.

C. Structural and Biochemical Properties

During the last decades, numerous studies have been undertaken to unravel the molecular, structural, and biophysical properties of gap junctions and their identification by immunocytochemical methods. In short, gap junctions are established by fusion of two hemichannels (also called *connexons*) in apposed membrane areas of two adjacent cells. Each connexon consists of six individual subunits that belong to an extended family of 12 polypeptides with approximately 59% sequence homology called *connexins*. The different polypeptides are identified by their specific molecular weight, for example, Cx43 refers to a connexin with a molecular weight of 43 kDa (Hertzberg, 1985; Bennett *et al.,* 1991; Kumar and Gilula, 1992; Wolburg and Rohlmann, 1995; Goodenough *et al.,* 1996).

Connexins are characterized by the similarity in their core amino acid sequence, which results in an almost identical tertiary structure of all known subtypes. All connexins possess four helical transmembrane segments connected by two extracellular (E1 and E2) and one intracellular loop; the N and C termini both protrude into the intracellular space (Milks *et al.,* 1988; cf. Yeager and Glilula, 1992). The third transmembrane domain (M3) is amphipathic and thus anchors the entire protein in the membrane, which, through its bilipid character, stabilizes the complex (Malewicz *et al.,* 1990).

Distinct intra- or extracellular regions confer specific properties to eachconnexin. The intracellular side of the connexin forms the "pore" of the gap junction and determines the permeability of the *nexus*. Gating properties appear to be mediated by a "blocking structure" (Makowski, 1985) contained within the C-terminal arm (Peracchia and Girsch, 1985; Ek-Vitorín *et al.,* 1996). Hemichannel docking occurs between the extracellular loops through noncovalent bonds at highly conserved cysteine motives (Peracchia and Peracchia, 1985; Milks *et al.,* 1988; Goodenough *et al.,* 1988; John and Revel, 1991). Channel diameter and the ionic charge of the amino acids constituting the intracellular loop and both termini set the limitation and selectivity for permeability of charged solutes (Veenstra *et al.,* 1995).

The connexin diversity may also underpin a diversity in their response to pharmacological regulation and thereby allow selective responses to second messengers within single cells. For example, the homocellular gap junctions between retinal A II amacrine cells are uncoupled by cAMP, whereas the heterocellular gap junctions between the same A II amacrine cell and a cone bipolar cell are closed by cGMP (Hampson *et al.,* 1992; Mills and Massey, 1995). The structural compatibility of the extracellular domains, and particularly of domains E2, appears to be the critical determinator for the ability of connexins to dock onto an apposing counterpart (Bruzzone *et al.,* 1993; White *et al.,* 1994, 1995; Elfgang *et al.,* 1995). Depending on compatibility among each other and subsequent sequence similarities in their extracellular domains, Kumar and Gilula (1992) proposed a nomenclature that divides all known connexins into two families, α and β connexins, respectively. Generally, within each of the two families, there appears to be a fair degree of compatibility, whereas the compatibility between α and β connexins is only marginal.

On consideration of their molecular diversity and quaternary structure, gap junctions could be built according to four different blueprints: (1) homomeric homotypic (all six subunits of a connexon are identical and both connexons are identical); (2) heteromeric homotypic (connexons are identical but composed of different subunits); (3) homomeric heterotypic (the channel consists of two different connexons); or (4) heteromeric heterotypic. Heteromeric connexons have been recently identified in a number of tissues and cells (Sosinsky, 1995; Stauffer, 1995; Konig and Zamphigi, 1995; Jiang and Goodenough, 1996). In addition, many studies have generated heterotypic junctions in model systems and characterized their unique properties, including electrical assymetry (Werner *et al.,* 1989; Barrio *et al.,* 1991). At this point, however, there appears to be no hard evidence for the natural occurrence of heterotypic gap junctions (Kumar and Gilula, 1996).

Degradation of most gap junctions follows a conventional lysosomal pathway after endocytosis of entire junctional plaques, which includes fractions of the cytoplasm of the apposing cell (Vaughan and Lasater, 1992). The exception to the rule appears to be Cx43, the degradation of which is mediated through an ubiquitin–proteasomal pathway (Laing and Beyer, 1995).

III. GAP JUNCTIONS OF THE CILIARY BODY

A. Electron Microscopy

In the ciliary body epithelium, membrane specializations conforming to the picture of gap junctions were already described by Bairati and Orzalesi (1966) as *zonae occludentes,* according to the old nomenclature of Farquhar and Palade (1963). In retrospect, however, it is clear that Bairati and Orzalesi indeed observed and documented gap junctions connecting the two epithelial layers. In subsequent studies (Rentsch, 1970; Raviola, 1971, 1974; Smith and Rudt, 1973; Ohkuma and Nishiura, 1974; Kogon and Pappas, 1975; Hirsch *et al.,* 1977; Dabagian *et al.,* 1979; Ober and Rohen, 1979; Ganieva and Dabagian, 1980; Okinami *et al.,* 1980), more gap junctions were documented in various aspects of the ciliary body epithelium, however, for some time, there was still no clear agreement as to their exact distribution among the individual layers. In particular, gap junctions between adjacent NPE cells appeared to be difficult to observe (Reale and Spitznas, 1975) or were not found altogether (Kogon and Pappas, 1975). This inability to demonstrate homocellular NPE gap junctions may have been caused by their rapid endocytosis under anoxic conditions (see later discussion). Nevertheless, in two systematic studies, Reale and Spitznas (1975) and Raviola and Raviola (1978) laid the foundations of our current understanding of the gap junctional connectivity of the ciliary body epithelium (CBE). The outcome of both studies demonstrated the ubiquitous interconnection of the entire CBE, which includes homocellular junctions between adjacent PE cells as well as NPE cells and, in addition, heterocellular junctions between PE and NPE cells. Junctional complexes between the NPE cells appear to have a uniform density throughout all portions of the ciliary body if one allows for quantitative changes dictated by the varying degree of basal and lateral infoldings and regional changes in cell morphology. Between pigmented cells, gap junctional particles are often found in an unusual arrangement, concentrated in parallel rows near the apical surface. This particular pattern is found in all parts of the ciliary body from the pars pliccata to the transitional zone between the CBE and the retina (Shabo and Maxwell, 1973, referred to by the authors as *zomulae occludentes*) and extends farther into the retinal pigmented epithelium (Hudspeth and Yee, 1973; Reale *et al.,* 1974).

Of particular interest are the gap junctions at the NPE–PE interface since their location in the apical membrane seems to ignore the rules that, in other epithelial cells, tie gap junctions to the basolateral membranes. In addition, their plentitude appears to exceed the number of homocellular gap junctions between either PE or NPE cells (Raviola and Raviola, 1978). Furthermore, this particular class of gap junctions is easily accessible to freeze-fracture experiments because of the fairly high straight apical surfaces of both epithelia. Accordingly, the heterocellular interface has been favored for the study of gap junctions of the ciliary epithelium and, among the three different types, the heterocellular gap junctions are probably the best described. In their morphology, as revealed by freeze etching, they strongly resemble the homocellular gap junctions between the PE cells, and there appears to be an increase in size of individual junctional plaques and thus of the total number of channels toward the ora serrata (Reale and Spitznas, 1975; Ober and Rohen, 1979). In addition, in the pars pliccata, local gradients in the number of heterocellular gap junctions have been reported by Ober and Rohen (1979) with peak values at the ridges and the lowest number in the valleys. Similarly, in the spiny dogfish, Flügel *et al.* (1989) reported an increase of gap junctions toward pars plana. In addition to the plasma membrane, gap junction plaques have been found to be contained in partially digested vesicles within the PE cytosol and, less frequently, within the NPE of both rabbit and rat (Tenkova and Chaldakov, 1990).

B. Connexin Distribution in the Ciliary Body

Coca-Prados *et al.* (1992) were the first to conduct a systematic investigation regarding the molecular identity of the connexins involved in the formation of gap junctions in the ciliary body epithelium. The studies were carried out in fresh human and bovine tissue as well as in cultured cells. Out of four transcripts probing for the α_1, α_3, β_1, and β_2 (Cx43, Cx46, Cx32, and Cx26, respectively) gene products in Northern blots, only the α_1 transcript was identified within the whole tissue. The same transcript was also easily identified in primary cell cultures of human and bovine PE, whereas in cultures of human nonpigmented cells, an α_1 mRNA signal was detectable only after prolonged development of the blots. In addition, the transcript identified in NPE cells showed a molecular weight slightly below that detected in the intact tissue and the PE cell cultures. Immunocytochemistry using an antibody against Cx43 revealed that most of the putative connexin was accumulated at the PE–NPE border of the bovine ciliary epithelium, as well as the lateral flanks of the PE cells proximal to the heterocellular interface. In primary confluent monolayers of PE cells, Cx43-like immunoreactivity revealed an abundance of large *puncta* at cell-to-cell contacts. In contrast, in NPE cell cultures, only a tenuous stain consisting of minute *puncta* was discernible in some cells. These results led the authors to believe that the α_1 protein may be present in both cells, with high levels of mRNA and Cx43 protein in the PE cells and only marginal levels of both in the NPE. It also was necessary for the authors to postulate exclusive confinement of the protein in the NPE to the apical (PE-facing) membrane domain. The spatial constraint against diffusion was thought to be provided by the tight junctional ring girdling the apicolateral shoulder of NPE cells. The presence of Cx43 at the PE–NPE border was also observed in rat (Shin *et al.,* 1996).

Several considerations recently led us to reexamine the connexin distribution within the CBE. First, as described under Section III,A, electron microscopy of the CBE cells indicates numerous gap junctional plaques in intracellular structures on an endosomal–lysosomal protein degradation pathway. Yet as mentioned in Section II,C, Cx43 is removed from the membrane through ubiquitinylation and subsequent protesoal degradation that bypasses endocytosis of membrane plaques. Thus, since Cx43 is unlikely to be part of the gap junctional plaques observed in the NPE endosomes, the presence of other connexins could be suspected. A second consideration relates to the specific manner in which hemichannels assemble into functional gap junctions. From studies of cells experimentally brought into contact, it has become clear that the localization of the hexameric connexon to a particular membrane domain is controlled primarily by lock and key interactions with compatible connexons present in the apposing membrane (Swenson *et al.,* 1987). Given that the study by Coca-Prados *et al.* demonstrated minimal levels of Cx43 protein in the NPE cells, the majority of the intense stain observed at the NPE–PE interface is likely to represent primarily Cx43 of PE cells. It is not clear which mechanism maintains this excess of Cx43 tightly localized at the PE–NPE interface.

In a survey of other connexins in the CBE, we confirmed the highly localized distribution of Cx43 in rabbit and rat and discovered the presence of copious levels of Cx50 (α_8 connexin) in the NPE of both species (Wolosin *et al.,* 1997). Furthermore, examination of samples in which the two epithelial layers had spontaneously separated during cryohistological processing (Fig. 3) revealed a mutually exclusive distribution of the two connexins, with Cx43 expression only in the PE, and Cx50, in turn, only in the NPE. In sections of the intact tissue, however, spatial overlap between the two connexins at the PE–NPE interface was a prevalent feature of double-stained immunofluorescent samples examined by high-resolution confocal

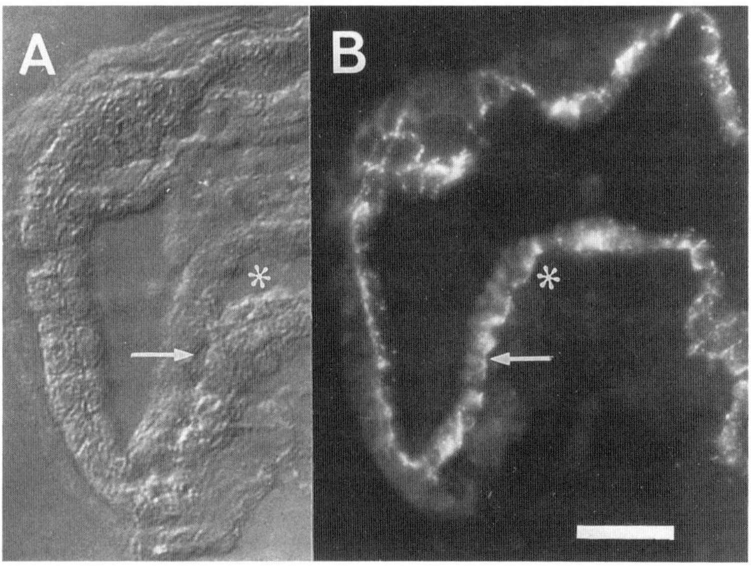

FIGURE 3 Details of a rabbit ciliary body cryosection immunostained for Cx43. (A) Differential interference contrast micrograph. (B) Epifluorescence. At two locations, one indicated by the horizontal arrow, the other by the asterisk, the two cell layers have spontaneously separated from each other. The immunostaining demonstrates that the Cx43 present at the NPE–PE interface is located in the PE cell exclusively.

microscopy. These data lead us to suggest that the heterocellular gap junctions connecting the PE and NPE may assemble from two different connexons consisting of Cx43 and Cx50, respectively; that is, they form a heterotypic gap junction (Fig. 4). This notion is somewhat conflicting with data provided by White *et al.* (1994) who showed that in transfected oocyte systems the very same connexins (α_1 and α_8) are unable to establish viable gap junctions, despite the fact that both belong to the same subfamily. However, posttranslational protein modification may vary substantially between an intact mammalian tissue and an experimental system based on amphibian oocytes. Hence, it would be unwise to draw final conclusions from one system for the other. The indisputable fact is that both Cx43 and Cx50 accumulate in opposing membranes at the heterocellular interface. Whether these two proteins, indeed, interact with each other or, alternatively, with other yet to be identified connexins poses a challenging enterprise for further studies. In this regard, it is noteworthy that neither Cx32 nor Cx26 (β_1 or β_2 connexin) could be identified in the epithelial layers of the CBE (Wolosin *et al.*, 1997; Shin *et al.*, 1996), confirming the earlier data of Coca-Prados *et al.* (1992).

IV. FUNCTIONAL STUDIES OF JUNCTIONAL COMMUNICATION

A. Dye Microinjection

Green *et al.* (1985) microinjected Lucifer Yellow (LY) into the rabbit ciliary epithelium by iontophoresis performed intermittently over a 20-min period. The extent of dye diffusion beyond the impaled cell was determined after fixation of the tissue by examination of microtome sections under the fluorescent microscope. LY was identified in both nonpigmented and

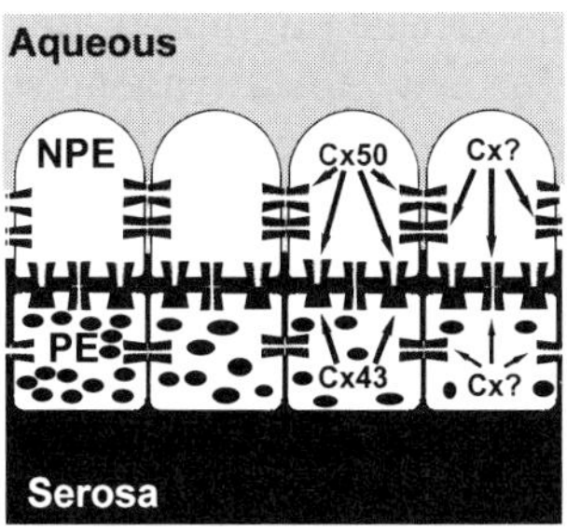

FIGURE 4 Connexin distribution within the CBE. In addition to the already identified Cx43 and Cx50, other connexins (represented by Cx?) may be present.

pigmented epithelia. From a point of maximal fluorescence visible in an NPE–PE pair, likely to contain the impaled cell, the dye was seen to spread laterally within both cell layers, decreasing gradually in concentration, over an area covering 5–10 cells.

Oh *et al.* (1994) performed similar experiments, except that the iontophoretic dye injection was limited to 1 min and the examination of fluorescent labels was carried out by confocal microscopy to avoid the tedious process of tissue sectioning. Their results confirmed the data obtained by Green *et al.* (1985) in that pronounced dye spreading from the impaled NPE cell to neighboring NPE and PE cells was observed. Interestingly, despite the large difference in time available for dye diffusion (20 vs. 1 min) the extent of spreading observed in both studies was comparable. Dye spreading was markedly decreased when the pH of the bathing solution was lowered, consistent with the high sensitivity of gap junctional permeability to pH.

Edelman *et al.* (1994) studied dye transference in bovine NPE–PE couplets obtained during cell dissociation protocols. LY was microinjected into the couplet's pigmented cell using a pressure-driven volume delivery device. Simultaneously the cells were observed under fluorescent illumination. Within seconds after the injection, fluorescence was seen to appear in the NPE cell. Taken together, the dye studies clearly indicate the physiological coupling of the CBE and define it as syncytium with a high capacity to share and propagate ions and metabolic signals between cells belonging to the same layer as well as between individual cells of the two distinct layers.

B. Microelectrode Studies

The complex anatomy of the ciliary body presents severe obstacles to microelectrophysiological techniques for the study of intracellular potentials in the individual cell layers and in particular by dual microelectrode techniques used to study gap junctional communications. Thus only a few relevant measurements have been attempted. Berggren (1960) and Miller and Constant (1960) performed experiments in isolated rabbit ciliary processes where glass microelectrodes were moved perpendicular to the surface of the tissue from aqueous to stroma either continuously or step by step, respectively. The basic results of both studies were essentially identical. Berggren determined values of -28 ± 7.9 mV for the NPE and -60 ± 8.2 mV for the PE ($n = 28$), respectively, against the aqueous humor as reference; Miller and Constant (1960) reported values of -27.8 ± 3.1 mV and -55.6 ± 4.4 mV, respectively. Thus, these early studies suggested that (1) the two cell types exhibit very different ionic exchange with their respective outside environment and (2) either there is no layer-to-layer communication or the level of communication is restricted enough to prevent cell potential equalization.

It is unfortunate that these investigations were not repeated in more recent times using more advanced techniques for microelectrode fabrication and data acquisition. The only notable explanation is a study by Candia *et al.* (1986), which has been reported only in abstract form. The experiments were performed with rabbit ciliary bodies mounted in modified Ussing-type chambers so that it became possible to affect selectively the membrane permeability or transport functions of one cell layer or the other. Under control conditions cell potentials in both cell layers were indistinguishable from each other (ca. −68 mV). This value for the NPE is consistent with other contemporary measurements of NPE potentials (Green *et al.*, 1985; Carré *et al.*, 1992) and may be interpreted as an indication that both cell layers exhibit a higher degree of commmunication. However, addition of the nonselective monovalent cation ionophore amphotericin B (see Section IV,D) to the hemichamber facing the aqueous side caused selective depolarization of the NPE, whereas the PE cell transmembrane potential was preserved. Similarly, addition of amphotericin B to the serosal hemichamber to abolish cation selectivity of the PE plasma membrane and, thus, depolarize the PE cell potential had no substantial effect on the contralaterally located NPE layer. Equivalent results were obtained by selective addition of ouabain to either the aqueous or the serosal side of the tissue to inhibit the Na^+ pump in the NPE or PE, respectively. In each case the gradual loss of cell potential expected from the loss of transmembrane ionic gradients was observed only in the layer modified by the inhibitor. Thus, even though these more recent results are partially in discord with the results obtained more than two decades earlier, they strongly support the notion of a restricted or absent NPE–PE ion exchange mechanism; in a highly connected syncytial arrangement, one would expect that any unilateral depolarization caused by amphotericin B or sodium pump inhibitors would be reflected immediately in the resting potential of the opposing cell layer.

With respect to the discrepancy between earlier and late studies, Green *et al.* (1985) hypothesized that their finding of a much more hyperpolarized NPE resting potential "undoubtedly reflects the improvements in microelectrode recording techniques that occurred since the 1960s" (p. 378). That is, reduced NPE transmembrane potentials measured in the earlier studies may have resulted from the use of low resistance electrodes (10–20 MΩ) that caused cellular damage. Notwithstanding the high likelihood of this explanation, it should be noted that the measurement of two distinct transmembrane potentials in the two distinct epithelial layers in the studies by Berggren and by Miller and Constant is fully consistent with restricted or even absent interlayer communication; otherwise electrode-induced damage to the NPE should have been reflected in the PE measurements as well.

The study of Green *et al.* (1985) stands alone in the technical approach of dual electrode recordings from the CBE. Along the longitudinal axis of the dual-layered epithelium, propagation of potentials was recorded up to 300 μm, that is, as much as 30 cells away from the stimulus electrode.

C. Ionic Equilibrium and Calcium Signal Transfer

An indirect method for the evaluation of the extent of NPE–PE communications in a physiologically relevant setting employs simultaneous measurements of the concentration of diffusable cellular components controlled by cellular activity. Bowler *et al.* (1996) measured the concentration of intracellular Cl^- using ion microprobe analysis. Since Cl^- is actively extruded from the NPE, the authors reasoned that a rate-limiting barrier maintained by the heterocellular gap junctional path should cause a higher concentration in the PE as the site of ion uptake, than in the NPE, the site of efflux. Since in both cells essentially identical Cl^- concentrations were found it was concluded that the two cells must share a common intracellular ionic milieu, that is, uptake and secretion across the cellular membranes of either cell must occur at a rate that is slower than the rate for gap junction-mediated PE–NPE Cl^- transfer.

Studies of calcium regulation have led to a different view. Given the intrinsic inaccuracy involved in the calibration of intracellular [Ca^{2+}], absolute [Ca^{2+}]$_i$ values in the layers could not be used as a valid criterion for comparative studies. Rather, the comparison depended on examining whether pharmacologic stimuli or other ionic perturbations could generate different calcium mobilization response patterns in both cell layers. For the performance of this study (Schütte *et al.*, 1996), preestablished differences in adrenoceptor agonist-induced calcium mobilization patterns between the PE and NPE cells were highly instrumental. That is, a saturating α_1-adrenergic stimulus (phenylephrine) causes an approximately threefold, sustained, slowly rising, increase in [Ca^{2+}]$_i$ of the isolated PE but has no measurable effect in the isolated NPE [Ca^{2+}]$_i$. Conversely, α_2-adrenergic stimulation (brimonidine), in combination with acetylcholine (ACH) causes a dramatic, transient [Ca^{2+}]$_i$ increase in the NPE (Farahbakhsh and Cillufo, 1994; Schütte *et al.*, 1996), but fails to induce Ca^{2+} mobilization in the isolated PE, beyond the twofold increase elicited by the cholinergic stimulus itself.

These observations raised several questions regarding the fate of intracellular [Ca^{2+}] within the complete, undissociated epithelium. To obtain data from the inner PE, ciliary processes had to be examined in a slice preparation exposing both NPE and PE cells simultaneously for optical data acquisition. This experimental approach was accomplished by either manual slicing of isolated processes (Fig. 5a) or by optical sectioning of the intact tissue using a confocal microscope (Schütte and Wolosin, 1996). Introduction of 3 m*M* heptanol (the best known inhibitor of gap junctional permeability; Bernardini et al., 1984) resulted, in most cases, in a small increase in PE $[Ca^{2+}]_i$ concomitant with a more substantial (ca. 50%) decrease of NPE $[Ca^{2+}]_i$. On removal of heptanol this divergence rapidly reversed. The sim-

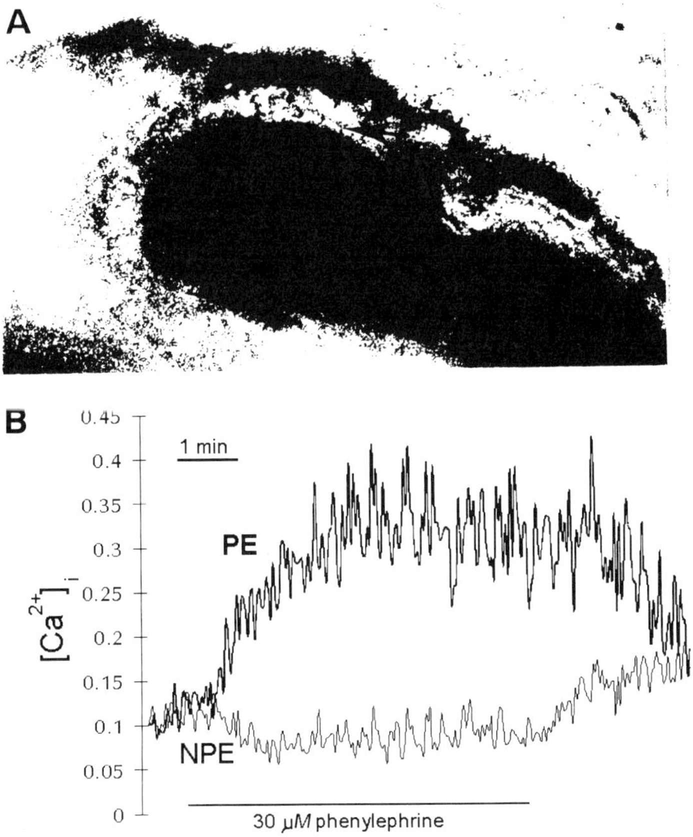

FIGURE 5 Simultaneous measurements of $[Ca^{2+}]_i$ in adjacent NPE and PE cells in CBE tissue slices. (A) Differential interference contrast of a living ciliary process. The arrow points to the NPE–PE interface. (B) Opposite direction effects of phenyeprine on adjacent NPE and PE cells of the intact CBE as measured by fura-2 ratio imaging fluorometry.

plest interpretation of these results is that under steady-state conditions there is constant net movement of calcium from the PE to the NPE. Jacob (1991), predicted this process on the basis of the finding that only the PE cells contain Ca^{2+} channels. L-type Ca^{2+} channels have since been observed in NPE cells (Cooper *et al.*, 1991).

The next experimental phase concerned the differential effects of α_1- versus α_2-adrenoceptor agonists. In the PE of the intact epithelium, phenylephrine caused the same sustained stimulation as observed in the isolated layer. Phenylephrine effects on the NPE, however, were variable. That is, in about 90% of the experiments, there was either a small decrease in NPE $[Ca^{2+}]_i$ or no measurable change (Fig. 5b). In only a few experiments did phenylephrine elicit a NPE response similar to that seen in the PE. The generation of a sustained three- to sixfold difference between PE and NPE $[Ca^{2+}]_i$ does not necessarily imply that calcium is unable to move down along the concentration gradient from one cell to another. It does indicate, though, that the flow rate from PE to NPE must be substantially smaller than the rate at which the PE can accumulate Ca^{2+} upon stimulation. Furthermore, one may conclude that the rate of Ca^{2+} extrusion/sequestration in the NPE exceeds the rate of inflow across the PE–NPE heterocellular path. Nevertheless, in view of the effects of heptanol described earlier, one would expect an increase in NPE $[Ca^{2+}]_i$ in all experiments, rather than the small decrease observed frequently. This apparent inconsistency can be explained, however, by the fact that, as described in Section IV,D, α_1-adrenergic stimulation elicits a large inhibition of the PE–NPE heterocellular junctional permeability in addition to an increase in PE $[Ca^{2+}]_i$.

In contrast to the state of affairs with respect to the PE-initiated α_1-adrenergic-dependent calcium increase, Ca^{2+} increases elicited in the NPE through the synergetic interactions between ACh and the α_2-adrenoceptor agonist brimonidine were transmitted rapidly (<2 sec) and with high efficiency to the PE cells (Schütte and Wolosin, 1996). It was clear from the mobilization patterns, though, that in this case one is dealing with transference of cytosolic components that initiate large Ca^{2+} mobilization responses within the PE itself rather than with plain transference of Ca^{2+} from cell to cell. Within the NPE layer, the Ca^{2+} mobilization responses were extremely synchronous; temporal differences between adjacent and even distant cells, if any, were well below the 1-sec resolution of the measuring technique employed. Such a high synchronism is consistent with the presence of a large area of cell–cell membrane contact between the interdigitating lateral membranes and the high density of Cx50 containing gap junctions in these membranes.

Elimination of junctional communications by tissue preincubation in Ca^{2+}-free medium as described by Wolosin *et al.*, (1993) resulted in total loss of Ca^{2+} signal transmission to the PE as well as moderate asynchronism in the response of the individual NPE cells. Reduction of gap junctional permeability with heptanol resulted in varied attenuation of the transmission of the signal to the PE, ranging from increased delay of onset (>10 sec) to full inhibition.

The rectificatory pattern where calcium signal propagation occurs only from the NPE to the PE but not in the opposite direction may originate in a number of factors, including (1) differences in the rate and magnitude of accumulation of intracellular relevant messages in the originating cell (e.g., it was calculated that the initial calcium rise in the NPE caused by the synergistic stimulation is 20- to 100-fold faster than that elicited by phenylephrine in the PE; these rate differences may reflect comparable rate differences for the accumulation of the second messengers eliciting the $[Ca^{2+}]_i$ change); (2) differences in the spatial localization of receptors, such as for IP_3, relative to the localization of heterocellular gap junctions; and/or (3) an intrinsic rectifactory behavior of the NPE–PE gap junctions that facilitates the transfer of relevant messenger molecules, such as the negatively charged IP_3 (Hansen *et al.*, 1995), in one preferential direction.

The few isolated cases in which the NPE displayed calcium mobilization in response to phenylephrine may represent a state where either the baseline cell-to-cell communication level between both layers and/or the degree/rate of the initial response in the PE allowed signals for calcium mobilization to diffuse to the NPE to an extent sufficient to initiate Ca^{2+} mobilization responses. In these cases, PE-to-NPE signal transmission may have succeeded only at a single cell pair and then, through the abundant coupling of NPE cells, could have rapidly spread throughout the entire layer.

D. A Macroelectrophysiological Approach to the Study of the NPE–PE Junctional Path

When the rabbit ciliary body is mounted under zero voltage-clamp conditions in Ussing–Zerhan chambers, a positive aqueous-to-serosa short circuit (I_{SC}) current can be measured. This current primarily represents the energy-dependent transport of chloride and bicarbonate ions from the serosa to the aqueous humor. This current has been interpreted to reflect the overall transport processes that drive aqueous humor secretion, even though the magnitude of the I_{SC} measured in the *in vitro* conditions can account for only a small fraction of the *in vivo* fluid secretory rates.

As discussed in the introduction, ion and solute absorption at the serosal side of the tissue may, in principle, involve either the apical membrane of the NPE or the whole PE cell or a combination of both. Bilateral addition of 3 mM heptanol to inhibit junctional permeabilities causes a >90% reduction of the I_{SC} (Wolosin et al., 1997), indicating an almost complete dependence of the secretory process on the passage of solutes through the heterocellular gap junctions and, by extension, the lack of participation of the apical NPE membrane in ion uptake processes.

To further characterize the heterocellular junctional path, the chamber-mounted CBE was modified in ways that allow assessment of the junctional permeability relative to the transport capacities of the cell's basolateral membranes (Shi *et al.,* 1996). The method employed is based on two critical facts. The first is the presence of Na^+ pumps (Na^+-K^+- ATPases) in both CBE domains, that is, in the basolateral membranes of both NPE and PE (Pesin and Candia, 1983). The second is the uncanny ability of both the monovalent cation ionophore amphotericin B (Pesin and Candia, 1983) and the cholesterol-dependent detergent digitonin (Malinowska *et al.,* 1981; Morita *et al.,* 1985; Sokol *et al.,* 1990) to permeabilize exposed cell plasma membranes without directly affecting either intracellular organelles or contralaterally located plasma membranes. In the case of digitonin the permeabilization is not selective. Thus, to temporarily maintain cell viability, the experiments were performed with low, controlled concentrations so that critical intracellular components were partially retained over the experiment's duration. Additionally, the side of digitonin addition was kept Ca^{2+} free to prevent serious cell damage secondary to a massive $[Ca^{2+}]_i$ elevation (Wolosin, 1991). Addition of these agents to the aqueous side hemichamber results in a large increase in I_{SC}. The following observations were made: (1) The maximal I_{SC} attained with either treatment show high similarity; (2) three junctional permeability inhibitors (heptanol, α-18-glycyrrhetinic acid, and dieldrin) as well as ouabain on the serosal side elicit complete I_{SC} inhibition; (3) reduced aqueous-side pH ($I_{C50} \sim$ pH = 6.7) reduces I_{SC}; (4) the I_{SC} is fully dependent on sodium; 5) the I_{SC} is not sensitive to either aqueous-side ouabain or anion composition in the case of permeabilization with digitonin; and (6) the I_{SC} is abolished by aqueous-side Ca^{2+}. These observations led to the conclusion that the permeabilization-induced current reflects a three-step process that includes (1) unrestricted Na^+ flow from the aqueous to the NPE through its permeabilized basolateral membrane, (2) sodium translocaton from the PE to the NPE through the heterocellular gap junctions, and, finally, (3) Na^+ extrusion into the serosal environment by the Na^+,K^+-ATPase. A schematic description of this process is depicted in Fig. 6a. When amphotericin B is added to the serosal instead of to the aqueous side, and the polarity of the I_{SC} reverses,consistent with the establishment of a mirror-image process in which the last step is active sodium extrusion into the aqueous compartment by the Na^+,K^+ATPase of the NPE. Of course, in either case, given the $3Na^+$ for $2K^+$ stoichiometry

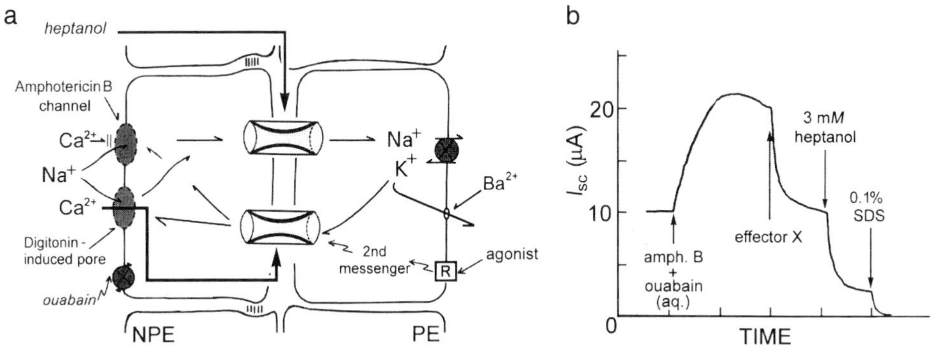

FIGURE 6 A primer for the methodology employed to study electrophysiologically the amphotericin B or digitonin-modified CBE mounted in Ussing–Zehran chambers. (a) Diagram of the ion flows established by the permeabilization of the NPE basolateral membrane with either amphotericin B or digitonin. Na^+ freely enters the NPE cytosol from the aqueous compartment through either path, diffuses down its concentration gradient to the PE via the NPE–PE gap junctions, and from there it is transported into the serosa in a 3-for-2 electrogenic exchange for K^+ by the Na^+,K^+-ATPase. PE K^+ can flow toward the aqueous side via the gap junctions or diffuse back to the serosa through the PE K^+ channels. The difference between the net flows of Na^+ from aqueous to serosa K^+ from serosa to aqueous can be measured as an I_{SC}. Inhibition of PE K^+ channels by serosal side Ba^{2+} is expected to increase transmural K^+ flow and thus decrease the I_{SC}. The current will also be modified if the junctional permeability is changed by inhibitors such as heptanol or modulated otherwise by intracellular events initiated by an agonist-cell receptor (R) interaction. In the case of digitonin the aqueous-side solution needs to be kept Ca^{2+} free. Otherwise, massive Ca^{2+} influx will close gap junctions. Finally, addition of ouabain to the aqueous side to block the NPE Na^+,K^+-ATPase has no effect on the I_{SC} because the NPE is essentially short circuited but it will reduce NPE adenosine triphosphate consumption. (b) Schematic graph of changes in I_{SC} on the subsequent addition of amphotericin B and ouabain to the aqueous; any undefined effector (effector x; e.g., an agonist); 3 m*M* heptanol to attain maximal gap junctional inhibition and SDS to destroy the epithelial barrier.

of the ATPase K^+ will be moving in the opposite direction of Na^+. These artificially imposed I_{SC}'s were employed to investigate (1) the rate limitation for the entire transmural movement of sodium and (2) possible pharmacologic modulation of the heterocellular gap junctional permeability (Fig. 6b).

Stepwise bilateral addition of heptanol in the 0.2–1.2 m*M* range resulted in near-proportional decreases of the I_{SC} induced by the addition of amphotericin B to the aqueous side. Considering the steps involved in I_{SC} generation, heptanol may, as expected, act on either junctional ion translocation and/or, though less likely, on the Na^+ pump. A parallel study with the simple epithelium of the rabbit gall bladder, in which an apex-to-serosal Na^+,K^+-ATPase-dependent I_{SC} was established by permeabilization of the apical membrane with amphotericin B showed, however, no effect of

3.5 m*M* heptanol, demonstrating the complete insensitivity of the rabbit enzyme activity to the alkanol. Thus, the I_{SC} inhibition in the CBE can only reflect the inhibition of junctional permeability, and the near linearity of the dose–response relationship of heptanol-induced inhibition of I_{SC} indicates that the heterocellular junctional translocation step plays a major role in limiting the overall transmural ion flow. Otherwise, one would expect to observe no effect of heptanol on the I_{SC}, at least at low concentrations causing only partial decrease of junctional permeability, similar to that shown previously in the unmodified rabbit corneal epithelium, where the rate-limiting step falls squarely on the apical membrane (Wolosin, 1991).

The apparently rate-limiting role of the junctional translocation step was also demonstrated with respect to the potassium permeability of the basolateral PE membranes. If this permeability were exceedingly large, all potassium pumped into the PE by its Na^+,K^+-ATPase would diffuse back into the serosa leading to the net transference of three charges ($3Na^+$) from aqueous to serosa for each mole of adenosine triphosphate (ATP) consumed. At the other extreme, if the potassium membrane permeability is null, the 2 mol of potassium pumped per mole of ATP will diffuse toward the NPE and aqueous. In that case, each mole of ATP consumed will translate in the net translocation of only one positive charge ($3Na^+$-$2K^+$). Addition of 5 m*M* Ba^{2+} to the serosal side to inhibit K^+ membrane channel permeability in the PE caused a 52% reduction of I_{SC}, indicating that prior to the addition of Ba^{2+}, a large fraction of the potassium pumped into the PE was diffusing back into the serosa via Ba^{2+}-sensitive channels.

The extent of inhibition can be used to obtain a rough estimate of the gap junction/PE K^+ channel permeability ratio. For this purpose I_{SC} is arbitrarily defined as

$$I_{SC} = C\left\{3 - [2P_K{}^{Gj}/(P_K{}^{Gj} + P_K{}^{Ch})]\right\}$$

where $P_K{}^{Gj}$ and $P_K{}^{Ch}$ are the permeabilities of the NPE–PE gap junctions and PE cell membranes, respectively; the numbers 3 and 2 reflect the number of Na^+ and K^+ ions transported per cycle by the Na^+K^+ pump; and C is the turnover rate of the Na^+ pump in unity of μA/tissue. If one assumes that (1) $P_K{}^{Ch} = 0$ after the introduction of 5 m*M* Ba^{2+} to the serosal side and (2) the Na^+ pump rate is not changed by Ba^{2+} inhibition of K^+ permeability, then the 52% I_{SC} decrease implies $P_K{}^{Gj}/P_K{}^{Ch} = 0.85$.

Pharmacological studies revealed that cholinergic and α_1-adrenoceptor agonists caused about a 50% decrease of the I_{SC}'s generated by addition of amphotericin B to either the aqueous or serosal side (Shi *et al.,* 1996). The only common step for these currents of opposite polarity is the translocation of Na^+ through the NPE–PE heterocellular interface. Therefore, it is rational to assume that the I_{SC} inhibitions reflected inhibition of this rate-limiting step and not of the Na^+ pump. The inhibitions were fully dependent on the presence of serosal calcium (irrespective of whether calcium was present or absent from the aqueous solution) indicating that they depend on pharmacological events occurring in the PE that involve opening of Ca^{2+} channels.

Cholinergic and α_1-adrenergic inputs elicit two to threefold increases in $[Ca^{2+}]_i$ in the PE (Schütte *et al.,* 1996; Schütte and Wolosin, 1996). These increases are roughly additive irrespective of the order of application. Yet with respect to the I_{SC} inhibition, each acetylcholine and phenylephrine alone was able to elicit the maximally attainable inhibition, suggesting that, while the pharmacological phenomenon requires Ca^{2+} elevation, it is not directly proportional to its concentration. This pattern decreases the likelihood of a direct interaction of Ca^{2+} with the gap junctions. Of many agents able to interface with intracellular kinases and phosphatases, only the calmodulin inhibitors trifluoperazine and W7 were able to block the adrenergic- or cholinergic-induced I_{SC} reductions.

The combination of α_2-adrenergic and cholinergic agonists added to the NPE-side hemichamber caused a moderate (30%) I_{SC} inhibition (Fig. 7)

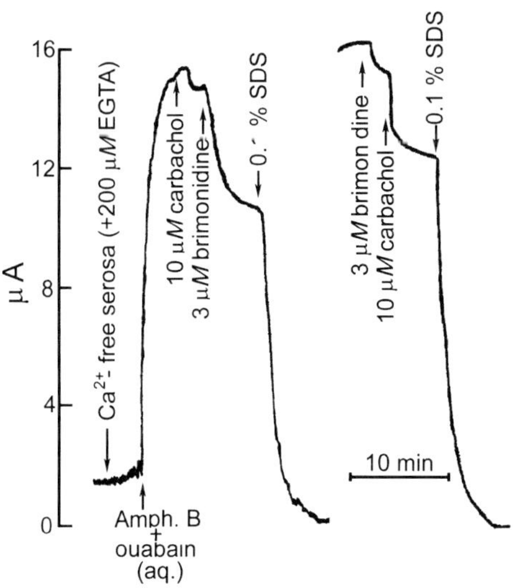

FIGURE 7 Effect of agonists at α_2-adrenergic and cholinergic receptors on the amphotericin B-induced I_{SC}. Experiments were carried out in the absence of serosal Ca^{2+}. Under these conditions, all effects are due to NPE-initiated events. Note the potentiation of the carbachol effect by brimonidine and vice versa. To generate this figure, text was wrapped around actual chart recorder traces that were scanned into a graphic program.

resembling in its cooperative pattern the synergistic interaction leading to calcium mobilization described earlier in Section IV,C. It should be noted, though, that given the rapid transmission of the calcium mobilization signal from NPE to PE, it is not possible to determine whether this effect is related directly to phenomena within the NPE or PE cells.

The overall macroelectrophysiologic studies with amphotericin B or digitonin-modified CBE indicate that the NPE–PE junctional path presents a substantial permeability barrier to ionic diffusion in the context of whole transmural ion flow and that this path is under adrenergic and cholinergic control. It could be argued or suspected, though, that the induced perturbances in the membrane permeability cause metabolic changes leading to reductions of the junctional permeability. Nevertheless, when heptanol was added to unmodified tissues, its inhibitory effect was nearly indistinguishable from the effect seen on the I_{SC} of the permeabilized tissues (Fig. 8), suggesting that here also the heterocellular permeability barrier has a certain rate-limiting role. Likewise, halothane (Tang *et al.*, 1991), a known gap junction inhibitor (Burt and Spray, 1989), as well as α_1-adrenoceptor agonists acting from the serosal side (Krupin *et al.*, 1991) have been shown to inhibit the I_{SC} generated by the intact unmodified rabbit CBE. Given the effect of heptanol as described earlier it is intriguing to consider whether this pharmacological effect on the normal I_{SC} may originate primarily from inhibition of the permeability of the heterocellular path, rather than from modification of ion-specific cell membrane transporters.

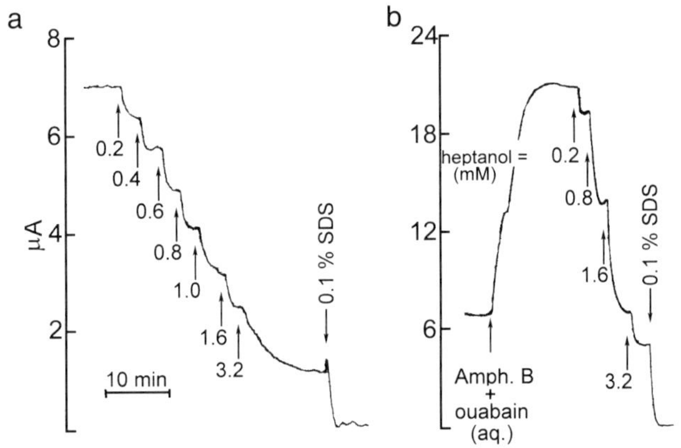

FIGURE 8 Effect of the gap junction permeability inhibitor heptanol on the I_{SC} of the control and amphotericin B modified CBE. (a) Control tissue. (b) Amphotericin B-modified tissue. Note similarity of dose-response in both cases. To generate this figure, text was wrapped around actual chart recorder traces that were scanned into a graphic program.

V. SUMMARY

Fluid and solute secretion by the ciliary body depends on complex interactions between the transporters and regulatory mechanisms present in the two distinct constituent cell layers. These interactions are fully dependent on the gap junctions present at the heterocellular apex-to-apex interface. The structural immunochemical, electrophysiological, and pharmacological observations described indicate that these junctions may play a central role in the control of the secretory activity.

Acknowledgments

Supported by Public Health Service Grant RO1 EY09074 and an unrestricted grant from Research to Prevent Blindness to the Department of Ophthalmology. J. W. W is the recipient of a Senior Scientific Investigators Award from Research to Prevent Blindness.

References

Bairati, A., and Orzalesi, N. (1966). The ultrastructure of the epithelium of the ciliary body: A study of the junctional complexes and of the changes associated with the production of plasmoid aqueous humour. *Zeitschrift Zellforschung* **69,** 635–658.

Barrio, L. C., Suchyna, T., Bargiello, T., Xu, L. X., Roginski, R. S., Bennett, M. V., and Nicholson, B. J. (1991). Gap junctions formed by connexins 26 and 32 alone and in combination are differently affected by applied voltage. *Proc. Natl. Acad. Sci. U.S.A.* **88,** 8410–8414.

Bastide, B., Hervé, J. C., Cronier, L., and Délèze, J. (1995). Rapid onset and calcium independence of the gap junction uncoupling induced by heptanol in cultured heart cells. *Eur. J. Physiol.* **429,** 386–393.

Bennett, M. V. L. (1966). Physiology of electronic junctions. *Ann. N.Y. Acad. Sci.* **137,** 509–539.

Bennett, M. V. L., Barrio, L. C., Bargiello, T. A., Spray, D. C., Hertzberg, E. L., and Saze, J. C., (1991). Gap junctions: New tools, new answers, new questions. *Neuron* **6,** 305–320.

Berggren, L. (1960). Intracellular potential measurements from the ciliary processes of the rabbit eye *in vivo* and *in vitro. Acta Physiol. Scand.* **48,** 461.

Bernardini, G., Peracchia, C., and Peracchia, L. (1984). Reversible effects of heptanol on gap junction structure and cell-to-cell electrical uncoupling. *Eur. J. Cell Biol.* **34,** 307–312.

Boitano, S., Dirksen, E. R., and Sanderson, M. J. (1992). Intercellular propagation of calcium waves mediated by inositol trisphosphate. *Science* **258,** 292–295.

Bowler, J. M., Pert, D., Purves, R. D., Carré, D. A., MacKnight, A. D. C., and Civan, M. M. (1996). Electron probe x-ray microanalysis of rabbit ciliary epithelium. *Exp. Eye Res.* **62,** 131–139.

Brightman, M. W., and Reese, T. J. (1969). Junctions between intimately apposed cell membranes in the vertebrate brain. *J. Cell Biol.* **40,** 648–677.

Bruzzone, R., Haefliger, J.-A., Gimlich, R. L., and Paul, D. L. (1993). Connexin40, a component of gap junctions in vascular endothelium, is restricted in its ability to interact with other connexins. *Mol. Biol. Cell* **4,** 7–20.

Burt, J. M., and Spray, D. C. (1989). Volatile anesthetics block intercellular communication between neonatal rat myocardial cells. *Circ. Res.* **65,** 829–837.

Candia, O. A., Iizuka, S., and Chu, T.-C. (1986). Intracellular recordings in the isolated ciliary epithelium. *Invest. Ophthalmol. Vis. Sci.* **27,** (suppl), 178.

Carré, D. A., Tang, C. S., Krupin, T., and Civan, M. M. (1992). Effect of bicarbonate on intracellular potential of rabbit ciliary epithelium. *Curr. Eye Res.* **11,** 609–624.

Coca-Prados, M., Ghosh, S., Gilula, N. B., and Kumar, N.M. (1992). Expression and cellular distribution of the alpha 1 gap junction gene product in the ocular pigmented ciliary epithelium. *Curr. Eye Res.* **11,** 113–122.

Cooper, K., Rae, J. L., and Dewey, J. (1991). Inwardly rectifying potassium current in mammalian lens epithelial cells. *Am. J. Physiol.* **261,** C115–C123.

Dabagian, N. V., Ganieva, L. K., and Eshkind, L. G. (1979). Electron microscope study of the intercellular junctions during the ciliary epithelial differentiation of the eye of the common frog. *Ontogenez* **10,** 576–585.

Dewey, M. M. and Barr, L. (1962). Intercellular connection between smooth muscle cells: The nexus. *Science* **137,** 670–672.

Edelman, J. L., Sachs, G., and Adorante, J. S. (1994). Ion transport asymmetry and functional coupling in bovine pigmented and nonpigmented ciliary epithelial cells. *Am. J. Physiol.* **266,** C1210–C1221.

Ek-Vitorín, J. F., Calero, G., Morley, G. E., Coombs, W., Taffet, S. M., and Delmar, M. (1996). pH regulation of connexin43: Molecular analysis of the gating particle. *Biophys. J.* **71,** 1273–1284.

Elfgang, C., Eckert, R., Lichtenberg-Fraté, H., Butterweck, A., Traub, O., Klein, R. A., Hülser, D. F., and Willecke, K. (1995). Specific permeability and selective formation of gap junction channels in connexin-transfected HeLa cells. *J. Cell. Biol.* **129,** 805–817.

Farahbakhsh, N. A., and Cilluffo, M. C. (1994). Synergistic effect of adrenergic and muscarinic receptor activation on $[Ca^{2+}]_i$ in rabbit ciliary body epithelium. *J. Physiol.* **447,** 215–221.

Farquhar, M. G., and Palade, G. E. (1963). Junctional complexes in various epithelia. *J. Cell Biol.* **17,** 375–412.

Fawcett, D. W. (1958). Specializations of the cell surface. In "Frontiers in Cytology" (S. L. Palay, ed.), pp. 19–45. Yale University Press, New Haven, CT.

Flügel, C., Drecoll, E. L., Zadunaisky, J. A., and Wiederholt, M. (1989). Regional differences in the morphology and enzyme distribution of the spiny dogfish (*Squalus acanthias*) ciliary epithelium. *Exp. Eye Res.* **49,** 1097–1111.

Furshpan, E. J., and Potter, D. D. (1959). Transmission at the giant motor synapses of crayfish. *J. Physiol.* **145,** 289–325.

Ganieva, L. K., and Dabagian, N. V. (1980). Ultrastructure of the ocular ciliary epithelium of the common frog. *Tsitologiia* **22,** 744–750.

Giaume, C., and Korn, H. (1983). Bidirectional transmission at the rectifying, electrotonic synapse: A voltage-dependent process. *Science* **220,** 84–87.

Giaume, C., and Korn, H. (1984). Voltage-dependent dye coupling at a rectifying electrotonic synapse of the crayfish. *J. Physiol.* **356,** 151–167.

Goodenough, D. A., Paul, D. L., and Jesaitis, L. (1988). Topological distribution of two connexin32 antigenic sites in intact and split rodent hepatocyte gap junctions. *J. Cell. Biol.* **107,** 1817–1824.

Goodenough, D. A., Goliger, J. A., and Paul, D. L. (1996). Connexins, connexons, and intercellular communication. *Annul. Rev. Biochem.* **65,** 475–502.

Green, K., Bountra, C., Georgiu, P., and House, C. R. (1985). An electrophysiological study of rabbit ciliary body epithelium. *Invest. Ophthalmol. Vis. Sci.* **26,** 371–381.

Hampson, E. C., Vaney, D. I., and Weiler, R. (1992). Dopaminergic modulation of gap junction permeability between amacrine cells in mammalian retina. *J. Neurosci.* **12,** 4911–4922.

Hansen, M., Boitano, S., Dirksen, E.R., and Sanderson, J. (1995). A role for phospholipase C activity but not ryanodine receptors in the initiation and propagation of intercellular calcium waves. *J. Cell Sci.* **108,** 2583–2590.

Hertzberg, E. L. (1985). Antibody probes in the study of gap junctional communication. *Annu. Rev. Physiol.* **47,** 305–318.
Hirsch, M., Montcourrier, P., and Renard, G. (1977). Ultrastructure of the blood–aqueous barrier in normal condition and after paracentesis: A freeze fracture study in the rabbit. *Albrecht Von Graefes Arch. Klin. Exp. Ophthalmol.* **203,** 169–179.
Hudspeth, A. J., and Yee, A. G. (1973). The intercellular junctional complexes of retinal pigment epithelia. *Invest. Ophthalmol.* **12,** 354–365.
Jacob, T. J. C. (1991). Identification of a low-threshold T-type calcium channel in bovine ciliary epithelial cells. *Am J. Physiol.* **261,** C808–C813.
Jiang, J. X., and Goodenough, D. A. (1996). Heteromeric connexons in lens gap junction channels. *Proc. Natl. Acad. Sci. U.S.A.* **93,** 1287–1291.
John, S. A., and Revel, J.-P. (1991). Connexon integrity is maintained by non-covalent bonds: Intramolecular disulfide bonds link the extracellular domains in rat connexin 43. *Biochem. Biophys. Res. Commun.* **178,** 1312–1318.
Johnston, M. F., Simon, S. A., and Ramón, F. (1980). Interaction of anaesthetics with electrical synapses. *Nature* **286,** 498–500.
Kogon, M., and Pappas, G. D. (1975). Atypical gap junctions in the ciliary epithelium of the albino rabbit eye. *J. Cell Biol.* **66,** 671–676.
Konig, N., and Zampighi, G. A. (1995). Purification of bovine lens cell-to-cell channels composed of connexin44 and connexin50. *J. Cell Sci.* **108,** 3091–3098.
Krupin, T., Wax, W. B., Carré, D. A., Moolchandani, J., and Civan, M.M. (1991). Effects of adrenergic agents on transepithelial electrical measurements across the isolated iris–ciliary body. *Exp. Eye Res.* **53,** 709–716.
Kumar, N. M., and Gilula, N. B. (1992). Molecular biology and genetics of gap junction channels. *Semin. Cell Biol.* **3,** 3–16.
Kumar, N. M., and Gilula, N. B. (1996). The gap junction communication channel. *Cell* **84,** 381–388.
Kurz-Isler and Wolburg, H. (1986). Gap junctions between horizontal cells in the cypramid fish alter rapidly their structure during light and dark adaptation. *Neurosci. Lett.* **67,** 7–12.
Laing, J. G., and Beyer, E. C. (1995). The gap junction protein connexin43 is degraded via the ubiquitin proteasome pathway. *J. Biol. Chem.* **270,** 26399–26403.
Loessberg-Stauffer, P. L., Zhao, H., Luby-Phelps, K., Moss, R. L., Star, R. A., and Muallem, S. (1993). Gap junction communication modulates $[Ca^{2+}]_i$ oscillations and enzyme secretion in pancreatic acini. *J. Biol. Chem.* **268,** 19769–19775.
Loewenstein, W. R. (1967). Cell surface membranes in close contact. Role of calcium and magnesium ions. *J. Colloid Interface Sci.* **25,** 34–46.
Loewenstein, W. R. (1981). Cell-to-cell communication. *Physiol. Rev.* **61,** 829–913.
Loewenstein, W. R., and Kanno, Y. (1964). Studies on an epithelial (gland) cell junction. I. Modifications of surface membrane permeability. *J. Cell Biol.* **22,** 565–586.
Makowski, L. (1985). Structural domains in gap junctions: Implications for the control of intercellular communication. *In* "Gap Junctions" (M. V. L. Bennett and D. C. Spray, eds.), pp. 5–12. Cold Spring Harbor Laboratory Press, Cold Spring Harbor, NY.
Malewicz, B., Kumar, V. V., Johnson, R. G., and Baumann, W. J. (1990). Lipids in gap junction assembly and function. *Lipids* **25,** 419–427.
Malinowska, D. H., Koelz, H. R., Hersey, S. J., and Sachs, G. (1981). Properties of the gastric proton pump in unstimulated permeable gastric glands. *Proc. Natl. Acad. Sci. U.S.A.* **79,** 5908–5912.
Margiotta, J. F., and Walcott, B. (1983). Conductance and dye-permeability of a rectifying electrical synapse. *Nature* **305,** 52–56.

McNutt, N. S., and Weinstein, R. S. (1970). The ultrastructure of the nexus. A correlated thin-section and freeze-cleave study. *J. Cell Biol.* **47,** 666–688.

McNutt, N. S., and Weinstein, R. S. (1973). Membrane ultrastructure at mammalian intercellular junctions. *Prog. Biophys. Mol. Biol.* **26,** 45–101.

Milks, L. C., Kumar, N. M., Houghton, R., Unwin, N., and Gilula, N. B. (1988). Topology of the 32-kD liver gap junction protein determined by site-directed antibody localizations. *EMBO J.* **7,** 2967–2976.

Miller, J. E., and Constant, M. A. (1960). The measurement of rabbit ciliary epithelial potentials *in vitro. Am. J. Physiol.* **50,** 855.

Mills, S. L., and Massey, S. C. (1995). Differential properties of two gap junctional pathways made by AII amacrine cells. *Nature* **377,** 734–737.

Morita, K., Levine, M., Heldman, E., and Pollard, H. B. (1985). Ascorbic acid and catecholamine release from digitonin treated chromafin cells. *J. Biol. Chem.* **260,** 15112–15116.

Murakami, M., Miyachi, E.-I., and Takahashi, K.-I. (1995). Modulation of gap junctions between horizontal cells by second messengers. *Prog. Retinal Eye Res.* **14,** 197–221.

Murray, A. W., and Gainer, H. S. C. (1989). Regulation of gap junctional communication by protein kinases. *In* "Cell Interactions and Gap Junctions" (N. Sperelakis and W. C. Cole, eds.), pp. 97–106. CRC Press, Boca Raton, FL.

Ober, M., and Rohen, J. W. (1979). Regional differences in the fine structure of the ciliary epithelium related to accommodation. *Invest. Ophthalmol. Vis. Sci.* **18,** 655–664.

Oh, J., Krupin, T., Tang, L. Q., Sveen, J., and Lahum, R. A. (1994). Dye coupling of rabbit ciliary epithelial cells *in vitro. Invest. Ophthalmol. Vis. Sci.* **35,** 2509–2514.

Ohkuma, M., and Nishiura, M. (1974). Intercellular junctions of the ciliary epithelium as revealed by freeze-fracture and lanthanum tracer technique. *Acta Soc. Ophthalmol. Japonica* **78,** 1419–1430.

Okinami, S., Matsumura, M., and Ohkuma, M. (1980). Freeze-fractured and freeze-etched replicas of tight junctions and gap junctions in the ciliary epithelium of rhesus monkeys. *Acta Soc. Ophthalmol. Japonica* **84,** 510–521.

Oliveira-Castro, G. M., and Loewenstein, W. R. (1971) Junctional membrane permeability: Effects of divalent cations. *J. Membr. Biol.* **5,** 51–77.

Pappas, G. D., Asada, Y., and Bennett, M. V. L. (1971). Morphological correlates of increased coupling resistance at an electrotonic synapse. *J. Cell Biol.* **49,** 173–188.

Perracchia, C. (1989). Control of gap junction permeability and calmodulin like proteins. *In* "Cell Interactions and Gap Junctions" (N. Sperelakis and W. C. Cole, eds.), pp. 125–142. CRC Press, Boca Raton, FL.

Peracchia, C., and Girsch, S. J. (1985). Functional modulation of cell coupling: Evidence for a calmodulin-driven channel gate. *Am J. Physiol.* **248,** H765–H782,

Perracchia, C., and Perracchia, L .L. (1980). Gap junction dynamics: Reversible effects of hydrogen ions. *J. Cell Biol.* **87,** 719–727.

Peracchia, C., and Peracchia, L. L. (1985). Bridges linking gap junction particles extracellularly. A freeze-etching rotary-shadowing study of split junctions. *Eur. J. Cell Biol.* **36,** 286–293.

Pesin, S. R., and Candia, O. A. (1983). Na^+ and Cl^- fluxes and effect of pharmacological agents on the short circuit current of the isolated iris–ciliary body. *Curr. Eye Res.* **1983,** 815–898.

Piccolino, M., Neyton, J., and Gerschenfeld, H. M. (1984). Decrease of the gap-junction permeability induced by dopamine and cyclic 3′-5′ adenosine-monophosphate in horizontal cells of the turtle retina. *J. Neurosci.* **4,** 2477–2488.

Raviola, G. (1971). The fine structure of the ciliary zonule and ciliary epithelium. *Invest. Ophthalmol.* **10,** 851–869.

Raviola, G. (1974). Effects of paracentesis on the blood–aqueous barrier: An electron microscope study on *Macaca mulatta* using horseradish peroxidase as a tracer. *Invest. Ophthalmol.* **13,** 828–858.

Raviola, G., and Raviola, E. (1978). Intercellular junctions in the ciliary epithelium. *Invest. Ophthalmol. Vis. Sci.* **17,** 958-981.

Raviola, E., Goodenough, D.A., and Raviola, G. (1980). Structure of rapidly frozen gap junctions. *J. Cell Biol.* **87,** 273-279.

Reale, E., and Spitznas, M. (1975). Freeze-fracture analysis of junctional complexes in human ciliary epithelia. *Albrecht von Graefes Arch. Klin. Exp. Ophthalmol.* **195,** 1–16.

Reale, E., Luciano, L., and Spitznas, M. (1974). Introduction to freeze-fracture method in retinal research. *Albrecht von Graefes Arch. Ophthalmol.* **192,** 73–87.

Rentsch, F. J. (1970). Elektronenmikroskpische Untersuchungen über die intracellulären Verbindungen des unpigmentierten Ciliarepithels in den Hauptverankerungsgebieten des Zonularapparates. *Albrecht von Graefes Arch. Klin. Exp. Ophthalmol.* **180,** 113–133.

Revel, J.-P., and Karnovsky, M. J. (1967). Hexagonal array of subunits in intercellular junctions of the mouse heart and liver. *J. Cell Biol.* **33,** C7–C12.

Revel, J.-P., and Sheridan, J. D. (1968). Electrophysiological and ultrastructural studies of intercellular junctions in brown fat. *J. Physiol.* (*London*) **194,** 34P–35P.

Revel, J.-P., Olson, W., and Karnovsky, M. J. (1967). A twenty-angström gap junction with a hexagonal array of subunits in smooth muscle. *J. Cell Biol.* **35,** 112A.

Revel, J.-P., Yee, A. G., and Hudspeth, A. J. (1971). Gap junctions between electronically coupled cells in tissue culture and in brown fat. *Proc. Natl. Acad. Sci. U.S.A.* **68,** 2924–2927.

Robertson, J. D. (1963). The occurrence of a subunit pattern in the unit membranes of club endings in Mauthner cell synapses in goldfish brains. *J. Cell Biol.* **19,** 201–221.

Rose, B., and Loewenstein, W. R. (1975). Permeability of cell junction depends on local cytoplasmic calcium activity. *Nature* (*London*) **254,** 250–252.

Sáez, J. C., Spray, D. C., Nairn, A. C., Hertzberg, E., Greengard, P., and Bennett, M. V. L. (1986). cAMP increases junctional conductance and stimulates phosphorylation of the 27-kDa principal gap junction polypeptide. *Proc. Natl. Acad. Sci. U.S.A.* **83,** 2473.

Sáez, J. C., Connor, J. A., Spray, D. C., and Bennett, M. V. L. (1989). Hepatocyte gap junctions are permeable to the second messenger, inositol 1,4,5-trisphosphate, and to calcium ions. *Proc. Natl. Acad. Sci. U.S.A.* **86,** 2708–2712.

Schütte, M., and Wolosin, J.M. (1996). Ca^{2+} mobilization and interlayer signal transfer in the heterocellular bilayered epithelium of the rabbit ciliary body. *J. Physiol.* **496,** 25–37.

Schütte, M., Diadori, A., Wang, C., and Wolosin, J. M. (1996). Comparative adrenocholinergic Ca^{2+} pharmacology in the dissociated rabbit ciliary epithelial cell layers. *Invest. Ophthalmol. Vis. Sci.* **37,** 212–220.

Shabo, A. L., and Maxwell, D. S. (1973). Electron microscope localization of occluding junctions in the primate ciliary process and pars plana. *Invest. Ophthalmol.* **12,** 863–865.

Shi, X.-P., Wolosin, J. M., and Candia, O. A. (1996). Adrenocholinergic modulation of junctional communication between the pigmented and nonpigmented layers of the ciliary body epithelium. *Invest. Ophthalmol. Vis. Sci.* **37,** 1037–1046.

Shin, B.-C., Suzuki, T., Tanaka, S., Kuraoka, A., Shibata, Y., and Takata, K. (1996). Connexin 43 and the glucose transporter, GLUT1, in the ciliary body of the rat. *Histochem. Cell Biol.* **106,** 209–214.

Smith, R. S., and Rudt, L. A. (1973). Ultrastructural studies of the blood–aqueous barrier. II. The barrier to horseradish peroxidase in primates. *Am. J. Ophthalmol.* **76,** 937–947.

Sokol, J. L., Masur, S. K., Asbell, P. A., and Wolosin, J. M. (1990). Layer-by-layer desquamation of corneal epithelium and maturation of tear-facing membranes. *Invest. Ophthalmol. Vis. Sci.* **31,** 294–304.

Sosinsky, G. (1995). Mixing of connexins in gap junction membrane channels. *Proc. Natl. Acad. Sci. U.S.A.* **92,** 9210–9214.

Spray, D. C., Stern, J. H., Harris, A. L., and Bennett, M. V. L. (1982). Gap junctional conductance: Comparison of sensitivities to H and Ca ions. *Proc. Natl. Acad. Sci. U.S.A.* **79,** 441.

Stauffer, K. A. (1995). The gap junction proteins β_1-connexin (connexin-32) and β_2-connexin (connexin-26) can form heteromeric hemichannels. *Proc. Natl. Acad. Sci. U.S.A.* **270,** 6768–6772.

Swenson, K. I., Jordan, J. R., Beyer, E. C., and Paul, D. L. (1987). Formation of gap junctions by expression of connexins in *Xenopus* oocyte pairs. *Cell* **57,** 145–155.

Tang, L. Q., Krupin, T., Milner, M., Woods, D., Miller, K., Carré, D. A., and Civan, M. M. (1991). Halogenated inhalation anesthetic agents decrease transepithelial electrical measurements across the isolated iris–ciliary body. *Invest. Ophthalmol. Vis. Sci.* **32,** 1912–1915.

Tenkova, T., and Chaldakov, G. N. (1990). Internalized gap junctions in ciliary epithelium of rabbit and rat. A transmission electron-microscopic study. *Cell Tissue Res.* **261,** 205–210.

Vaughan, D. K., and Lasater, E. M. (1992). Acid phosphatase localization in endocytosed horizontal cell gap junctions. *Vis. Neurosci.* **7,** 77–81.

Veenstra, R. D., Wang, H.-Z., Belbo, D. A. Chilton, M. G., Harris, A. L., Beyer, E. C., and Brink, P. R. (1995). Selectivity of connexin-specific gap junctions does not correlate with channel conductance. *Circ. Res.* **77,** 1156–1165.

Werner, R., Levine, E., Rabadan-Diehl, C., and Dahl, G. (1989). Formation of hybrid cell-cell channels. *Proc. Natl. Acad. Sci. U.S.A.* **86,** 5380–5384.

White, T. W., Bruzzone, R., Wolfram, S., Paul, D. L., and Goodenough, D. A. (1994). Selective interactions among the multiple connexin proteins expressed in the vertebrate lens: The second extracellular domain is a determinant of compatibility between connexins. *J. Cell Biol.* **125,** 879–892.

White, T. W., Paul, D. L., Goodenough, D. A., and Bruzzone, R. (1995). Functional analysis of selective interactions among rodent connexins. *Mol. Biol. Cell* **6,** 459–470.

Wolburg, H., and Rohlmann, A. (1995). Structure–function relationships in gap junctions. *Int. Rev. Cytol.* **157,** 315–373.

Wolosin, J. M. (1991). Gap junctions in rabbit corneal epithelium: Limited permeability and inhibition by cAMP. *Am. J. Physiol.* **261,** C857–C864.

Wolosin, J. M., Candia, O. A., Peterson-Yantorno, K., Civan, M. M., and Shi, X.-P. (1997). Effect of heptanol on the short circuit current of cornea and ciliary body demonstrates rate limiting role of heterocellular gap junctions in active ciliary body transport. *Exp. Eye Res.* **64,** 945–952.

Wolosin, J. M., Chen, M., Gordon, R. E., Stegman, A., and Butler, G. A. D. (1993). Separation of the rabbit ciliary body epithelial layers in viable form: Identification of differences in bicarbonate transport. *Exp. Eye Res.* **56,** 401–409.

Wolosin, J. M., Schütte, M., and Chen, S. (1997). Connexin distribution in the rabbit and rat ciliary body: A case for heterotypic epithelial gap junctions. *Invest. Ophthalmol. Vis. Sci.* **38,** 341–348.

Yeager, M., and Gilula, N. B. (1992). Membrane topology and quarternary structure of cardiac gap junction ion channels. *J. Mol. Biol.* **223,** 929–948.

Zimmermann, K. W., and von Palczewska, I. (1910). Über den Bau der Herzmuskulatur. *Arch. Microscop. Anat.* **75,** 41–65.

CHAPTER 7

The Trabecular Meshwork and Aqueous Humor Reabsorption

Michael Wiederholt and Friederike Stumpff
Institut für Klinische Physiologie, Universitätsklinikum Benjamin Franklin, Freie Universität Berlin, 12200 Berlin, Germany

I. INTRODUCTION

The precise mechanism of outflow regulation is not fully understood. It has been generally accepted that contraction of the ciliary muscle fibers

1063-5823/98 $25.00

extending into the trabecular meshwork influences outflow of aqueous humor and hence lowers intraocular pressure (Kaufman, 1984; Lütjen-Drecoll and Rohen, 1989). Contraction of the ciliary muscle via its insertion at the scleral spur and its tendons may ultimately expand and spread the trabecular lamellae and thus increase the filtration area of the cribriform meshwork (Rohen, 1964; Rohen *et al.,* 1981). The cribriform layer accounts for most of the outflow resistance (Mäepea and Bill, 1992). The concept of the direct contribution of the ciliary muscle to aqueous humor reabsorption is undisputed and has been supported by morphological observations showing that pilocarpine reduced outflow resistance by ciliary muscle contraction (Bárány, 1967; Rohen and Lütjen-Drecoll, 1982) and modified the intercellular spaces of the cribriform meshwork (Grierson *et al.,* 1978; Lütjen-Drecoll *et al.,* 1981). The experimental model of disinsertion of the ciliary muscle in monkey eyes also supported the concept that pilocarpine increased aqueous humor outflow by drug-induced contraction of the ciliary muscle (Kaufman and Bárány, 1976; Kaufman, 1984). Thus, the trabecular meshwork was considered to have no direct effect on aqueous humor reabsorption.

However, morphological studies have presented evidence that muscarinergic agonists such as pilocarpine may directly modify trabecular meshwork geometry (Flocks and Zweng, 1957; Bárány, 1962; Holmberg and Bárány, 1966; Lütjen-Drecoll *et al.,* 1981) and change the shape of cultured trabecular meshwork cells (Tripathi and Tripathi, 1980; Zadunaisky and Spring, 1995) suggesting direct contractile properties of the trabecular meshwork. It has already been demonstrated that trabecular meshwork tissue and cultured meshwork cells of the eye of humans and various animals contain contractile filaments that are consistent with a smooth muscle-like function of the meshwork (Ringvold, 1978; Tripathi and Tripathi, 1980; Grierson *et al.,* 1986; Coroneo *et al.,* 1991a; Flügel *et al.,* 1991; De Kater *et al.,* 1992). In this chapter, our recent research is summarized demonstrating a direct effect of the trabecular meshwork on aqueous humor reabsorption. Thus, the concept of the ciliary muscle being the exclusive modulator of aqueous humor regulation has to be extended.

II. ELECTROPHYSIOLOGY OF CULTURED TRABECULAR MESHWORK CELLS

A. *Resting Membrane Voltage*

Because *in vivo* access by microelectrodes is difficult, we used cell culture techniques to facilitate membrane voltage recordings from cultured trabecular meshwork (TM) cells. Primary cultures of bovine cells were grown by established techniques. At least two different trabecular cell types were identified. Phase-contrast microscopy revealed the presence of epithelial-like and spindle-shaped cell types. These and the mixed types were further characterized by electron microscopy and immunofluorescence methods. The spindle cells in particular showed parallel alignment of intracellular smooth muscle specific α-isoactin filaments and had long cytoplasmic processes and abundant intermediate filaments and microfilaments (Coroneo *et al.*, 1991a). Human TM cells were generated from five donors of different ages including a cell line from a donor with reported glaucoma. Human TM cells are more difficult to grow and proliferate more slowly in culture. However, all cell lines grew in the absence of exogenously added growth factors (except fetal calf serum) and extracellular matrix component. When compared to bovine TM cells, which reached confluency within 2 weeks, the human cells needed 6–8 weeks to reach confluency after passaging at a splitting factor of 1:2 (Lepple-Wienhues *et al.*, 1994; Coroneo *et al.*, 1991b). In these TM cells, the classical method of puncturing cells with glass microelectrodes (tip resistance of 60–140 MΩ) was used. Stable membrane voltage recordings could be obtained from individual cells for periods of up to 20 min.

In Table I, a summary of the membrane voltages recorded in cultured bovine and human TM cells is given. The frequency distribution of mem-

TABLE I

Membrane Voltage and Relative Potassium Permeability (Apparent Transference Number) in Cultured Trabecular Meshwork Cells

Species	Mean voltage[a] (mV)	Relative K^+ permeability (%)	Ref[b]
Bovine			
Epithelial cells	−49.7 ± 0.8 (143)	50	1
Spindle cells	−70.9 ± 1.9 (48)	71	1
Mixed cells	−55.0 ± 1.0 (191)	—	3, 4
Human			
Primary culture	−62.0 ± 5.0 (17)	—	3
Cell line, H_1 TM	−63.1 ± 1.0 (46)	65	2
Cell line, BER 5	−58.5 ± 1.6 (80)	—	5
Cell line, ALC 4	−47.4 ± 3.8 (17)	—	5
Cell line, ALC 10	−44.0 ± 1.4 (71)	—	5
Cell line, ALC 2 (glaucomatous eye)	−58.7 ± 1.2 (79)	—	5

[a] X ± SEM, number of experiments in parentheses.

[b] 1, Coroneo *et al.*, 1991a; 2, Coroneo *et al.*, 1991b; 3, Lepple-Wienhues *et al.*, 1991c; 4, Lepple-Wienhues *et al.*, 1992a; 5, Lepple-Wienhues *et al.*, 1994.

brane voltage and morphological arguments both point to the presence of at least two different cell populations in cultured bovine TM. The smooth muscle-like spindle cells had higher membrane voltages than the epithelial cells, which revealed the characteristics of metabolically active cells in electron microscopy studies. In addition to the experiments performed on bovine cultures, similar studies were conducted using a variety of different TM cell cultures of human origin. The cell lines showed a spectrum of different potential levels. However, in principle, the membrane voltages in human cells were similar to those obtained in bovine cells. The membrane potential of a cell line derived from a glaucomatous eye fell within the range observed for the other cells. No significant morphological differences could be observed. Furthermore, recent work has demonstrated that the amount of contractile filaments in human trabecular meshwork is similar in glaucomatous and nonglaucomatous eyes (Flügel et al., 1992).

Inspection of Table I indicates that all membrane voltages measured are more positive than the calculated potassium equilibrium potential, indicating some cell conductivity for other ions, that is, for sodium, calcium, or chloride. An overall relative potassium permeability can be calculated by measuring the membrane voltage changes induced by changing extracellular potassium concentration. The calculated apparent transference number (t_K) indicates that the potassium ion only determines approximately 50–71% of the total membrane permeability. Again, there is no principal difference in relative K^+ permeability between cultured bovine and human TM cells.

B. Excitability

One of the key steps in electromechanical coupling in smooth muscle cells is the depolarization of membrane voltage and the appearance of bursts of calcium-dependent voltage changes. Smooth muscle cells frequently exhibit either spontaneous oscillations of their membrane voltage or induced excitability. In contractile smooth muscle cells, oscillating membrane voltages ("abortive action potentials") may be induced by such maneuvers as applying extracellular Ba^{2+} (Korbmacher *et al.,* 1989). In bovine (Coroneo *et al.,* 1991a) and in human TM cells (Lepple-Wienhues *et al.,* 1994), such spontaneous fluctuations could be observed, and depolarization by application of extracellular Ba^{2+} induced voltage spikes (Table II). A typical recording is shown in Fig. 1. These voltage oscillations were dependent on the presence of extracellular Ca^{2+} and could be blocked by calcium channel blockers such as nifedipine, but were not inhibitable by tetrodotoxin (an inhibitor of fast Na^+ channels). Abortive action potentials of smooth muscle cells are typically insensitive to tetrodotoxin.

TABLE II

Characterization of Excitability (≈ Voltage Spikes, ≈ Abortive Action Potentials) in Cultured Trabecular Meshwork Cells

Experimental procedure	Human TM cells	Bovine TM cells	Ref.[a]
Spontaneous fluctuations	Rare	≈10% of all cells	1, 2
Voltage spikes			
Induced by Ba^{2+}	5–10 mmol/liter	1–10 mmol/liter	
Sensitive to extracellular Ca^{2+}	+	+	1, 2
Inhibitable by nifedipine	+	+	1, 2
Insensitive to tetrodotoxin	+	+	1, 2

[a] 1, Coroneo *et al.,* 1991a; 2, Lepple-Wienhues *et al.,* 1994.

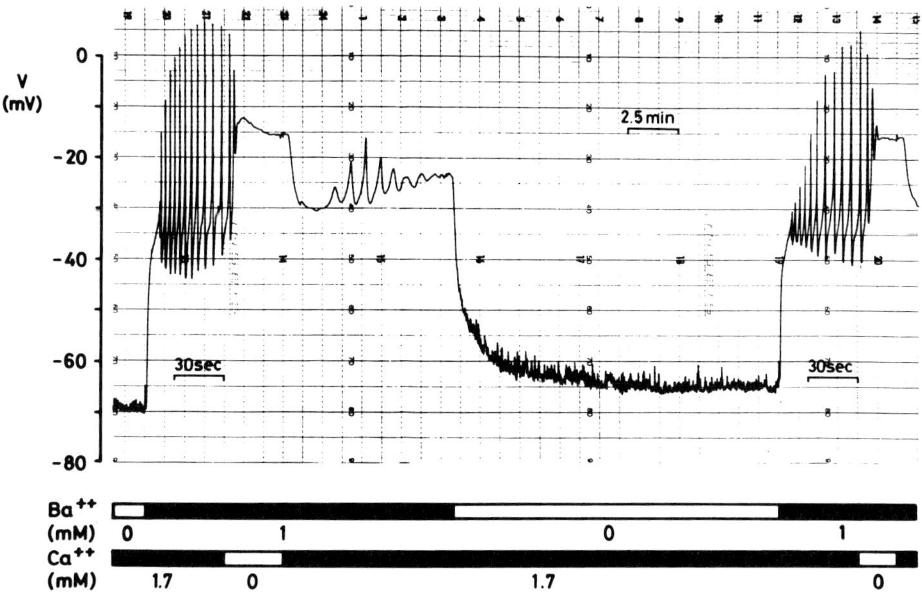

FIGURE 1 Repetitive overshooting action potentials in a cultured bovine TM cell induced by 1 mmol/liter Ba^{2+}. The action potentials are abolished in the absence of extracellular Ca^{2+}. Only voltage fluctuations are seen on readdition of extracellular Ca^{2+}. After a period of superfusion with the control solution, Ba^{2+}-induced action potentials can again be elicited. (From Coroneo *et al., Exp. Eye Res.* **52,** 375–388, 1991, Fig. 8, with permission of the publisher.)

The sequence of events is generally thought to be the following: (1) The membrane voltage is depolarized by a blockade of Ba^{2+}-sensitive K^+ channels; (2) the depolarization opens voltage-dependent Ca^{2+} channels, which are sensitive to dihydropyridine (nifedipine); (3) the influx of Ca^{2+} (or Ba^{2+}) ions drives the membrane voltage toward the equilibrium potential for Ca^{2+} (Ba^{2+}), which is well in the positive range; (4) by increasing the intracellular Ca^{2+} concentration contractions may be induced; and finally (5) the overshooting depolarization is then stopped and reversed by Ca^{2+}-sensitive K^+ channels, which are activated by the increasing intracellular Ca^{2+} concentration. Again, the presented arguments support the assumption that human as well as bovine TM cells are equivalent to excitable smooth muscle cells and are functionally an independent contractile element in the aqueous humor outflow pathway.

C. *Effect of Various Drugs on Membrane Voltage*

Table III summarizes the various substances and drugs we used in human and bovine trabecular meshwork cells. As expected, ouabain depolarized the membrane voltage, indicating the electrogenic component of a Na^+,K^+-ATPase. Also typical for most mammalian cells is the presence of a large K^+ conductance with Ba^{2+}-sensitive K^+ channels. We already discussed the induced excitability, which could be largely inhibited by nifedipine and verapamil, both of which typically block voltage-dependent Ca^{2+} channels of the L type.

An electrogenic Na^+–HCO_3^- cotransport has been described in bovine corneal endothelium and human ciliary muscle (Jentsch et al., 1984; Stahl et al., 1992). In contrast to a simple Na^+ conductance, this symport mediates a net flux of negative charge (HCO_3^-) coupled to the flux of sodium so that Na^+ removal should induce a flux of HCO_3^- out of the cell. Like many anion exchangers, this symport is sensitive to DIDS. Although bovine cells do not seem to express an electrogenic Na^+–HCO_3^- transport, all the different human cell lines studied showed evidence for the presence of this symport. Besides having an impact on membrane voltage, such transporters are involved in the regulation of intracellular pH (Jentsch et al., 1988).

Muscarinic receptors have been described in human trabecular meshwork cells (WoldeMussie *et al.*, 1990). In our studies with human TM cells, acetylcholine evoked a membrane voltage response typical for muscarinic receptors coupled to phospholipase C (PLC) (Lepple-Wienhues *et al.*, 1994). The voltage changes induced by acetylcholine were highly sensitive to atropine. Activation of such muscarinic receptors generally evokes Ca^{2+} release from intracellular stores and an influx of Ca^{2+}. This, in turn, activates Ca^{2+}-sensitive K^+ channels as a possible explanation for the transient hyperpolarization observed. The consecutive depolarization could be mediated

TABLE III

Effect of substances and Drugs on Membrane Voltage of Cultured Human and Bovine Trabecular Meshwork Cells

Substance/drug	Human TM	BovineTM	Mechanisms involved	Refs.[a] Human TM	Refs.[a] Bovine TM
Ouabain	Depolarization	Depolarization	Na,K^+-ATPase	2	1
High K^+	Depolarization	Depolarization	K^+ conductance	2	1
Ba^{2+}	Depolarization	Depolarization	Ba^{2+}-sensitive K^+ channels	2, 3, 5	1, 4
Low Ca^{2+}	⊥Excitability[b]	⊥Excitability	Dihydropyridine-sensitive	2	1
Nifedipine	⊥Excitability	⊥Excitability	Ca^{2+} channels (L-type)	3, 5	1, 4
Verapamil	∅[c]	⊥Excitability		—	4
Tetrodotoxin	No effect	No effect	Lack of fast Na^+ channels	2, 3, 5	1, 4
Low Na^+ (± DIDS)	Depol. DIDS	Depol. DIDS	Electrogenic Na^+/HCO_3		
Low HCO_3^- (± DIDS)	Depol. sensitive	Depol. insensitive	symporter (only in humans)	5	1
Acetylcholine (± atropine)	Trans. hyperpolarization followed by depolarization	∅	Muscarinic receptors	2, 5	—
Isoproterenol (± metipranolol)	Depolarization	∅	β-Adrenergic receptors	5	—
Endothelin-1	Depolarization	Depolarization	ET_A receptors, nonselective cation channels	3, 5	4

[a] 1, Coroneo *et al.*, 1991a; 2, Coroneo *et al.*, 1991b; 3, Lepple-Wienhues *et al.*, 1991c; 4, Lepple-Wienhues *et al.*, 1992b; 5, Lepple-Wienhues *et al.*, 1994.

[b] ⊥ inhibition of excitability.

[c] Not tested.

by opening of nonselective cation channels permeable to Ca^{2+} (Berridge, 1993; Sims and Janssen, 1993).

Opening of cation channels leading to membrane depolarization is also a well-known effect following activation of β-adrenergic and endothelin receptors (Resink *et al.*, 1990; Lepple-Wienhues *et al.*, 1992a). The dose-dependent effect of isoproterenol on membrane voltage and its sensitivity to metipranolol indicates the presence of β-adrenergic receptors in cultured human TM cells. β-adrenergic receptors have been described in human TM cells using autoradiographic localization (Jampel *et al.*, 1987). The receptors are mainly of the β_2 subtype (Wax *et al.*, 1989).

The first report of an effect of endothelin on the eye was the endothelin-mediated membrane depolarization and elevation of intracellular Ca^{2+} in a human ciliary muscle cell line (Korbmacher *et al.,* 1989). As in ciliary muscle cells, endothelin dose dependently depolarized membrane voltage in cultured human and bovine TM cells. This indicates the presence of ET_A receptors in both tissues. Endothelin was initially supposed to act directly on L-type calcium channels (Yanagisawa *et al.,* 1988). However, subsequent studies have shown an involvement of the PLC-mediated hydrolysis of phosphoinositoles and release of intracellular calcium (Resink *et al.,* 1990). Nonselective cation channels have also been shown to mediate endothelin-evoked calcium entry and contraction (van Renterghem *et al.,* 1989; Pollock *et al.,* 1995).

The description of transporters, channels, and functional presence of receptors (Table III) is just a first step in analyzing the complex transport properties across the TM membrane.

D. Relevance of Membrane Voltage Measurements

Membrane voltage is an important factor in determining contractile cell tone and thus contractility of the trabecular meshwork. In arterial smooth muscle cells, the height of membrane voltage, which is mainly determined by K^+ channels, is an important regulator of contractile tone (Nelson and Quayle, 1995). In most arterial smooth muscle cells, membrane voltage hyperpolarization leads to vasodilation/relaxation, and voltage depolarization leads to vasoconstriction/contraction. However, there are important exceptions from this common feature of smooth muscle cells. Furthermore, beside electromechanical coupling, pharmacomechanical coupling is an important mechanism in the regulation of contraction in some smooth muscle cells (Somlyo and Somlyo, 1968) including the ciliary muscle (Ito and Yoshitomi, 1986). Thus, the functional relevance of the described channels and transporters has to be tested directly by measuring contractility.

E. Characterization of a Calcium-Dependent Maxi-K^+ Channel by Patch-Clamp Techniques

So far, no patch-clamp data exist that describe the properties of TM cells. We were interested in studying channels known to contribute to smooth muscle contraction/relaxation. Among the various K^+ channels present in TM cells, we concentrated on the maxi-K^+ channel (K_{Ca}), which is an important regulator of the balance depolarization and hyperpolarization and thus contraction and relaxation in smooth muscle cells (Nelson and Quayle, 1995).

The whole-cell and single-channel configurations of the patch-clamp technique were used to study membrane conductance (Stumpff et al., 1996). Table IV summarizes the characteristics of this maxi-K^+ channel. The channel exhibited voltage-dependent activation and inactivation could be blocked by extracellular charybdotoxin, tetraethyl ammonium ions (TEA^+), and internal Ba^{2+}, while extracellular Ba^{2+} and intracellular TEA^+ had little effect. Increase of intracellular Ca^{2+}, adenosine triphosphate (ATP) (1 mmol/liter), and cGMP (8-bromo-cGMP 10^{-3} mol/liter) activated the channel. These are typical characteristics of the maxi-K^+ channel observed in almost every type of smooth muscle cells (Nelson and Quayle, 1995). The data presented are, in consequence, additional arguments for TM cells being smooth muscle cells functionally. It is of importance that elevation of intracellular Ca^{2+} in the physiological range and depolarization increases the open probability of this channel. Activation of the maxi-K^+ channel leads to an increase of K^+ efflux, which counteracts depolarization. Substances that block the maxi-K^+ channel should depolarize and increase contractility, while openers of the channel should hyperpolarize and relax trabecular meshwork cells. Figure 2 is a schematic representation of the postulated role of the maxi-K^+ channel for contraction/relaxation in TM cells. Recent evidence indicates that cGMP-dependent protein kinases are able to activate the maxi-K^+ channel in smooth muscle cells (Archer *et al.*, 1994; Nelson and Quayle, 1995) in addition to a direct effect of NO on this channel (Bolotina *et al.*, 1994). The maxi-K^+ channel appears to serve as a negative-feedback mechanism in regulating membrane voltage and hence the balance between contraction and relaxation. Thus, it is an important

TABLE IV

Properties of a Ca^{2+}-Dependent Maxi-K^+ Channel (K_{Ca}) in Cultured Bovine Trabecular Meshwork Cells[a]

Whole-cell configuration
Strong outward current, voltage dependent
Charybdotoxin (10^{-7} mol/liter) blocks the outward current by 35%
TEA^+ blocks the outward current dose dependently
Ionomycin (10^{-5} mol/liter) increases outward current by 650%
cGMP stimulates the current to 290% of the original level, 44% of which is blockable by charybdotoxin
Single-channel configuration
High conductance for potassium (>300 pS), negligible conductance for sodium (<1 pS)
Open probability (NP_o) increases with both voltage and calcium concentration
ATP increases the open probability and shifts the voltage dependence toward more negative values

[a] Data from Stumpff *et al.* (1996).

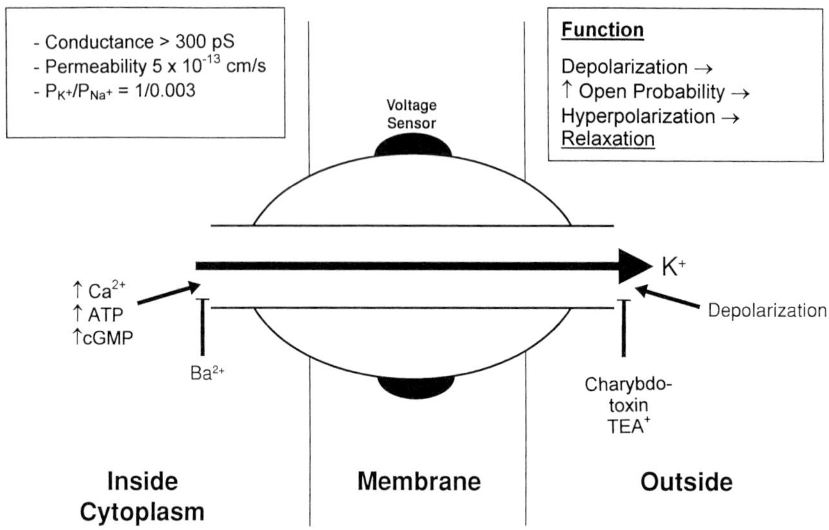

FIGURE 2 Schematic representation of the maxi-K^+ channel (K_{Ca}, BK) in trabecular meshwork cells. Increases in intracellular calcium, ATP, and cGMP increase the open probability of the channel and thus stimulate the charybdotoxin-sensitive outward current of potassium. This efflux leads to hyperpolarization of the membrane voltage and might induce relaxation of smooth muscle cells like the TM cell.

target for vasoactive substances that could modify TM contractility and consequently modulate aqueous humor reabsorption. Concerning the regulation of intraocular pressure it is most interesting that in arterial smooth muscle cells changes in pressure interfere with the effects of substances that block the maxi-K^+ channel (Nelson and Quayle, 1995).

III. INTRACELLULAR CALCIUM

There is general agreement about the fact that the concentration of intracellular Ca^{2+} is a most important regulator of smooth muscle contraction (Berridge, 1993; Somlyo and Somlyo, 1994). Using fura-2-loaded bovine TM cells, we measured the concentration of intracellular calcium by recording the fluorescence ratio of calcium-bound to calcium-free dye (Lepple-Wienhues *et al.*, 1992a). Similar to observations on cultured ciliary muscle cells (Stahl *et al.*, 1991), we observed an agonist-induced biphasic increase in intracellular calcium in cultured trabecular meshwork cells. The basal resting calcium concentration was 40–80 nmol/liter. Application of endothelin resulted in an initial $[Ca^{2+}]_i$ peak followed by a recovery phase and a $[Ca^{2+}]_i$ plateau. In smooth muscle cells, the initial calcium peak is due to release of intracellular calcium stores, whereas the calcium plateau is the consequence of entry of extracellular calcium (Stahl *et al.,* 1991). Since endothelin as well as muscarinic acetylcholine receptors are coupled to the PLC system, both agonists induce a calcium release from inositol trisphosphate (IP_3)-sensitive intracellular stores (Berridge, 1993; Sims and Janssen, 1993; Somlyo and Somlyo, 1994). The calcium entry induced by both agonists also seems to be mediated by common mechanisms, which include entry through receptor-operated and voltage-operated calcium channels. Furthermore, as with acetylcholine, an endothelin-induced inward current through a nonselective cation channel has been demonstrated in smooth muscle cells (van Renterghem *et al.,* 1989). Such a nonselective cation channel may thus serve as an important common calcium entry mechanism in smooth muscle cells such as TM cells.

The initial transient hyperpolarization induced by acetylcholine and endothelin (see Section II.C) has a time course similar to the transient $[Ca^{2+}]_i$ peak induced by endothelin. This transient hyperpolarization can be explained by a calcium-dependent potassium conductance that is also present in smooth muscle cells (Stahl *et al.,* 1991; Nelson and Quayle, 1995).

Recently, intracellular calcium measurements were performed in cultured human TM cells (Kohmoto *et al.,* 1994; Shade *et al.,* 1996). Various muscarinic agonists (carbachol, aceclidine, pilocarpine, and oxotremorine-M) increased intracellular calcium dose dependently. The absolute concentration of steady-state intracellular calcium and the agonist-induced calcium increase with a transient peak and a plateau phase were almost identical to the measurements obtained in cultured bovine TM cells (Lepple-Wienhues *et al.,* 1992a). Furthermore, the use of muscarinic receptor subtype antagonists showed that the M_3 receptor subtype is most important for the effect of carbachol. As in other smooth muscle cells, carbachol stimulated phosphoinositide production as an indication for the coupling between the muscarinic receptor and the PLC system. Various neuropeptides (neuropeptide Y, substance P, bombesin, calcitonin gene-related peptide, VIP) increased intracellular calcium in cultured bovine TM cells (Ohuchi *et al.,* 1992). Neuropeptide Y was the most potent peptide, and its effect on calcium was accompanied by an increase in phosphoinositide turnover.

The measurement of intracellular calcium in cultured TM cells, and its elevation by substances known to have an effect on smooth muscle contractility, is an important argument for documenting the smooth-like properties of trabecular meshwork. However, it is of preeminent importance to measure contractility *directly.*

IV. REGULATION OF INTRACELLULAR pH

Regulation of intracellular pH (pH_i) is an essential feature of cell homeostasis to which many cellular processes are sensitive. Transporters involved in pH_i regulation participate in many cell types in transmembrane transport, transepithelial transport, growth factor activation, and cell proliferation (Roos and Boron, 1981). It is possible that these transporters may play a similar role in TM function.

Cytoplasmic pH was continuously monitored using the pH-sensitive absorbance of 5(6)carboxy-4′,5′-dimethylfluorescein in cultured bovine TM cells (Coroneo *et al.*, 1992). In standard HCO_3^- Ringer's solution, pH_i averaged 7.02 and was significantly lower in HCO_3^--free solution (Table V). Intracellular buffering capacity was evaluated by incubating cells for brief periods in Ringer's solutions containing HCO_3^- or in HCO_3^--free Ringer's containing acetate. Acetate results in an initial fall in pH_i as a result of nonionic diffusion of acetic acid, which dissociates intracellularly into acetate and H^+. From the amplitude of the pH_i changes, the intracellular buffering capacity can be estimated. The measured values were different in HCO_3^- and HCO_3^--free media (Table V). However, the buffering capacity measurements are only valid for the pH_i range at which they were obtained.

We have demonstrated the presence of three mechanisms involved in pH_i regulation (Table VI). Replacement of extracellular Na^+ resulted in an intracellular acidification of 0.35 pH units/min. pH_i recovery after an acute intracellular acid load was Na^+ dependent and completely blocked by amiloride or reversed by extracellular Na^+ replacement in the absence of HCO_3^-. These findings are compatible with the presence of a Na^+–H^+ exchanger. This exchanger is the main mechanism responsible for pH_i

TABLE V

Intracellular pH in Cultured Bovine Trabecular Meshwork Cells[a]

Experimental procedure	Steady state (pH_i)		Intracellular buffering capacity (mmol H^+/pH unit)
Ringer's solution (pH 7.4) with HCO_3^-	7.02 ± 0.03 (40)[b]		89.4
		$P < 0.001$	
Ringer's solution (pH 7.4) without HCO_3^-	6.89 ± 0.01 (39)		56.6

[a] Data from Coroneo *et al.* (1992).

[b] $X \pm$ SEM, number of experiments in parentheses.

TABLE VI

Ion Transporters Involved in Regulation of Intracellular pH in Bovine Trabecular Meshwork Cells[a]

Transporter	Characteristics
Na^+–H^+	Na^+ dependent H^+ dependent Amiloride sensitive (1 mmol/liter)
Na^+-dependent Cl^-–HCO_3^- exchanger (Na^+–HCO_3^- symporter)	Na^+ dependent HCO_3^- (Cl^-) dependent DIDS-sensitive (1 mmol/liter) Pyridoxal 5′-phosphate-sensitive (10 mmol/liter)
Na^+-independent Cl^-–HCO_3^- exchanger	HCO_3^- (Cl^-) dependent DIDS sensitive (1 mmol/liter) Ethacrynic acid sensitive (0.1 mmol/liter)

[a] Data from Coroneo *et al.* (1992).

regulation after an acid load, and was also described by Chu *et al.* (1992) in their preparation of bovine trabecular meshwork.

In HCO_3^- containing media, the pH_i recovery was only partially blocked by amiloride alone but completely blocked by amiloride + DIDS or + pyridoxal 5′-phosphate, suggesting participation of anion (HCO_3^-) transporters.

Replacement of extracellular Cl^- induced an intracellular alkalization rate of 0.42 pH units/min. This alkalization was dependent on the presence of extracellular HCO_3^-. Replacement of extracellular Cl during pH_i recovery from an alkaline load reversed pH_i back-regulation, implying the presence of a Cl^-/HCO_3^- exchange. Further experiments using sodium replacement revealed that part of this exchange was sodium independent, and could be blocked by ethacrynic acid, an effective blocker of the Na^+-independent Cl^-/HCO_3^- exchanger (Madshus and Olsnes, 1987). Thus, three independent transporters can be postulated for regulating the intracellular pH of TM cells: The Na^+–H^+ exchanger, the Na^+-independent Cl^-–HCO_3^- exchanger, and the Na^+-dependent Cl^-–HCO_3^- exchanger.

It is not our aim in this review to discuss the available information on these transporters in general. However, we should mention that in other cell systems of the eye such as cornea and ciliary epithelium similar transporters involved in cytoplasmic pH_i regulation have been described in detail (Jentsch *et al.,* 1988; Helbig *et al.,* 1989). Although the kinetics of the transporters may vary between different cells and species, the principal properties are very similar. In Fig. 3 the importance of the described transporters for pH_i regulation in trabecular meshwork cells is summarized

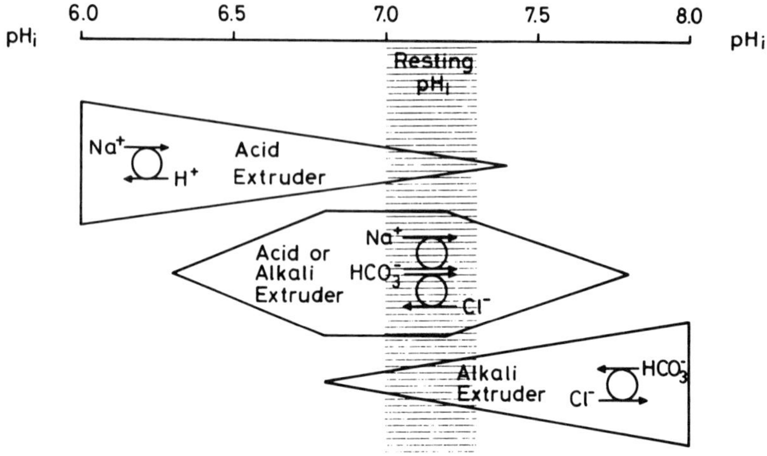

FIGURE 3 Relative activity of ionic transporters involved in regulation of pH_i in TM cells. Resting pH_i is 7.0–7.2 in TM cells. Na^+–H^+ exchange is activated at more acidic pH_i, while HCO_3^-–Cl^- exchange is stimulated at more alkaline pH.

schematically. The amiloride-sensitive Na^+–H^+ exchanger is functionally an acid extruder and is activated by intracellular acidification. At resting pH_i this transporter has only small activity; the internal modifier site of this transporter is mainly activated at pH_i lower than 6.9. However, the Cl^-/HCO_3^- exchanger is an alkali extruder and seems to be effective at more alkaline cytoplasmic pH_i. This exchanger is also active at resting pH_i and expresses a small activity also in the more acidic range. The Na^+-coupled HCO_3^- transporter is the dominant force at resting pH_i and functions as an acid extruder at normal intracellular pH of 6.8–7.1 (by taking up HCO_3^- and Na^+ in exchange for Cl^-). In alkali-loaded cells this transporter can be reversed to become an alkali extruder. In bovine TM cells this transporter seems to be electrically neutral while in human TM cells an electrogenic Na^+–HCO_3^- symporter was described (see Table III).

In conclusion, we have described three important mechanisms involved in pH_i regulation in TM cells. Their functional role, besides their capability to deal with acid and alkali load, remains to be elucidated. An important aspect is the fact that TM cells as well as most other cells respond to growth factors and can be stimulated to proliferate, both processes being pH_i dependent. Endothelin but not angiotensin II increased intracellular pH in cultured bovine TM cells (Kohmoto *et al.*, 1994).

V. DIRECT MEASUREMENT OF CONTRACTILITY OF ISOLATED TRABECULAR MESHWORK AND CILIARY MUSCLE STRIPS

A. Method

The contractility of isolated ciliary muscle strips has been measured in several mammalian species including man (Ito and Yoshitomi, 1986; Lograno and Reibaldi, 1986). Measurements of contractility of trabecular meshwork had not been performed before we attempted to obtain direct contractility measurements in isolated TM strips and compared them with those in ciliary muscle strips. Because of the anatomical proximity of ciliary muscle and trabecular meshwork and the extension of ciliary muscle fibers into the trabecular meshwork in the eye of higher primates (Rohen, 1964; Tripathi, 1974), we chose the bovine eye as a model. In this species, the ciliary muscle is rudimentary and posteriorly located and can be easily separated from the trabecular meshwork, which in the bovine eye more closely resembles a reticular meshwork (Rohen, 1964; Flügel *et al.,* 1991). The bovine chamber angle has recently been examined in detail (Flügel *et al.,* 1991; Wiederholt *et al.,* 1995).

Strips of the trabecular meshwork with a width of approximately 0.5 mm and a length of 2–4 mm were dissected (Lepple-Wienhues *et al.,* 1991a). To obtain isometric recordings of very small forces in fragile elastic tissue strips, we developed a force-length transducer similar to that described by Brutsaert *et al.* (1988). In principle, contractions of the isolated strips are electrically counterbalanced via a lever/magnetic coil system and an optoelectronic device. This results in a position clamp (isometric contraction) of the tissue with negligible length changes of less than 10 μm. It was possible to measure forces in the range of 0.5–2000 μN (equivalent to forces of approximately 0.5–200 mg). The average tensions measured in the trabecular meshwork were in the range of 50–500 μN and 200–2000 μN in ciliary muscle strips of similar length but larger diameter. After an equilibration period of approximately 1 hr, stable and reproducible measurements could be obtained for several hours (Lepple-Wienhues *et al.,* 1991a; Wiederholt *et al.,* 1993).

B. Effect of Substances and Drugs on Contraction/Relaxation

Table VII is a summary of the measurements obtained in TM strips. The relevant literature is also given in the table.

TABLE VII

Characterization of Contractile Properties of Isolated Bovine Trabecular Meshwork Strips by Isometric Force Measurements

Substance/drug	Contraction[a]	Relaxation[a]	Comment	Refs.[b]
High external K^+	+		Blocked by atropine	1, 3
Cholinergic agents				
Acetylcholine	++		Blocked by atropine	1, 3, 6
Pilocarpine	++		EC_{50}[c] 2×10^{-6} mol/liter	1, 3, 6
Aceclidine	++			1, 3
Carbachol	+++		EC_{50} 2×10^{-7} mol/liter Blocked by $M_3 \gg M_1$ antagonists	2, 3, 6
Adrenergic agents				
Phenylephrine (α_1)	+		Blocked by prazosin	6
Brimonidine (α_2)	++		Blocked by yohimbine	6
Isoproterenol (β)		−	Blocked by metipranolol	3, 6
Epinephrine (α,β)	+			3, 6
Epinephrine + metipranolol	++			6
Low external Ca^{2+}		−−	~60% of carbachol-induced force	2
Ca^{2+} channel blockers				
Nifedipine		−		2
Verapamil		−−		7
Ni^{2+}		−−		2
Endothelin				
ET-1	++		EC_{50} 5×10^{-9} mol/liter	2
Nitric oxide system				
cGMP		−−		4
L-NAG	+		Inhibitor of NO synthase	4
Organic nitrovasodilators				
ISDN		−		4
5-ISMN		−		4

(*continues*)

continued

Substance/drug	Contraction[a]	Relaxation[a]	Comment	Refs.[b]
Non-nitrates				
SNP		−−		4
SNAP		−−		4
Prostaglandins				
$PGF_{2\alpha}$	No effect	No effect	FP receptor	5
17-Phenyl $PGF_{2\alpha}$	No effect	No effect	FP receptor	5
Sulprostone	+		$EP_3 > EP\ 1$ receptor	5
11-Deoxy PGE_1		−	EP receptor	5
AH 13205		−−	EP_2 receptor	5
U-46619	+++		TP receptor	5
Cyclo-oxygenase inhibitor				
Indomethacin	+		Carbachol-contraction augmented	7
Diuretics				
Furosemide	No effect	No effect		7
Hydrochlorothiazide	No effect	No effect		7
Ethacrynic acid		−−−		7

[a] Relative amount of contraction/relaxation: strong: +++/−−−; medium: ++/−−; small: +/−.

[b] 1, Lepple-Wienhues *et al.*, 1991a; 2, Lepple-Wienhues *et al.*, 1991b; 3, Wiederholt *et al.*, 1993; 4, Wiederholt *et al.*, 1994; 5, Krauss *et al.*, 1997; 6, Wiederholt *et al.*, 1996; 7, Wiederholt *et al.*, 1997.

[c] EC_{50}, half-maximal effective concentration of agonists.

1. High External K^+

Depolarization of membrane voltage induced by raising extracellular potassium concentration is a classical mechanism for contracting smooth muscle cells (Somlyo and Somlyo, 1994). A reproducible biphasic contractile response was evoked by 120 mmol/lK^+. This response could be markedly reduced by atropine. Compared to the maximal contraction induced by acetylcholine (100%), the contraction induced by depolarization was only 19%. Because atropine partly inhibited the contraction, a significant fraction of force development must be due to a depolarization-induced acetylcholine release from cholinergic nerve terminals rather than resulting from depolarization of the muscle cell membrane per se. The small atropine-resistant contraction might be due to the release of nonmuscarinic transmitters and/or a direct effect of depolarization. Futhermore, the data are evidence for the typical smooth muscle behavior where contraction can be induced by electromechanical and/or pharmacomechanical coupling.

2. Cholinergic Agents

Various cholinergic agonists were tested that showed dose-dependent contractions. A typical original recording is shown in Fig. 4. The contrac-

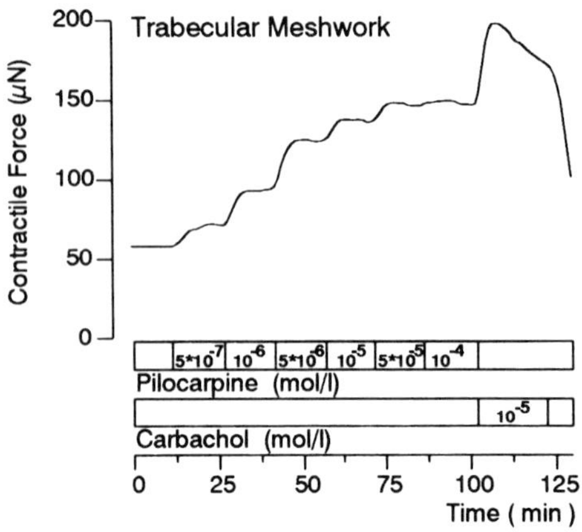

FIGURE 4 Original recording of isometric force measurement of isolated bovine TM strip. Dose-dependent contractions are induced by pilocarpine and a maximal concentration of carbachol.

tions induced by acetylcholine, pilocarpine, and aceclidine could be completely inhibited by atropine demonstrating the presence of muscarinic rather than nicotinic receptors in TM strips. When the reversible anticholinesterase agent physostigmine was given, the maximal force induced by acetylcholine increased, indicating acetylcholine esterase activity in bovine TM. Thus, in most experiments cholinergic agonists like pilocarpine and carbachol that are resistant to hydrolysis by esterases were used. The relative potency of the cholinergic agonists was carbachol $\gg$ aceclidine $>$ pilocarpine $>$ acetylcholine. As compared with pilocarpine (100%), carbachol evoqued a maximal response of 165% in TM. The half-maximal effective concentration, EC_{50}, was an order of magnitude lower for carbachol than for pilocarpine.

Pharmacological and molecular biological evidence has been provided for the existence of at least five muscarinic receptor subtypes (Dörje *et al.*, 1991). In our studies, the muscarinic antagonists pirenzepine and 4-DAMP were used. Pharmacological data indicate that pirenzepine is an M_1-receptor antagonist. 4-DAMP has equal affinity for M_1 and M_3 receptors (Dörje *et al.*, 1991). In isolated strips of TM, 4-DAMP was the most potent antagonist, displaying a selectivity that was approximately 100-fold that of pirenzepine; that is, 4-DAMP at 10^{-8} mol/liter was as effective as pirenzepine at 10^{-6} mol/liter in inhibiting maximal carbachol-induced contractions. Thus, the data suggest the presence of functional muscarinic receptors mainly of the M_3 subtype in bovine TM cells. Muscarinic receptors, mainly of the M_3 subtype, have also been described in human TM cells (WoldeMussie *et al.*,

1990; Gupta *et al.*, 1994). Receptor subtype antagonists are modulators of the aqueous humor outflow (Gabelt and Kaufman, 1992).

3. Adrenergic Agents

Adrenergic agonists, especially epinephrine, have been shown to reduce outflow resistance and thus increase outflow facility through direct actions on the trabecular meshwork and via the uveoscleral route (Kaufman, 1984, 1986; Alvarado *et al.*, 1990; Robinson and Kaufman, 1990; Erickson-Lamy *et al.*, 1992). β-adrenergic receptors have been described in human TM cells (Jampel *et al.*, 1987; Wax *et al.*, 1989) and are mainly of the β_2 subtype. However, it remains an open question whether the effects on outflow are mediated via α- or β-adrenergic receptors. α_1- (phenylephrine) and α_2-agonists (brimonidine) contracted the TM strips with approximately 20% of the potency of carbachol. The effects of both agonists could be completely blocked by the specific antagonists (prazosin, yohimbine). The effect of the α_2-agonist was more pronounced than the effect of the α_1-agonist, indicating that TM cells possess functional α_2- > α_1-adrenergic receptors.

In contrast to α-adrenergic agonists, which contracted the TM, a β-agonist such as isoproterenol significantly relaxed the tissue precontracted by carbachol. The isoproterenol-mediated relaxation could be blocked by metipranolol. Metipranolol per se had no effect on precontracted tissues.

The interpretation of effects of epinephrine is more complicated since epinephrine is a nonspecific α- and β-agonist. It has been widely accepted that epinephrine lowers intraocular pressure by increasing the aqueous humor outflow (Kaufman, 1984; Alvarado *et al.*, 1990). In the isolated bovine strips, high concentrations (10^{-3}–10^{-4} mol/liter) of epinephrine (and dipivefrin) induced significant contractions. The concentration of epinephrine in the aqueous humor must reach such a high level for an effective reduction in intraocular pressure to take place (Kaufman, 1984). In our experiments, further contraction was induced in the trabecular meshwork when epinephrine and metipranolol were given, indicating that the β-component of epinephrine (relaxation) was blocked. Thus, the effect of epinephrine depends on the balance between the activity of the α- (contraction) and the β-adrenergic component (relaxation). The data we obtained with isolated strips are consistent with the assumption of functional α- and β-adrenergic receptors, which may act on outflow facility by direct modification of the TM contractility. There appears to be a relaxant response to β-adrenergic agonists that is sensitive to β-blockers.

4. Low External Ca^{2+} in Tissues Precontracted by Carbachol or Endothelin

Total removal of external Ca^{2+} led to a fast and reversible relaxation of contraction (see Table IX in a later section). It is most interesting that in trabecular meshwork, 42% of the carbachol-induced and 23% of the endothelin-induced force response remained after removal of extracellular calcium. In isolated ciliary muscle strips, removal of external calcium led to complete relaxation of contractions induced by carbachol and endothelin (see Table IX in a later section). In both tissues, readdition of Ca^{2+} resulted in an immediate recovery of the tension. Thus, a calcium-sensitive pathway and additional mechanisms independent of external calcium are involved in the effects of carbachol and endothelin on TM contractility. Contractions partially independent of extracellular calcium have been shown in vascular smooth muscle tissues (Kodama *et al.*, 1989; Marsault *et al.*, 1990). A possible stimulation of calmodulin-independent protein kinases regulating myosin activity has been postulated (Kodama *et al.*, 1989).

5. Ca^{2+} Channel Blockers

In the human eye, the Ca^{2+} channel blocker verapamil decreases outflow resistance by an unknown mechanism (Erickson *et al.*, 1995). Two well-described calcium blockers, extensively used in clinical medicine, and the inorganic calcium blocker Ni^{2+} (10^{-3} mol/liter) were tested on meshwork strips precontracted either by carbachol or endothelin (Table VII and Table VIII). The effects were dose dependent. In a low concentration, nifedipine had only a slightly relaxing (10%), whereas verapamil, given at higher doses, relaxed a significant fraction of the agonist-induced contraction (76–92%). The relaxing effect of Ni^{2+} is more pronounced with endothelin (86%) than with carbachol (41%) as a contracting agonist. Ni^{2+} is an inorganic calcium blocker that has been shown to inhibit low-threshold (T-type) Ca^{2+} channels and Na^{+}–Ca^{2+} exchange (Tang *et al.*, 1989; Kaczorowski *et al.*, 1989). The fast and reversible actions of Ni^{2+} support the hypothesis of Ca^{2+}-entry

TABLE VIII

Effect of Calcium Channel Blockers on Contractility of Isolated Meshwork Strips[a]

		Relative relaxation in precontracted tissues maximally stimulated (=100%) by	
Substance	Dose (mol/liter)	Carbachol (%)	Endothelin (%)
Nifedipine	10^{-5}	10	5
Verapamil	10^{-4}	92	Not tested
	5×10^{-5}	76	
Ni^{2+}	10^{-3}	40	86

[a] Data from Lepple-Wienhues *et al.* (1991b) and unpublished observations.

inhibition leading to relaxation of the TM cells. At low doses nifedipine-(dihydropyridine)-sensitive Ca^{2+}-influx (L-type) channel is only a small pathway for pharmacomechanic coupling. The activation of a nonselective cation channel was suggested as an additional mode of calcium entry (Sims and Janssen, 1993; Pollock *et al.*, 1995) and is currently being investigated by us in trabecular meshwork. It is interesting that in smooth muscle cells, verapamil is a rather nonspecific Ca^{2+} channel blocker that blocks a variety of channels, including Ca^{2+}-activated K^{+} channels (Nelson et al., 1990). Our measurements of contractility can only indicate that in the action of carbachol and endothelin, Ca^{2+}-sensitive pathways and additional mechanisms independent of external calcium are involved.

6. Endothelin

As summarized in Tables VII and IX endothelin-1 is a potent contracting agent for the trabecular meshwork with a half-maximal effective concentration of 5×10^{-9} mol/liter. The maximal force evoked by endothelin was 73% of the maximal carbachol response, and 77% of the endothelin-induced contraction was sensitive to extracellular calcium, indicating that the release of intracellular calcium stores is an important component in mediating the effect of endothelin. The range of effective endothelin concentration is in good agreement with those reported by other authors (Pollock *et al.*, 1995). It seems, however, that our calculation of the Hill coefficient near 4 does not resemble a simple kinetic model and may represent two or more cooperative intracellular mechanisms, coupling the endothelin receptor to contractile filaments (Lepple-Wienhues *et al.*, 1991b). Our data on measurements of membrane voltage, intracellular calcium, and contractility are in line with the present model of the effect of endothelin on contractility (van Renterghem *et al.*, 1989; Nelson and Quayle, 1995; Pollock *et al.*, 1995): Endothelin-induced contractions are partly dependent on extracellular calcium; L-type Ca^{2+} channels and outward Ca^{2+}-activated K^{+} currents (initial transient effect) are involved; in addition, activation of nonspecific cation channels that are also permeable to Ca^{2+} are important; furthermore, increase of intracellular calcium and thus contractility must be mediated in part by other mechanisms.

The functional significance of endothelin in regulation of aqueous humor outflow can only be mentioned briefly (Erickson-Lamy *et al.*, 1991; for review, Wiederholt *et al.*, 1993). Besides the effect of endothelin on membrane voltage, intracellular calcium, and contraction of trabecular meshwork and ciliary muscle, we demonstrated for the first time that endothelin-like immunoreactivity in aqueous humor of human and bovine eyes is two to three times higher than the corresponding plasma level (Lepple-Wienhues *et al.*, 1992b; Wiederholt *et al.*, 1993). Furthermore, we showed that human nonpigmented ciliary epithelial cells express a high potential for releasing endothelin-like immunoreactivity. Immunocytochemically, an intensive staining for endothelin was found in cultured human nonpigmented ciliary epithelial cells and in ciliary epithelial cells of donor eyes (Lepple-Wienhues *et al.*, 1992b). Taking the data together, this suggests an important role for endothelin in the regulation of aqueous humor secretion and reabsorption. We postulated that a constant secretion of endothelin sets the basic tone for contractile elements in the eye, which include trabecular meshwork. In addition to being an important circulating vasoregulatory hormone, endothelin may represent a local humoral factor involved in regulation of aqueous humor dynamics (Wiederholt *et al.*, 1993). In this context, it is important to note that release of endothelin can be increased by stretch and by stimulation of fluid flow rate (Rubanyi and Polokoff, 1994). Thus endothelin release could be changed by modification of aqueous humor secretion/reabsorption and/or intraocular pressure. Endothelin in aqueous humor could act as a counterbalance by influencing outflow facility via modulation of the contractile elements of the ciliary muscle and trabecular meshwork. It may be speculated on the dysfunction of such a feedback system in a disease such as primary open-angle glaucoma. In fact, we found that endothelin-like immunoreactivity in aqueous humor of patients with primary open-angle glaucoma was significantly higher than in age-matched controls with normal intraocular pressure (Noske *et al.*, 1997).

7. Nitric Oxide System

In TM strips precontracted by carbachol, the membrane-permeable cGMP evoked a relaxation of 41% as compared to the maximal contraction. Inhibition of NO-synthase by L-NAG increased the carbachol-induced contraction significantly (Tables VII and IX). The organic nitrovasodilators ISDN and 5-ISMN produced significant relaxation. The nonnitrates SNP and SNAP were the most potent relaxants (65 and 67%, respectively). ISDN and SNP had also significant relaxing activity in tissues without carbachol-induced precontraction, indicating that besides the presence of the inducible NO-synthase there is also a continuous release of NO at resting conditions in trabecular meshwork. Thus, nitric oxide is a cotransmitter of smooth muscle relaxation in the chamber angle and may be involved in the regulation of aqueous humor reabsorption. NO-synthase could be detected in the outflow pathway of the bovine (Geyer *et al.*, 1993) and human eye (Nathanson and McKee, 1995a), and sodium nitroprusside increased intracellular cGMP in transformed human TM cells (Pang *et al.*, 1994). The NO-synthase immunoreactivity was reduced in patients with primary open-angle glaucoma (Nathanson and McKee, 1995b).

8. Prostaglandins

Prostaglandins (PGs) represent a new class of topically effective ocular antihypertensive drugs. Mediated by receptors, prostaglandins have been reported to enhance uveoscleral outflow (Bito *et al.*, 1983). $PGF_{2\alpha}$ and its analogues have some effect on outflow facility (Nilsson *et al.*, 1989; Gabelt and Kaufman, 1990; Woodward *et al.*, 1993). The precise mechanism of action of the prostanoid receptor agonists remains speculative.

By measuring the contractility of isolated trabecular meshwork strips we functionally identified the prostanoid receptor subtypes using receptor-selective agonists (Tables VII and IX). $PGF_{2\alpha}$ and 17-phenyl $PGF_{2\alpha}$ had no effect on contractility; sulprostone contracted the meshwork. The nonselective EP-agonist (11-deoxy PGE_1) and the specific EP-agonist AH 13205 significantly relaxed the precontracted tissue by 16 and 21%, respectively. The thromboxane-mimetic U-46619 elicited a strong dose-dependent contraction of the trabecular meshwork with the highest concentration (10^{-6} mol/liter) being almost twice as efficient (187%) as the maximal carbachol concentration. The contraction induced by the agonists could be totally blocked with a potent and selective TP-receptor antagonist (SQ 29548). It is important to note that the TP-agonist had no effect at all on contractility of the ciliary muscle (Table IX). The studies suggest the existence of TP and EP_2 receptors in the trabecular meshwork.

Thromboxane mimetics and EP_2-agonists have opposing activities and may modulate trabecular outflow in a functionally antagonistic manner (Krauss *et al.*, 1997).

The presence of TP receptors in the outflow pathway is possibly clinically relevant. Intraocular inflammation, surgical and pharmacological maneuvers, and reflux of blood into Schlemm's canal from the intrascleral veins (Hamanaka and Bill, 1994) may activate the TP receptors in the outflow pathway and thus may induce an increase in ocular pressure.

9. Cyclo-Oxygenase Inhibitor

On muscarinic stimulation, arachidonic acid and endogenous prostaglandins are released and thus contractility of smooth muscle cells is modulated via second messengers such as cAMP, IP_3, and diacyl glycerol (DAG). This cascade of events has been well described in the ciliary muscle (Yousufzai *et al.*, 1994; Abdel-Latif, 1996). We tested the effect of the cyclo-oxygenase inhibitor indomethacin (5×10^{-6} mol/liter) and could show that the carbachol-induced contractions (= 100%) were greater than those observed in the absence of the inhibitor (TM: 138 ± 5%, $n = 7$; CM: 137 ± 6%, $n = 6$; Tables VII and IX, unpublished observations). Results obtained from these studies suggest that carbachol (and probably also endothelin) releases relaxing prostaglandins, which partially diminish the effect of both substances on contraction of the trabecular meshwork (and the ciliary muscle). In the monkey eye, the effect of decreasing outflow resistance is partly inhibited by indomethacin (Crawford *et al.,* 1996).

10. Diuretics

Diuretics are extensively used for treatment of hypertension and edema and their mechanism of action on epithelial and nonepithelial cells has been well described (Greger and Wangemann, 1987). "Loop" diuretics such as furosemide and bumetanide are relative specific inhibitors of the Na^+–$2Cl^-$–K^+ cotransporter, whereas hydrochlorothiazide diuretics are inhibitors of a Na^+–Cl^- cotransporter. The effect of ethacrynic acid is more complex and includes inhibition of the Na^+–$2Cl^-$–K^+ cotransporter, inhibition of Na^+-independent anion transporters, and modulation of the cytoskeleton (Erickson-Lamy *et al.,* 1992). Systemic or local application of various diuretics does not have an effect on aqueous humor dynamics and intraocular pressure, whereas local application of ethacrynic acid increases outflow facility in the human eye (Liang *et al.,* 1992). It was recently shown that in human and bovine TM cells a Na^+–$2Cl^-$–K^+ cotransporter can be demonstrated that is sensitive to bumetanide and ethacrynic acid (O'Donnell *et al.,* 1995). As in almost every cell, this transporter is involved in volume regulation when cells are exposed to changes of osmolality. Because bumetanide induced a decrease in trabecular meshwork cell volume and consequently increased the permeability of cultured TM cells, a prominent role for the Na^+–$2Cl^-$–K^+ cotransporter on regulation of outflow facility was postulated (O'Donnell *et al.,* 1995).

We tested high doses of diuretics on contractility of isolated TM and ciliary muscle (CM) strips (Table VII). Furosemide and hydrochlorothiazide had no effect on baseline contractility and did not modify the contraction induced by carbachol in both tissues. However, ethacrynic acid was able to relax precontracted tissues totally (Wiederholt *et al.,* 1997). Because both furosemide and ethacrynic acid are effective blockers of the Na^+–$2Cl^-$–K^+ cotransporter, the relaxing effect of ethacrynic acid seems to be independent of the cotransporter and a prominent role of this transporter on regulation of aqueous humor reabsorption seems to be very unlikely. This is in agreement with the observation that bumetanide has no effect on outflow facility in the living monkey (Gabelt *et al.,* 1996).

C. Trabecular Meshwork versus Ciliary Muscle

Table IX summarizes a comparison of our contractility measurements of isolated TM, and CM strips. Concerning electro- and pharmacomechanical coupling, there are important similarities and differences of the various substances on the contractility of both tissues. The most prominent differences follow:

1. The absolute force generated by TM strips is much smaller than the force generated by CM strips of similar length. However, in our preparations CM strips are much thicker than TM strips and absolute contractility per smooth muscle fiber may be similar in both tissues.
2. Atropine reduced the contractile response of the TM and inhibited completely the response of the CM to depolarization induced by high extracellular K^+.
3. Aceclidine was more effective in eliciting contractions in TM than in CM.
4. The α_2-agonist brimonidine induced contractions only in TM.
5. In ciliary muscle, the tension induced by carbachol or endothelin was completely dependent on extracellular calcium. In trabecular meshwork, a significant fraction of the endothelin- and carbachol-induced force was independent of extracellular calcium. This indicates different intracellular mechanisms mediating the action of endothelin and carbachol. There seems to be a calcium- and nickel-sensitive pathway in both tissues, and additional mechanisms independent of external calcium in TM.
6. When the NO/cGMP system was modulated, the effects on relaxation or contraction were stronger in trabecular meshwork than in ciliary muscle.
7. The ciliary muscle does not appear to be a major target for prostaglandins. The prostaglandins tested had only modest effects on the CM compared to cholinergics. However, the contractility studies suggest the existence of TP and EP_2 receptors in the trabecular meshwork and only a small number of these receptors in the ciliary muscle. A TP-agonist elicited a very strong contraction of the TM, whereas the effect on CM was negligible. Thus, TP-agonists may have a powerful effect on aqueous humor reabsorption by directly stimulating the contractility of TM cells.

VI. MEASUREMENT OF CONTRACTION OF CULTURED TRABECULAR MESHWORK CELLS

Zadunaisky and Spring (1995) and Zadunaisky *et al.* (1996) induced changes of the area of cultured human and bovine TM cells by pharmacological agents. They were able to discriminate between regulatory volume decrease and a true contraction of the cells. By using quantitative optical microscopy, they showed that activation of muscarinic and α-adrenergic receptors contracted and activation of β-adrenergic receptors relaxed cul-

TABLE IX

Functional Similarities and Differences in Contractile Properties of Isolated Trabecular Meshwork and Ciliary Muscle Strips

	Trabecular meshwork	Ciliary muscle	Refs.[a]
Maximal isometric tension, F (μN)	50–500	500–2000	1–6
K^+-induced contractions (% of maximal acetylcholine response)	+[b] (19%)	++ (59%)	1
Relative contractions of aceclidine vs. pilocarpine (% of maximal pilocarpine response)	++ (173%)	+ (139%)	1
EC_{50} carbachol	TM = CM		2, 3, 6
EC_{50} pilocarpine	TM = CM		1, 3, 6
Muscarinic receptors, $M_3 \gg M_1$	TM = CM		6
Relative contraction induced by muscarinic agonists	carbachol > aceclidine > pilocarpine > acetylcholine TM = CM		1, 6
Relative potency of adrenergic agents on contraction (+) or relaxation (−)			3, 6
Phenylephrine (α_1)	+	(+)	
Brimonidine (α_2)	++	no effect	
Epinephrine (α, β)	+	(+)	
Epinephrine + metipranolol	++	++	
Isoproterenol (β)	−	−	
Isoproterenol (β) + metipranolol	+	+	
Ca^{2+} dependence of contractions evoked by carbachol or endothelin	58% 77%	100% 100%	2
Relative contraction induced by endothelin vs. carbachol (100%)	73%	52%	2
EC_{50} endothelin	TM = CM		2
Relative potency of calcium channel blockers on relaxation (−) in precontracted tissues			
Nifedipine (low dose)	(−)	(−)	2
Verapamil	− −	− −	7
Ni^{2+}	− −	−	2
Relative potency of substances interfering with the NO system on contraction (+) or relaxation (−)			4
8-bromo-cGMP	−− (41%)	− (13%)	
L-NAG	+ (19%)	(+) (9%)	
ISDN	− (15%)	(−) (12%)	
5-ISMN	− (20%)	− (10%)	
SNP	−− (65%)	−(−) (45%)	
SNAP	−−(67%)	−(−) (32%)	

(continues)

continued

	Trabecular meshwork	Ciliary muscle	Refs.[a]
Relative potency of prostanoids on contractility			5
$PGF_{2\alpha}$	no effect	no effect	
17-phenyl $PGF_{2\alpha}$	no effect	no effect	
Sulprostone	+ (10%)	no effect	
11-deoxy PGF_1	− (16%)	− (7%)	
AH 13205	− − (21%)	− (7%)	
U-46619	+++ (187%)	no effect	
Augmentation of carbachol-induced contractions by indomethacin	TM = CM		7
Relative potency of diuretics on contractility			7
Furosemide			
Hydrochlorothiazide	TM = CM		
Ethacrynic acid			

[a] 1, Lepple-Wienhues *et al.*, 1991a; 2, Lepple-Wienhues *et al.*, 1991b; 3, Wiederholt *et al.*, 1993; 4, Wiederholt *et al.*, 1994; 5, Krauss *et al.*, 1997; 6, Wiederholt *et al.*, 1996; 7, Wiederholt *et al.*, 1997.

[b] Relative amount of contraction/relaxation: strong: +++/−−−; medium: ++/−−; small: +/−.

tured TM cells. Inhibition of the β-adrenergic component of epinephrine-induced contraction. There seems to be no difference between bovine and human TM cells in the contraction induced by pharmacological agents. The data confirm an older observation by Tripathi and Tripathi (1984). These authors observed area changes of cultured TM cells induced by epinephrine and postulated a contraction induced by the α-adrenergic agonist. In summary, the data obtained on cultured human and bovine TM cells support our measurements where contractility of TM strips was recorded directly.

VII. THE PERFUSED ANTERIOR SEGMENT

As shown in Sections II,V, and VI, the contractility of human and bovine TM cells can be modulated by an impressive number of substances. However, the effect of TM contractility per se on outflow facility remains unclear. Bárány (1962) was the first author who postulated that the TM cells are possibly contractile and that pilocarpine could influence outflow reabsorption directly. The morphological evidence for contractile filaments in the human and bovine chamber angle was reviewed in Section I.

To characterize the regulation of outflow facility, isolated perfused eyes of primates and nonprimates have been used for a long time (for review,

Wiederholt *et al.,* 1995). In most species, the ciliary muscle extends into the trabecular meshwork, making a dissociation of the effect of CM cells and TM cells on outflow regulation impossible. In the bovine eye, the ciliary muscle can easily be detached from the trabecular meshwork and does not extend into the outflow pathway (Flügel *et al.,* 1991; Wiederholt *et al.,* 1995). It could be shown that in the model of the perfused anterior segment of the bovine eye with detached ciliary muscle, TM cells maintained their morphological integrity (Erickson-Lamy *et al.,* 1988). In this model, effects of substances that are used in glaucoma therapy were not tested. In studies with perfused human or monkey eyes only the effects of adrenergic agents were tested on outflow facility (Kaufman, 1986; Robinson and Kaufman, 1990; Erickson-Lamy and Nathanson, 1992).

Similar to the model described by Erickson-Lamy *et al.* (1988), we perfused anterior segments of bovine eyes with detached iris, ciliary body, and ciliary muscle at a constant pressure of 8.8 mm Hg. Under these conditions a constant ouflow of 6–8 μL/min could be obtained for at least 3 hr. In seven experiments the following mean values were obtained: outflow, 7.66 $\pm$ 0.91 μL/min; outflow facility, 0.87 μL $\cdot$ mmHg/min; and outflow resistance, 1.15 mm Hg $\cdot$ min/μL. We did not observe a "washout" effect with time of perfusion. Light and electron microscopy showed that the ultrastructure of the entire filtering tissue appeared well preserved and that in the majority of specimens CM cells could not be detected (Wiederholt *et al.,* 1995).

To test the relation between flow rate and perfusate, the pressure in the perfused anterior chamber was changed. Outflow rate was an almost linear function of pressure in the range of 5–25 mm Hg. Only at higher perfusion pressure, the outflow resistance increased as first described for the perfused human eye (Grant, 1963). Brubaker (1975) presented convincing evidence that in the perfused human eye higher intraocular pressure induced an increase in resistance. From our model of perfused anterior segments without ciliary muscle it can be derived that an increase in intraocular pressure directly increases resistance of the outflow pathway, independent of the effect on the ciliary muscle.

In perfused anterior segments with intact morphology and reversible responsiveness to changes in perfusion pressure, we tested several drugs (Table X). The relative outflow was significantly reduced by carbachol, reaching a maximal inhibition of 37%. This effect could be completely blocked by atropine. The half-maximal effective concentration for carbachol was 3×10^{-8} mol/liter, which is in the same range as the value reported for the effect of carbachol on contractility (Table VII). Pilocarpine reduced outflow by 15%. Epinephrine at a concentration of 10^{-5} mol/liter reduced outflow, while at a lower concentration (10^{-6} mol/liter) it slightly increased

TABLE X
Effect of Drugs on Outflow Facility in Perfused Anterior Segments with Detached Iris, Ciliary Body, and Ciliary Muscle of Bovine Eyes[a]

Drugs	Concentration (mol/liter)	Change in outflow facility
Carbachol	2×10^{-6}	Decrease (37%)
Pilocarpine	10^{-5}	Decrease (15%)
Epinephrine	10^{-5}	Decrease (7%)
Epinephrine	10^{-6}	Increase (8%)
Epinephrine	10^{-6}	
+		Decrease (10%)
Metipranolol	10^{-5}	(18% vs. epinephrine 10^{-6})
Endothelin	2×10^{-9}	Decrease (23%)
Endothelin	2×10^{-8}	Decrease (31%)

[a] Data from Wiederholt *et al.* (1995).

outflow. This increase in outflow was fully blocked by the β-antagonist metipranolol. In the perfused human eye, 10^{-6} mol/liter epinephrine also increased outflow (Erickson-Lamy and Nathanson, 1992). Thus, the relative activity of α- and β-adrenergic stimulation determines whether epinephrine increases or decreases outflow resistance. These experiments are compatible with the dose-dependent effect of epinephrine on contraction and relaxation of the isolated meshwork strips. Endothelin-1 inhibited relative outflow dose dependently. Thus, substances that contract isolated TM strips like carbachol, pilocarpine, endothelin, and a high dose of epinephrine induced a reduction of outflow rate and an increase of outflow resistance of the anterior segment. However, substances that relax isolated meshwork strips like epinephrine in a low dosage and cytocholasin D increased the outflow rate (Wiederholt *et al.,* 1995, 1997). Thus, at least in the bovine eye, the trabecular meshwork per se is directly involved in the regulation of aqueous humor reabsorption. However, in the intact eye the balance between contractility of ciliary muscle and trabecular meshwork may determine the total outflow reabsorption.

VIII. SUMMARY OF CHANNELS, TRANSPORTERS, AND RECEPTORS IN THE TRABECULAR MESHWORK CELL

Figure 5 is a schematic representation of channels, transporters, and receptors involved in the regulation of transport properties of TM cells. Most of the arguments for the presence of channels, transporters, and receptors and the relevant literature have been discussed already in the

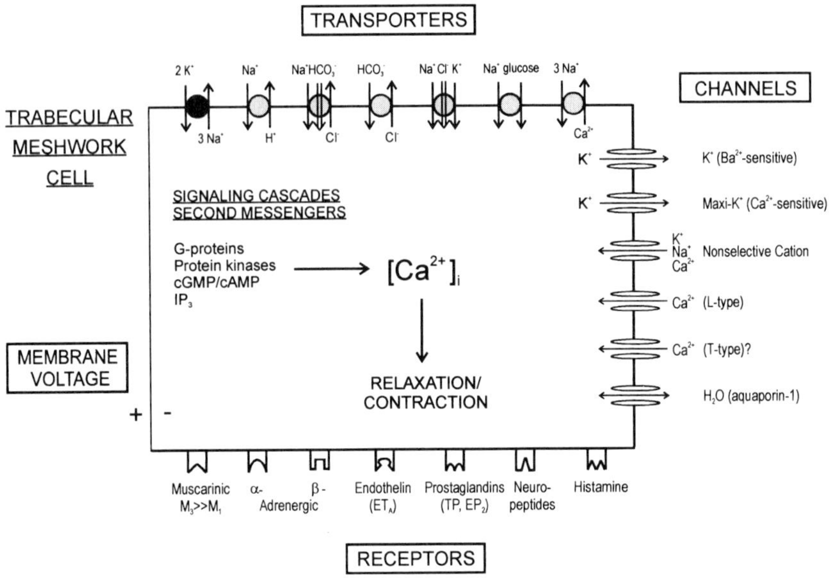

FIGURE 5 Summary of functional characterization of transporters, channels, and receptors in trabecular meshwork cells. *Transporters:* (a) primary active Na^+,K^+-ATPase (inhibitor: ouabain) which establishes the electrochemical gradient for the following secondary active pumps; (b) Na^+–H^+ antiporter (inhibitor: amiloride); (c) Na^+-dependent Cl^-–HCO_3^- exchanger (inhibitor: DIDS); (d) Cl^-–HCO_3^- exchanger (inhibitor: DIDS, ethacrynic acid); transporters b, c, and d are involved in regulation of intracellular pH; (e) Na^+–$2Cl^-$–K^+ cotransporter (inhibitors: furosemide, bumetanide, ethacrynic acid) involved in cell volume regulation; (f) Na^+–glucose symporter; (g) the 3 Na^+–Ca^{2+} exchanger is hypothetical, however, this transporter has been shown in all smooth muscle cells. *Channels:* (a) nonspecific K^+ channels (inhibitor: Ba^{2+}); (b) maxi-K^+ channel (= BK, K_{Ca}), Ca^{2+}-sensitive (inhibitors: charybdotoxin, TEA^+; activators: calcium, cGMP, ATP); the only channel which has been characterized by patch-clamp techniques; (c) nonselective cation channel (inhibitor: flufenamic acid; activators: carbachol, endothelin); (d) Ca^{2+} channel (voltage dependent, L-type; inhibitors: nifedipine, verapamil); (e) Ca^{2+} channel? (voltage dependent, T-type; inhibitor: Ni^{2+}); (f) water channel/aquaporin-1, not sensitive to antidiuretic hormone; (g) lack of fast Na^+ channel (inhibitor: tetrodotoxin). Presence of other channels (K_{ATP}, K_{IR}, various Na^+ and Cl^- channels) has not yet been tested. *Receptors:* All of these receptors have not been identified on a molecular level. The various signaling cascades and intracellular second messengers are not yet fully characterized. One of the most important final pathways in signal transduction is the intracellular Ca^{2+} concentration, which determines the balance between contraction and relaxation.

Sections II through VII. Not mentioned before are nonselective cation channels (Wiederholt *et al.,* 1997), a Na^+–glucose transporter (Kaulen *et al.,* 1991), a receptor for histamine (WoldeMussie and Ruiz, 1992), and the aquaporin-1 water channels (Stamer *et al.,* 1994, 1995). The presence of water channels in the human outflow pathway and in cultured human TM cells suggests that water channels may functionally be involved in aqueous humor reabsorption. The presence of aquaporin channels in trabecular meshwork is consistent with the observation that in cultured TM cells hydraulic conductivity can be changed by changing the perfusion pressure (Perkins *et al.,* 1988). Thus, transcellular and paracellular water flow is probably involved in the movement of water from the anterior chamber into Schlemm's canal.

In the model described in Fig. 5, the various signal transduction pathways and the intracellular second messengers are not mentioned. Besides the cyclic AMP (Pang *et al.,* 1994) and cyclic GMP system and their crosstalk with the polyphosphoinositide signaling cascade very little is known about signal transduction pathways in TM cells.

IX. FUNCTIONAL SYNERGISM/ANTAGONISM BETWEEN TRABECULAR MESHWORK AND CILIARY MUSCLE

The functional role of the contractility of trabecular meshwork for regulation of aqueous humor outflow is an open question. From a morphological and functional point of view it has been well established that pilocarpine increases aqueous humor outflow by affecting the contraction of the ciliary muscle, which then widens the functional spaces in the outflow pathway (Rohen, 1964; Rohen and Lütjen-Drecoll, 1982; Kaufman, 1984; Lütjen-Drecoll and Rohen, 1989). This well-established model has to be modified. In this paper we presented evidence that the trabecular meshwork per se is contractile. In Tables VII and X data are given that show that substances which contract isolated TM strips (muscarinic agonists, α-adrenergic agonists, endothelin) decrease aqueous humor outflow in the perfused eye model with an intact outflow pathway by increasing the resistance. However, maneuvers that relax isolated meshwork strips (reducing extracellular calcium, Ca^{2+} channel antagonists, β-adrenergic agonists, activation of the NO/cGMP system) may increase outflow by decreasing outflow resistance. Since it has been shown that cholinergic mimetics (Kaufman, 1984, 1986; Alvarado *et al.,* 1990; Robinson and Kaufman, 1990; Erickson-Lamy *et al.,* 1992) and endothelin (Erickson-Lamy *et al.,* 1991) increase the overall outflow in the intact eye and thus decrease intraocular pressure, it has to be postulated that the direct effect of both substances shown on contractility of the meshwork is functionally antagonistic to the direct effect of these substances on the ciliary muscle. As shown in Table IX, most substances tested for contractility of TM and CM strips affect both tissues (although sometimes in a different qualitative and quantitative manner). The overall effect of various drugs on outflow reabsorption is thus probably determined by the balance between the effects of drugs on the contractility of ciliary muscle and trabecular meshwork. Thus spaces within trabecular meshwork could be narrowed by contraction of TM fibers and widened by CM traction on the meshwork. Alternatively, contraction of trabecular meshwork could increase the rigidity of the outflow pathway, allowing CM traction on the meshwork to be more effective in altering the geometry of the outflow pathway. The concept of antagonism between ciliary muscle and trabecular meshwork (Fig. 6) has to be considered in the interpretation of the mechanism of action of currently used antiglaucoma drugs and in the search for new effective drugs. Contraction of the ciliary muscle and/or relaxation of the trabecular meshwork will determine the total effect of decreasing intraocular pressure (Fig. 6). In most conditions, contractility of the ciliary muscle probably dominates the overall effect on outflow reabsorption in the human and most likely also in the bovine eye. While the application of the bovine model to the human eye has yet to be tested, there are many similarities between the TM systems of the bovine and the human eye, especially when cultured cells are compared (Tables I and II). In the search for a new ideal antiglaucoma drug, hypothetically, a drug that only relaxes the trabecular meshwork without interfering with CM functions such as accommodation should be most beneficial in lowering intraocular pressure. In addition, such a substance would also have a relaxing/vasodilating effect on the microcirculation of the eye, potentially preventing the progression of nerve fiber damage. Mechanisms to prevent or delay nerve damage are

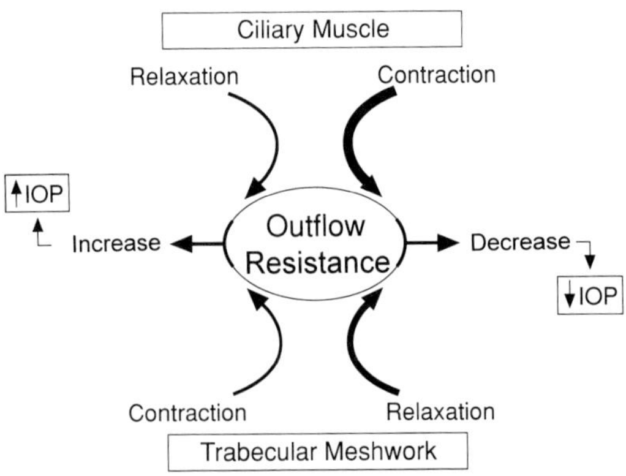

FIGURE 6 Model of functional synergism/antagonism between contractility of ciliary muscle and trabecular meshwork with regulation of outflow resistance and intraocular pressure (IOP).

key targets for development of new antiglaucoma drugs (Schumer and Podos, 1994).

In this concept of agonism/antagonism between ciliary muscle and trabecular meshwork the recent findings of Tamm *et al.* (1992, 1994, 1995) have to be considered. The authors described contractile cells in the human scleral spur that are probably functionally independent of the contractile elements of the ciliary muscle and the trabecular meshwork. The authors postulate that in the human scleral spur there are afferent mechanoreceptors and axons of parasympathetic origin, while the sympathetic innervation seems to be rare. The myofibroblast-like cells of the scleral spur in the human eye could resemble some characteristics of the myofibroblast-like cells of the bovine trabecular meshwork (Flügel *et al.,* 1991). Thus, at least in the human eye, three different contractile elements (ciliary muscle, scleral spur, trabecular meshwork) may modify outflow reabsorption, while in the bovine eye two contractile elements (ciliary muscle, trabecular meshwork) are involved in regulation of outflow.

X. SUMMARY

This review presents evidence for contractile properties of the trabecular meshwork and their effects on aqueous humor reabsorption.

Membrane voltage measurements and patch-clamp techniques were applied to cultured bovine and human TM and CM cells. Measurements of isometric tension were performed on isolated TM (and CM) strips. Anterior segments of bovine eyes with well-preserved TM were perfused to measure outflow rate.

1. Cultured bovine and human TM cells showed voltage spikes typical of smooth muscle cells which were inhibited by nifedipine, but insensitive to tetrodotoxin. The excitability of TM cells indicates that they function as contractile smooth muscle cells. There is no principal difference between human and bovine TM cells in terms of K^+ and Ca^{2+} channels, functional receptors for endothelin, and the effects of cholinergic and adrenergic agonists.

2. Direct measurements of contractility of isolated strips indicate the presence of muscarinic (M_3), α- and β-adrenergic, and endothelin receptors in the bovine TM (and CM). Cholinergic and α-adrenergic (mainly α_2) agonists produced contraction while β-agonists produced relaxation. Relaxation was induced by release of nitric oxide. The contractile properties of TM and CM are differently modulated by the various drugs.

3. Substances that produced contraction in TM strips induced a decrease in the outflow rate of the anterior segment. Relaxing substances induced an increase in the outflow rate.

4. Trabecular meshwork per se is a contractile element and is, at least in the bovine eye, directly involved in the regulation of aqueous humor outflow. The concept of functional antagonism between TM and CM has to be considered in the interpretation of mechanism of action of currently used antiglaucoma drugs and the search for new effective drugs.

Acknowledgment

Supported by the Deutsche Forschungsgemeinschaft, grant Wi 328/11.

References

Abdel-Latif, A. A. (1996). Cross talk between cyclic AMP and the polyphosphoinositide signaling cascade in iris sphincter and other nonvascular smooth muscle. *P.S.E.B.M.* **211,** 163–177.

Alvarado, J. A., Franse-Carman, L., McHolm, G., and Murphy, C. (1990). Epinephrine effects on major cell types of the aqueous outflow pathway: *In vitro* studies/clinical implications. *Trans. Am. Ophthalmol. Soc.* **88,** 267–282.

Archer, S. L., Huang, J. M. C., Hampl, V., Nelson, D. P., Shultz, P. J., and Weir, E. K. (1994). Nitric oxide and cGMP cause vasorelaxation by activation of a charybdotoxin-sensitive K channel by cGMP-dependent protein kinase. *Proc. Natl. Acad. Sci. U.S.A.* **91,** 7583–7587.

Bárány, E. H. (1962). The mode of action of pilocarpine on outflow resistance in the eye of a primate. *Invest. Ophthalmol.* **1,** 712–727.

Bárány, E. H. (1967). The immediate effect on outflow resistance of intravenous pilocarpine in the vervet monkey, *Cercopithecos ethiops. Invest. Ophthalmol.* **6,** 373–380.

Berridge, M. J. (1993). Inositol trisphosphate and calcium signaling. *Nature* **361,** 315–325.

Bito, L. Z., Draga, A., Blanco, J., and Camras, C. B. (1983). Long-term maintenance of reduced intraocular pressure by daily or twice daily topical application of prostaglandins to cat or rhesus monkey eyes. *Invest. Ophthalmol. Vis. Sci.* **24,** 312–319.

Bolotina, V. M., Najibi, S., Palacino, J. J., Pagano, P. J., and Cohen, R. A. (1994). Nitric oxide directly activates calcium-dependent potassium channels in vascular smooth muscle. *Nature* **368,** 850–853.

Brubaker, R. F. (1975). The effect of intraocular pressure on conventional outflow resistance in the enucleated human eye. *Invest. Ophthalmol. Vis. Sci.* **14,** 286–292.

Brutsaert, D., Meulemans, A., Sipido, K., and Sys, S. (1988). Effects of damaging the endocardial surface on the mechanical performance of isolated cardiac muscle. *Circ. Res.* **62,** 358–366.

Chu, T. C., Keith, C., and Green, K. (1992). Intracellular pH regulation by a Na^+/H^+ exchanger in cultured bovine trabecular cells. *Acta Ophthalmol.* **70,** 772–779.

Coroneo, M. T., Korbmacher, C., Stiemer, B., Flügel, C., Lütjen-Drecoll, E., and Wiederholt, M. (1991a). Electrical and morphological evidence for heterogenous populations of cultured bovine trabecular meshwork cells. *Exp. Eye Res.* **52,** 375–388.

Coroneo, M. T., Erickson-Lamy, K. A., and Wiederholt, M. (1991b). Membrane voltage recordings in a cell line (H_1TM) derived from human trabecular meshwork. *Invest. Ophthalmol. Vis. Sci.* **32,** 942.

Coroneo, M. T., Rössner, A., Lepple-Wienhues, A., and Wiederholt, M. (1992). Cytoplasmic pH regulation in cultured bovine trabecular meshwork cells. *Invest. Ophthalmol. Vis. Sci.* **33,** 1165.

Crawford, K. S., Gange, S. J., Gabelt, B. T., Heideman, W., Robinson, J. C., Hubbard, W. C., and Kaufman, P. L. (1996). Indomethacin and epinephrine effects on outflow facility and cyclic adenosine monophosphate formation in monkeys. *Invest. Ophthalmol. Vis. Sci.* **37,** 1348–1359.

De Kater, A. W., Shahsafaei, A., and Epstein, D. L. (1992). Localization of smooth muscle and nonmuscle actin isoforms in the human aqueous outflow pathway. *Invest. Ophthalmol. Vis. Sci.* **33,** 424–429.

Dörje, F., Wess, J., Lambrecht, G., Tacke, R., Mutschler, E., and Brann, M. R. (1991). Antagonist binding profiles of five cloned human muscarinic receptor subtypes. *J. Pharmacol. Exp. Ther.* **256,** 727–733.

Erickson, K. A., Schroeder, A., and Netland, P. A. (1995). Verapamil increases outflow facility in the human eye. *Exp. Eye Res.* **61,** 565–567.

Erickson-Lamy, K. A., and Nathanson, J. A. (1992). Epinephrine increases facility of outflow and cyclic AMP content in the human eye *in vitro*. *Invest. Ophthalmol. Vis. Sci.* **33,** 2672–2678.

Erickson-Lamy, K., Rohen, J. W., and Grant, W. M. (1988). Outflow facility studies in the perfused human ocular anterior segment. *Exp. Eye Res.* **7,** 799–807.

Erickson-Lamy, K. A., Korbmacher, C., Schuman, J. S., and Nathanson, J. A. (1991). Effect of endothelin on outflow facility and accommodation in the monkey eye *in vivo*. *Invest. Ophthalmol. Vis. Sci.* **32,** 492–495.

Erickson-Lamy, K. A., Schroder, A., and Epstein, D. L. (1992). Ethacrynic acid induces reversible shape and cytoskeletal changes in cultured cells. *Invest. Ophthalmol. Vis. Sci.* **33,** 2631–2640.

Flocks, M., and Zweng, H. C. (1957). Studies on the mode of action of pilocarpine on aqueous outflow. *Am. J. Ophthalmol.* **44,** 380–388.

Flügel, C., Tamm, E., and Lütjen-Drecoll, E. (1991). Different cell populations in bovine trabecular meshwork: An ultrastructural and immunohistochemical study. *Exp. Eye Res.* **52,** 681–690.

Flügel, C., Tamm, E., Lütjen-Drecoll, E., and Stefani, F. H. (1992). Age related loss of α-smooth muscle actin in normal and glaucomatous human trabecular meshwork of different age groups. *J. Glaucoma* **1,** 165–173.

Gabelt, B. T., and Kaufman, P. L. (1990). The effect of prostaglandin $F_{2\alpha}$ on trabecular outflow facility in cynomolgus monkeys. *Exp. Eye Res.* **51,** 87–91.

Gabelt, B. T., and Kaufman, P. L. (1992). Inhibition of outflow facility and accommodation and miotic responses to pilocarpine in rhesus monkeys by muscarinic receptor subtype antagonists. *J. Pharmacol. Exp. Ther.* **263,** 1131–1139.

Gabelt, B. T., Kiland, J. A., Peterson, J. A., and Kaufman, P. L. (1996). Effects of the Na/K/2Cl cotransport inhibitor bumetanide on aqueous humor formation and outflow facility in monkeys. *Invest. Ophthalmol. Vis. Sci.* **37,** S201.

Geyer, O., Podos, S. M., and Mittag, T. W. (1993). Nitric oxide synthase: Distribution and biochemical properties of the enzyme in the bovine eye. *Invest. Ophthalmol. Vis. Sci.* **34,** 826.

Grant, W. M. (1963). Experimental aqueous perfusion in enucleated human eyes. *Arch. Ophthalmol.* **69,** 783–801.

Greger, R., and Wangemann, P. (1987). Loop diuretics. *Renal Physiol.* **10,** 174–183.

Grierson, I., Lee, W. R., and Abraham, S. (1978). Effects of pilocarpine on the morphology of the human outflow apparatus. *Br. J. Ophthalmol.* **62,** 302–313.

Grierson, I., Millar, L., DeYong, J., Day, J., McKechnie, N., Hitchins, C., and Boulton, M. (1986). Investigations of cytoskleral elements in cultured bovine meshwork cells. *Invest. Ophthalmol. Vis. Sci.* **27,** 1318–1330.

Gupta, N., Drance, S. M., McAllister, R., Prasad, S., Rootman, J., and Cynader, M. S. (1994). Localization of M_3 muscarinic receptor subtype and mRNA in the human eye. *Ophthalmol. Res.* **26,** 207–213.

Hamanaka, T., and Bill, A. (1994). Platelet aggregation on the endothelium of Schlemm's canal. *Exp. Eye Res.* **59,** 249–256.

Helbig, H., Korbmacher, C., Stumpff, F., Coca-Prados, M., and Wiederholt, M. (1989). Role of HCO_3^- in the regulation of cytoplasmic pH in a cell clone derived from bovine pigmented ciliary epithelium. *Am. J. Physiol.* **257,** C696–C705.

Holmberg, A., and Bárány, E. H. (1966). The effect of pilocarpine on the endothelium formed on the inner wall of Schlemm's canal; Schlemm's canal, an electronmicroscopic study in the monkey. *Invest. Ophthalmol.* **5,** 53–58.

Ito, Y., and Yoshitomi, T. (1986). Membrane and contractile properties of the dog ciliary muscle. *Br. J. Pharmacol.* **88,** 629–638.

Jampel, H. D., Lynch, M. G., Brown, R. H., Kuhar, M. J., and DeSouza, E. B. (1987). β-Adrenergic receptors in human trabecular meshwork. Identification and autoradiographic localization. *Invest. Ophthalmol. Vis. Sci.* **28,** 772–779.

Jentsch, T. J., Koch, M., Bleckmann, H., and Wiederholt, M. (1984). Effect of bicarbonate, methazolamide and stilbenes on the intracellular potentials of cultured bovine corneal endothelial cells. *J. Membr. Biol.* **78,** 103–107.

Jentsch, T. J., Korbmacher, C., Janicke, I., Fischer, D. G., Stahl, F., Helbig, H., Cragoe Jr., E. J., Keller, S. K., and Wiederholt, M. (1988). The regulation of cytoplasmic pH of cultured bovine corneal endothelial cells in the absence and presence of bicarbonate. *J. Membr. Biol.* **103,** 29–40.

Kaczorowski, G., Slaughter, R. S., King, V. F., and Garcia, M. L. (1989). Inhibitors of sodium-calcium exchange: Identification and development of probes of transport activity. *Biochim. Biophys. Acta* **988,** 287–302.

Kaufman, P. (1984). Aqueous humor outflow. *Curr. Top. Eye Res.* **4,** 97–138.

Kaufman, P. L. (1986). Epinephrine, norepinephrine, and isoproterenol dose—outflow facility response relationships in cynomolgus monkey eyes with and without ciliary muscle retrodisplacement. *Acta Ophthalmol.* **64,** 356–363.

Kaufman, P., and Bárány, E. (1976). Loss of acute pilocarpine effect on outflow facility following surgical disinsertion and retrodisplacement of the ciliary muscle from the scleral spur in the cynomolgus monkey. *Invest. Ophthalmol. Vis. Sci.* **15,** 793–807.

Kaulen, P., Kahle, G., Keller, K., and Wollensak, J. (1991). Autoradiographic mapping of the glucose transporter with cytochalasin B in the mammalian eye. *Invest. Ophthalmol. Vis. Sci.* **32,** 1903–1911.

Kodama, M., Kanaide, H., Abe, S., Hirano, K., Kai, H., and Nakamura, M. (1989). Endothelin-induced calcium-independent contraction of the porcine coronary artery. *Biochem. Biophys. Res. Comm.* **160,** 1302–1308.

Kohmoto, H., Matsumoto, S., and Serizawa, T. (1994). Effects of endothelin-1 on $[Ca^{2+}]_i$ and pH_i in trabecular meshwork cells. *Curr. Eye Res.* **13,** 197–202.

Korbmacher, C., Helbig, H., Haller, H., Erickson-Lamy, K. A., and Wiederholt, M. (1989). Endothelin depolarizes membrane voltage and increases intracellular calcium concentration in human ciliary muscle cells. *Biochem. Biophys. Res. Comm.* **164,** 1031–1039.

Krauss, A. H.-P., Wiederholt, M., Sturm, A., and Woodward, D. F. (1997). Prostaglandin effects on contractility of bovine trabecular meshwork and ciliary muscle. *Exp. Eye Res.* **64,** 447–453.

Lepple-Wienhues, A., Stahl, F., and Wiederholt, M. (1991a). Differential smooth muscle-like contractile properties of trabecular meshwork and ciliary muscle. *Exp. Eye Res.* **53,** 33–38.

Lepple-Wienhues, A., Stahl, F., Willner, U., Schäfer, R., and Wiederholt, M. (1991b). Endothelin-evoked contractions in bovine ciliary muscle and trabecular meshwork: Interaction with calcium, nifedipine and nickel. *Curr. Eye Res.* **10,** 983–989.

Lepple-Wienhues, A., Stahl, F., Wunderling, D., Willner, U., Schneider, U., and Wiederholt, M. (1991c). Membrane properties of trabecular meshwork endothelial cells—action of barium and endothelin. *Pflügers Arch.* **418,** R76.

Lepple-Wienhues, A., Stahl, F., Wunderling, D., and Wiederholt, M. (1992a). Effects of endothelin and calcium channel blockers on membrane voltage and intracellular calcium in cultured bovine trabecular meshwork. *Ger. J. Ophthalmol.* **1,** 159–163.

Lepple-Wienhues, A., Becker, M., Stahl, F., Berweck, S., Hensen, J., Noske, W., Eichhorn, M., and Wiederholt, M. (1992b). Endothelin-like immunoreactivity in the aqueous humour and in conditioned medium from cultured ciliary epithelial cells. *Curr. Eye Res.* **11,** 1041–1046.

Lepple-Wienhues, A., Rauch, R., Clark, A. F., Grässmann, A., Berweck, S., and Wiederholt, M. (1994). Electrophysiological properties of cultured human trabecular meshwork cells. *Exp. Eye Res.* **59,** 305–311.

Liang, L. L., Epstein, D. L., de Kater, A. W., Shahsafaei, A., and Erickson-Lamy, K. A. (1992). Ethacrynic acid increases facility of outflow in the human eye *in vitro*. *Arch. Ophthalmol.* **110,** 106–109.

Lograno, M., and Reibaldi, A. (1986). Receptor responses in fresh human ciliary muscle. *Br. J. Pharmacol.* **87,** 379–385.

Lütjen-Drecoll, E., and Rohen, J. W. (1989). Morphology of aqueous outflow pathways in normal and glaucomatous eyes. *In* "The Glaucomas" (E. A. Klein, ed.), pp. 89–123. C. V. Mosby, St. Louis.

Lütjen-Drecoll, E., Futa, R., and Rohen, J. W. (1981). Ultrahistochemical studies on tangential sections of the trabecular meshwork in normal and glaucomatous eyes. *Invest. Ophthalmol. Vis. Sci.* **21,** 563–573.

Madshus, I. H., and Olsnes, S. (1987). Selective inhibition of sodium-linked and sodium-independent bicarbonate/chloride antiport in vero cells. *J. Biol. Chem.* **262,** 7486–7491.

Mäepea, O., and Bill, A. (1992). Pressures in the juxtacanalicular tissue and Schlemm's canal in monkeys. *Exp. Eye Res.* **54,** 879–883.

Marsault, R., Vigne, P., and Frelin, C. (1990). The effect of extracellular calcium on the contractile action of endothelin. *Biochem. Biophys. Res. Comm.* **171,** 301–305.

Nathanson, J. A., and McKee, M. (1995a). Identification of an extensive system of nitric oxide-producing cells in the ciliary muscle and outflow pathway of the human eye. *Invest. Ophthalmol. Vis. Sci.* **36,** 1765–1773.

Nathanson, J. A., and McKee, M. (1995b). Alterations of ocular nitric oxide synthase in human glaucoma. *Invest. Ophthalmol. Vis. Sci.* **36,** 1774–1784.

Nelson, M. T., and Quayle, J. M. (1995). Physiological roles and properties of potassium channels in arterial smooth muscle. *Am. J. Physiol.* **268,** C799–C822.

Nelson, M. T., Patlak, J. B., Worley, J. F., and Standen, N. B. (1990). Calcium channels, potassium channels, and voltage dependence of arterial smooth muscle tone. *Am. J. Physiol.* **259,** C3–C18.

Nilsson, S. F. E., Sperber, G. O., and Bill, A. (1989). The effect of prostaglandin $F_{2\alpha}$-isopropylester ($PGF_{2\alpha}$-1E) on uveoscleral outflow. *In* "The Ocular Effects of Prostaglandins and Other Eicosanoids" (L. Z. Bito and J. Stjernschantz, eds.), pp. 429–436. Alan R. Liss, New York.

Noske, W., Hensen, J., and Wiederholt, M. (1997). Endothelin-like immunoreactivity in aqueous humor of patients with primary open angle glaucoma and cataract patients. *Graefe's Arch. Clin. Exp. Ophthalmol.* **235.**

O'Donnell, M. E., Brandt, J. D., and Curry, F. R. E. (1995). Na-K-Cl cotransport regulates intracellular volume and monolayer permeability of trabecular meshwork cells. *Am. J. Physiol.* **268,** C1067–C1074.

Ohuchi, T., Tanihara, H., Yoshimura, N., Kuriyama, S., Ito, S., and Honda, Y. (1992). Neuropeptide-induced $[Ca^{2+}]_i$ transients in cultured bovine trabecular cells. *Invest. Ophthalmol. Vis. Sci.* **33,** 1676–1684.

Pang, I. H., Shade, D. L., Clark, A. F., Steely, H. T., and DeSantis, L. (1994). Preliminary characterization of a transformed cell strain derived from human trabecular meshwork. *Curr. Eye Res.* **13,** 51–63.

Perkins, T. W., Alvarado, J. A., Stilwell, L., Maglio, M., and Juster, R. (1988). Trabecular meshwork cells grown on filters: Conductivity and cytochalasin effects. *Invest. Ophthalmol. Vis. Sci.* **29,** 1836–1846.

Pollock, D. M., Keith, T. L., and Highsmith, R. F. (1995). Endothelin receptors and calcium signaling. *FASEB J.* **9,** 1196–1204.

Resink, T. J., Scott-Burden, T., and Bühler, F. R. (1990). Activation of multiple signal transduction pathways by endothelin in cultured human vascular smooth muscle cells. *Eur. J. Biochem.* **189,** 415–421.

Ringvold, A. (1978). Actin filaments in trabecular endothelial cells in eyes of the vervet monkey. *Acta Ophthalmol.* **56,** 217–227.

Robinson, J. C., and Kaufman, P. L. (1990). Effects and interactions of epinephrine, norepinephrine, timolol, and betaxolol on outflow facility in the cynomolgus monkey. *Am. J. Ophthalmol.* **109,** 189–194.

Rohen, J. (1964). Das Auge und seine Hilfsorgane. *In* "Handbuch der mikroskopischen Anatomie des Menschen", Bd. III/2, "Haut und Sinnesorgane" (W. v. Möllendorff and W. Bargmann, eds.), pp. 189–328. Springer Verlag, Berlin.

Rohen, J., and Lütjen-Drecoll, E. (1982). Biology of the trabecular meshwork. *In* "Basic Aspects of Glaucoma Research" (E. Lütjen-Drecoll, ed.), pp. 141–166. Schattauer Verlag, Stuttgart.

Rohen, J., Futa, R., and Lütjen-Drecoll, E. (1981). The fine structure of the cribriform meshwork in normal and glaucomatous eyes. *Invest. Ophthalmol. Vis. Sci.* **21,** 574–585.

Roos, A., and Boron, W. F. (1981). Intracellular pH. *Physiol. Rev.* **61,** 296–434.

Rubanyi, G. M., and Polokoff, M. A. (1994). Endothelins: Molecular biology, biochemistry, pharmacology, physiology, and pathophysiology. *Pharmacol. Rev.* **46,** 325–415.

Schumer, R. A., and Podos, S. M. (1994). The nerve of glaucoma! *Arch. Ophthalmol.* **112,** 37–44.

Shade, D. L., Clark A. F., and Pang, I. H. (1996). Effects of muscarinic agents on cultured human trabecular meshwork cells. *Exp. Eye Res.* **62,** 201–210.

Sims, S. M., and Janssen, L. J. (1993). Cholinergic excitation of smooth muscle. *NIPS* **8,** 207–212.

Somlyo, A. V., and Somlyo, A. P. (1968). Electromechanical and pharmacomechanical coupling in vascular smooth muscle. *J. Pharmacol. Exp. Therap.* **159,** 129–145.

Somlyo, A. V., and Somlyo, A. P. (1994). Signal transduction and regulation in smooth muscle. *Nature* **372,** 231–236.

Stahl, F., Lepple-Wienhues, A., Kuppinger, M., Schneider, U., and Wiederholt, M. (1991). Measurements of intracellular calcium and contractility in human ciliary muscle. *Pflügers Arch.* **418,** 531–537.

Stahl, F., Lepple-Wienhues, A., Kuppinger, M., and Wiederholt, M. (1992). Electrogenic sodium-bicarbonate cotransport in human ciliary muscle cells. *Am. J. Physiol.* **262,** C427–C435.

Stamer, W. D., Snyder, R. W., Smith, B. L., Agre, P., and Regan, J. W. (1994). Localization of aquaporin CHIP in the human eye: Implications in the pathogenesis of glaucoma and other disorders of ocular fluid balance. *Invest. Ophthalmol. Vis. Sci.* **35,** 3867–3872.

Stamer, W. D., Seftor, R. E. B., Snyder, R. W., and Regan, J. W. (1995). Cultured human trabecular meshwork cells express aquaporin-1 water channels. *Curr. Eye Res.* **14,** 1095–1100.

Stumpff, F., Strauss, O., Wagner, U., and Wiederholt, M. (1996). Cultured bovine trabecular meshwork cells possess maxi-K-channels in high density. *Invest. Ophthalmol. Vis. Sci.* **37,** S 205.

Tamm, E., Flügel, C., Stefani, F. H., and Rohen, J. W. (1992). Contractile cells in the human scleral spur. *Exp. Eye Res.* **54,** 531–543.

Tamm, E. R., Flügel, C., Stefani, F. H., and Lütjen-Drecoll, E. (1994). Nerve endings with structural characteristics of mechanoreceptors in the human scleral spur. *Invest. Ophthalmol. Vis. Sci.* **35,** 1157–1166.

Tamm, E. R., Koch, T. A., Mayer, B., Stefani, F. H., and Lütjen-Drecoll, E. (1995). Innervation of myofibroblast-like scleral spur cells in human and monkey eyes. *Invest. Ophthalmol. Vis. Sci.* **36,** 1633–1644.

Tang, C. M., Presser, F., and Morad, M. (1989). Amiloride selectivity blocks the low threshold (T) calcium channel. *Science* **240,** 213–215.

Tripathi, R. C. (1974). Comparative physiology and anatomy of the aqueous outflow pathway. *In* "The Eye" (H. Davson and L. Graham, eds.), Vol. 5, pp. 163–356. Academic Press, New York.

Tripathi, B. J., and Tripathi, R. C. (1980). Contractile protein alteration in trabecular endothelium in primary open-angle glaucoma. *Exp. Eye Res.* **31,** 721–724.

Tripathi, B. J., and Tripathi, R. C. (1984). Effect of epinephrine *in vitro* on the morphology, phagocytosis and mitotic activity of human trabecular endothelium. *Exp. Eye Res.* **39,** 731–744.

Tripathi, R. C., Borisuth, N. S., Kolli, S. P., and Tripathi, B. J. (1993). Trabecular cells express receptors that bind TGF-beta 1 and TGF-beta 2: A qualitative and quantitative characterization. *Invest. Ophthalmol. Vis. Sci.* **34,** 260–263.

van Renterghem, C., Vigne, P., Barhanin, J., Schmid Alliana, A., Frelin, C., and Lazdunski, M. (1989). Molecular mechanism of action of the vasoconstrictor peptide endothelin. *Biochem. Biophys. Res. Commun.* **157,** 977–985.

Wax, M. B., Molinoff, P. B., Alvarado, J., and Polansky, J. (1989). Characterization of β-adrenergic receptors in cultured human trabecular cells and in human trabecular meshwork. *Invest. Ophthalmol. Vis. Sci.* **30,** 51–57.

Wiederholt, M., Lepple-Wienhues, A., and Stahl, F. (1993). Contractile properties of trabecular meshwork and ciliary muscle. *In* "Basic Aspects of Glaucoma Research III" (E. Lütjen-Drecoll, ed.), pp. 287–306. Schattauer, Verlag, Stuttgart.

Wiederholt, M., Sturm, A., and Lepple-Wienhues, A. (1994). Relaxation of trabecular meshwork and ciliary muscle by release of nitric oxide. *Invest. Ophthalmol. Vis. Sci.* **35,** 2515–2520.

Wiederholt, M., Bielka, S., Schweig, F., Lütjen-Drecoll, E., and Lepple-Wienhues, A. (1995). Regulation of outflow rate and resistance in the perfused anterior segment of the bovine eye. *Exp. Eye Res.* **61,** 223–234.

Wiederholt, M., Schäfer, R., Wagner, U., and Lepple-Wienhues, A. (1996). Contractile response of the isolated trabecular meshwork and ciliary muscle to cholinergic and adrenergic agents. *Ger. J. Ophthalmol.* **5,** 146–153.

Wiederholt, M., Dörschner, N., and Groth, J. (1997). Effect of diuretics, channel modulators, and signal interceptors on contractility of the trabecular meshwork. *Ophthalmologica* **211,** 153–161.

WoldeMussie, E., and Ruiz, G. (1992). Effect of histamine on signal transduction in cultured human trabecular meshwork cells. *Curr. Eye Res.* **11,** 987–995.

WoldeMussie, E., Ruiz, G., and Feldmann, B. (1990). Muscarinic receptor subtype involved in signalling mechanisms in cultured human trabecular meshwork cells. *Invest. Ophthalmol. Vis. Sci.* **31,** 388.

Woodward, D. F., Lawrence, R. A., Fairbairn, C. E., Shan, T., and Williams, L. S. (1993). Intraocular pressure effects of selective prostanoid receptor agonists involve different receptor subtypes according to radioligand binding studies. *J. Lipid Med.* **6,** 454–553.

Yanagisawa, M., Kurihara, H., Kimura, S., Yazaki, Y., Goto, K., and Masaki, T. (1988). A novel potent vasoconstrictor peptide produced by vascular endothelial cells. *Nature* **332,** 411–415.

Yousufzai, S. Y. K., Zheng, P., and Abdel-Latif, A. A. (1994). Muscarinic stimulation of arachidonic acid release and prostaglandin synthesis in bovine ciliary muscle: Prostaglandins induce cyclic AMP formation and muscle relaxation. *Exp. Eye Res.* **58,** 513–522.

Zadunaisky, J. A., and Spring, K. R. (1995). TBM cells area changes induced by drugs. Is it contraction or cell volume regulation? *Invest. Ophthalmol. Vis. Sci.* **36,** S 194.

Zadunaisky, J. A., Croft, D. F., and Spring, K. S. (1996). Human trabecular meshwork cells receptors. *Invest. Ophthalmol. Vis. Sci.* **37,** S205.

CHAPTER 8

Circadian Rhythms in Aqueous Humor Formation

Jonathan Sears* and Marvin Sears†
*Emory University Eye Center, Atlanta, Georgia 30322; and †Department of Ophthalmology and Visual Science, Yale University, New Haven, Connecticut 00520

I. HISTORICAL SUMMARY OF INVESTIGATIONS

Ocular diurnal rhythms were noted by Sidler-Hugenin (1899, quoted by Katavisto, 1964), who made tactile measurements of eye pressure. Five years later, Maslenikow (1904) confirmed the original observations by applanation tonometry. Köllner (1916) suggested that diurnal variation had a special significance in patients afflicted with glaucoma. Thiel (1925) demonstrated a peak in intraocular pressure (IOP) during the interval from 5:00 to 7:00 A.M. Indeed, Goldmann (1955) would visit patients in their homes in the early morning, before they arose, to catch the early morning rise. Langley and Swanljung (1951) described five patterns of diurnal change. Drance (1960), De Venecia and Davis (1963), and Katavisto (1964) extended these observations to characterize populations of normal and glaucoma patients. Later, Henkind *et al.* (1973), and Kitazawa (1973) both determined that the lowest IOP in both populations occurred at about 3:00 a.m. The nightly decrease in IOP in humans undoubtedly reflects a decreased rate of formation of aqueous humor. Ericson (1958), using

1063-5823/98 $25.00

Rosengren's suction cup technique, found that inflow decreased markedly during sleep but remained constant throughout the day and that a decrease in inflow caused the low nocturnal pressure. Indeed, Breebart (1966) had described a 19-year-old girl with Reiger's anomaly with glaucoma whose diurnal pattern totally reversed with the switching of the night–day cycle. Other early observations of the circadian rhythm in IOP were those of Schmerl *et al.* (1955). They suspected that night–day (sleep–wake) cycles might produce a neurohumoral factor. They isolated what appeared to be active factors from the cerebrospinal fluid of rabbits, dubbed hyperpiesin and myopiesin, which caused increases and decreases in IOP. Hyperpiesin was believed to be produced in animals exposed to light and was converted to myopiesin during darkness. Further ideas and work about possible neurohumoral controls for aqueous formation were largely forgotten until pursued by two housestaff officers at the Wilmer Institute who performed cross-circulation experiments between two rabbits to show the influence of the circulating blood of one on the eye pressure of the other (Stone and Sears, 1958).

The work cited definitely implied that alterations in aqueous humor formation are responsible for the circadian changes in intraocular pressure. The concept was corroborated in the following way. Anjou (1966) showed that rabbits have a cyclical rhythm of intraocular pressure. The IOP was highest at night and lowest during the day. Anjou found a rhythm of aqueous flare (protein in the anterior chamber) that was 180° out of phase with the rhythm of IOP. Krakau (1962) and later others argued that the flare rhythm was a reflection of the flow rhythm. Using clearance techniques for fluorescein, Rowland *et al.* (1981) reported that rabbits entrained to a cycle of 12 hr of light and 12 hr of dark had a rhythm of IOP persisting in continuous dark and abolished by continuous light. Gregory *et al.* (1985) showed that the zeitgeiber for the phase of the rhythm of IOP is the light–dark cycle. Rowland *et al.* (1986) developed data to suggest that changes in aqueous flow are responsible for the circadian rhythm of IOP. A detailed study of the kinetics of this phenomenon by Smith and Gregory (1989) firmly established that the rhythm of aqueous flow was circadian and determined the circadian rhythm of IOP. Reiss *et al.* (1984) and Topper and Brubaker (1985) and co-workers, also using fluorimetric clearance techniques, took up these suggestions and applied these experimental studies to human eyes. Indeed, they found a circadian rhythm of flow rates that was only susceptible to reduction during the day and to increases during the night. This finding undoubtedly reflected the tone of the system as had been suggested (Sears, 1984).

The site of aqueous humor formation has been known since 1717, when Méry (quoted by Davson, 1956) suspected that the ciliary body was the source of aqueous humor. Two centuries later research indicated that the energy available from the cellular metabolism of the epithelia of the ciliary processes accounts for the secretion (extraction and formation) of aqueous humor. Thus the ciliary epithelium (Figs. 1 and 2) itself was an ideal place to look for controls of this process, and the circadian cycle a tool for uncovering these controls. Factors for the regulation of IOP can emerge from studies of diurnal variation in eye pressure because this variation undoubtedly reflects the existence of regulators. Knowledge of the molecular machinery was necessary to put the circadian rhythm on a sound foundation; that is, specific controls for the rate of aqueous humor formation require a signal transduction pathway.

Clues for a signal transduction pathway came from clinical sources. At the turn of the nineteenth century, several observations of the human eye indicated that a drop in ipsilateral pressure occurred after cervical sympathectomy. In particular, stellate ganglion block of the sympathetic input to the eye was effective in reducing intraocular pressure. Still later, observations from the extensive studies of the effect of degeneration release of norepinephrine on intraocular pressure, or the "ganglionectomy effect" as it was known, (Linner and Prijot, 1955; Sears and Bárány, 1960; Sears, 1975), stimulated investigations on a molecular level (briefly reviewed later) showing that a molecular mechanism for aqueous secretion required signals from the β-adrenergic receptor.

In this way studies of the adrenergic system became the first to demonstrate and report signal transduction for the ciliary epithelium. Bromberg *et al* (1980) showed the presence of β-adrenergic receptors in isolated ciliary processes. The adrenergic drug receptor relationship was quantified in a well-controlled, *in vitro,* cell-free system in which the drug concentration at the receptor was determined (Gregory *et al.,* 1985). β-Adrenergic receptors were studied in crude particulate preparations of the ciliary processes of rabbit eyes by a direct ligand-binding assay using ^{125}I-hydroxybenzylpindolol and by examining the kinetic and regulatory properties of adenylate cyclase linked to the β-adrenergic receptors. High-affinity binding sites for ^{125}I-hydroxybenzylpindolol were found in the same particulate membrane factions of homogenized ciliary processes as was adenylate cyclase activity. Stimulation of adenylate cyclase activity by catecholamines was completely blocked by several β-adrenergic antagonists but not by phenoxybenzamine, and α-blocker. The K_i was comparable to that for β-adrenergic receptors of other tissues. The K_{act} for stimulation of enzyme activity was of the order expected for a β-adrenergic receptor-linked adenylate cyclase. The K_i for inhibition of levoepinephrine stimulation was similar to binding constants for β-antagonists in other systems. Similar results were later obtained in membrane preparations from sheep, rabbit, monkey, and human eyes. The potency order of agonist activation indicated that ciliary processes contain a predominance of β_2-adrenergic receptors (Gregory *et al.*, 1981; Mittag and Tormay, 1985). Finally, binding constants determined by the direct ligand-binding technique and by the assay for adenylate cyclase agree, indicating that the two techniques measure the interaction between the β-adrenergic receptor of the ciliary processes (Gregory *et al.*, 1985). Obviously, a functionally coupled β-adrenergic receptor is located in the ciliary processes. Horio *et al.* (1996) later showed that these receptors were present in isolated ciliary epithelia.

Functional coupling is influenced by the "tone" of the system. Inhibitory effects are a consequence of receptor-mediated guanyl nucleotide binding protein (N_i input) negatively coupled to adenylate cyclase for compounds such as α_1- and α_2-agonists, adenosine agonists, opiates, or somatostatin. Stimulatory effects of cAMP-generating systems are mediated by N_s input, positively coupled to adenylate cyclase for epinephrine and its analogs. Thus, it was very satisfying to find and report that, indeed, α_2-adrenergic agonists modulated the β-adrenergic receptor in the ciliary epithelium; that is, the former, under conditions of high tone, caused a decrease in aqueous flow, and the latter, under conditions of low tone, caused an increase in aqueous flow (Bausher *et al.*, 1987; Bausher and Gregory, 1989).

a

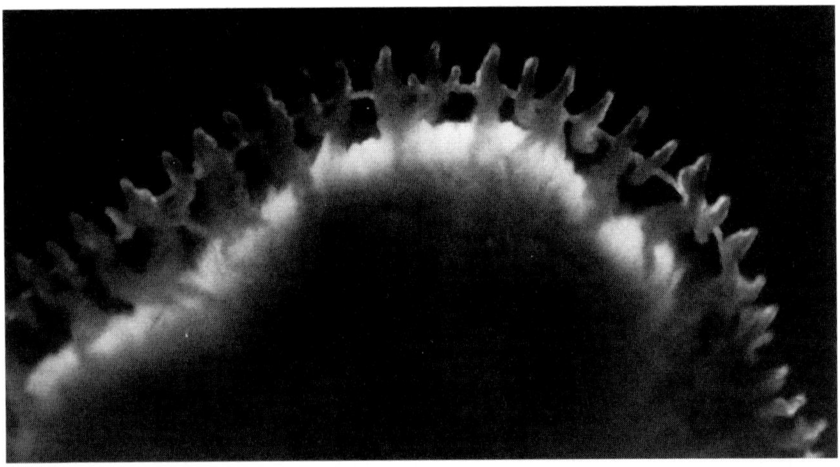

FIGURE 1 (a) The "digits" of the ciliary body, a silhouette of several plicated processes. (b) Scanning micrograph of the posterior structure of the rabbit iris showing the diverse morphology of the ciliary processes. Regional differences include the anteroposterior direction as well as between processes appearing side by side. (c) From inside the posterior chamber are seen thousands of microvilli, projections from the nonpigmented epithelium (NPE) basolateral membranes. [Reprinted with permission from Sears, M. L. (1994). Formation of aqueous humor. *In* "Principles and Practice of Ophthalmology" (D. M. Albert and F. A. Jakobiec, eds.), Chap. 11, pp. 182–206. W. B. Saunders, Philadelphia.]

b

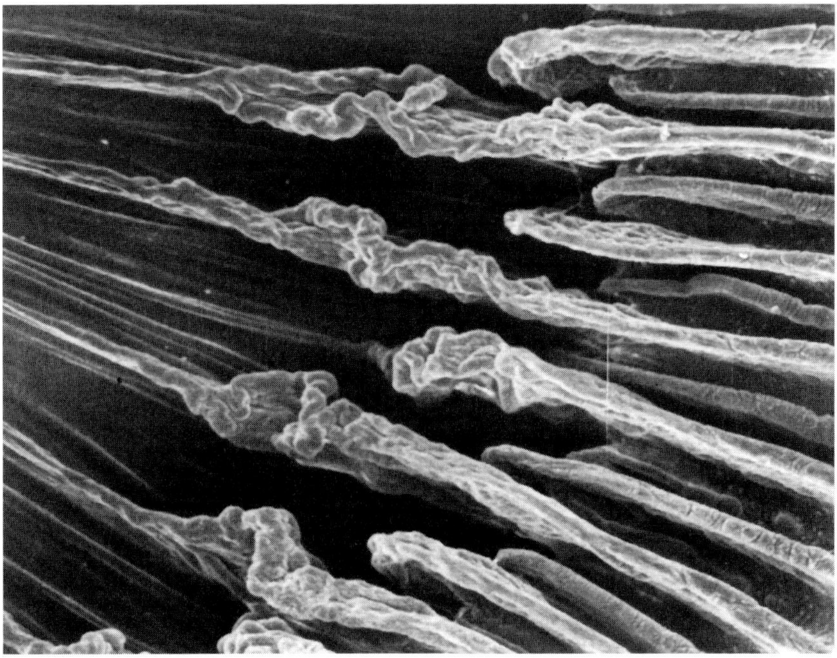

c

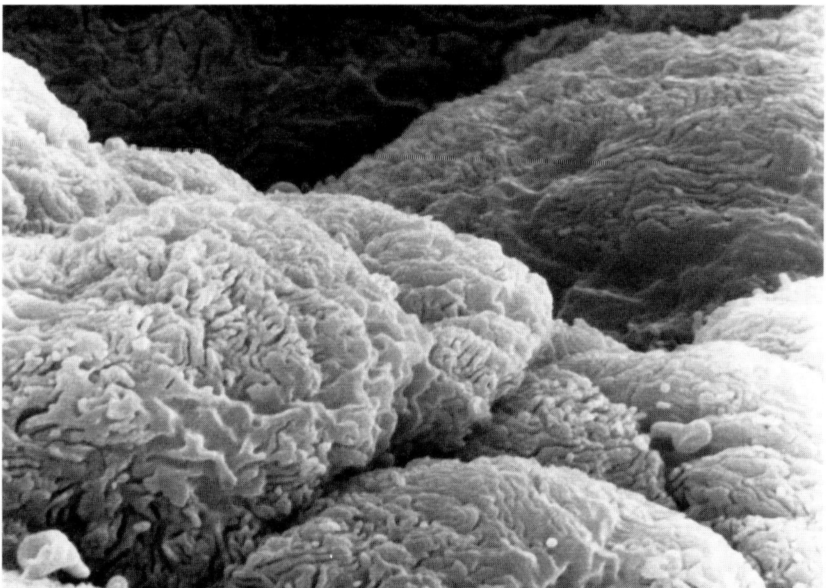

FIGURE 1—Continued

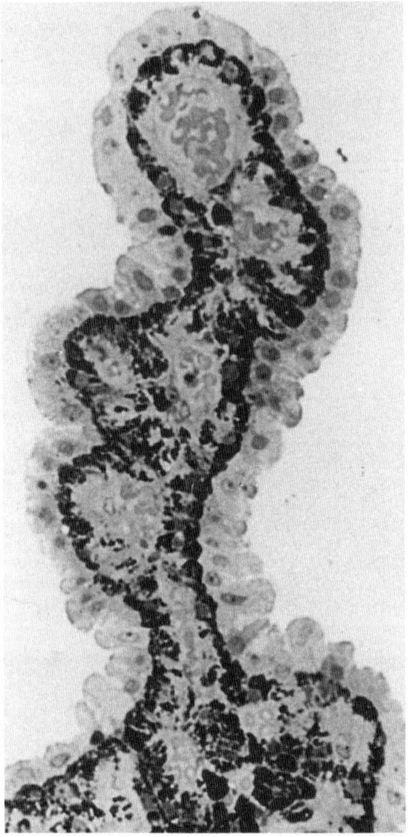

FIGURE 2 Typical ciliary process from the anterior or secretory region of the plicated ciliary processes, that is, the region of the embellished, extensively elaborated basolateral membranes of the villiform processes pictured in Fig. 1b.

In studies of circadian rhythms, intraocular pressure, and aqueous flow (Yoshitomi and Gregory, 1991; Yoshitomi *et al.*, 1991), it was found that intraocular pressure, aqueous flow, and the concentrations of norepinephrine and cAMP in the aqueous humor increase during the dark in rabbits entrained to 12 hr of light and 12 hr of dark. Depriving the eye of sympathetic input by excision of the superior cervical ganglion or preganglionic section of the cervical trunk did not eliminate, but only blunted slightly, the dark phase increases of IOP, flow, and aqueous norepinephrine and cAMP. Blockade of ocular β-adrenergic receptors with timolol decreased IOP and aqueous flow in the dark but not during the light phase, did not lower IOP during the dark phase in rabbits previously subjected to excision of the superior cervical ganglion or section of the sympathetic trunk, and abolished the dark phase increase of aqueous cAMP. These data are consistent with the idea that increased sympathetic input to the eye during the dark phase of the circadian cycle increases aqueous flow and IOP, and that these increases are mediated in part by ocular β-adrenergic receptors. We decided to continue these investigations by empowering the circadian cycle as a tool to uncover regulatory sources for aqueous secretion.

II. METHODS

Three primary methods were used in the current studies: (1) Isolation of a pure ciliary epithelial bilayer to exclude proteins and other mediators found in blood, blood elements, vasculature, and in contaminating stroma; (2) isolation of populations of mRNA particularized to intervals of the circadian cycle; and (3) the technique of subtractive hybridization employed during the circadian cycle. Our interest in light–dark variations in transcriptional programs of the ciliary epithelium began with the knowledge that peak and trough aqueous flow could be reproducibly predicted during the circadian cycle, thereby creating a model in which populations of mRNA could be segregated according to the functional state of the bilayered epithelium. We have proposed that this function is under transcriptional control and tested this hypothesis by examining β-arrestin and connexin 43 (Cx43), discovering that each was regulated in a different manner (see discussion later). Using the differential display of Liang and Pardee (1992), we were able to demonstrate differences in amplified cDNA sequences when comparing different populations of mRNA template, but we had difficulty in reproducibly amplifying these bands from experiment to experiment. Therefore, we sought to engineer a technique that would allow us to subtract mRNA populations reproducibly from a small amount of mRNA template using the polymerase chain reaction (PCR) method. NZW rabbits were entrained for at least 2 weeks in a 12-hr light/12-hour dark room. Animals were anesthetized with an intramuscular injection of ketamine hydrochloride, 25 mg/kg (Aveco Co., Fort Dodge, IA, USA) and Rompun, 25 mg/kg (xylazine; Haver, Shawnee, KS, USA). They were then intravenously anticoagulated with sodium heparin, 2500 U/kg (Elkins-Sinn, Cherry Hill, NJ, USA). After 2 min, the animals were killed with intravenous injection of 0.5 mL Beuthanasia (pentobarbital sodium, 390 mg/mL, and phenytoin sodium, 50 mg/mL; Schering-Plough Animal Health Corp., Kenilworth, NJ, USA) in accordance with the ARVO guidelines, as adopted from the rules of the National Institutes of Health. Eyes were enucleated, the long

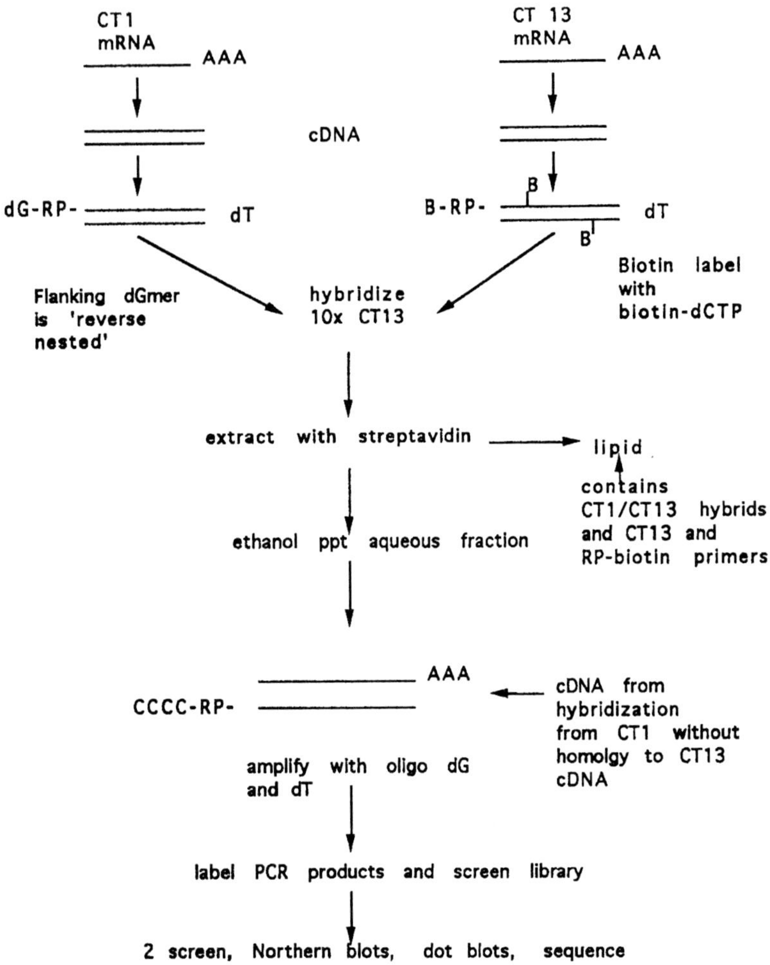

SCHEME 1

posterior ciliary arteries cannulated, and the secretory ciliary epithelial bilayer isolated by our previously described perfusion and microdissection technique (Sears *et al.,* 1991; Matsui *et al.,* 1996) (Figs. 3, 4, and 5), 1 hr after lights are on [circadian time (CT) 100] and 1 hr after lights are off (CT1300).

Total RNA was isolated by acid guanidinium thiocyanate-phenol-chloroform extraction, and polyadenylated RNA selected using oligo dT cellulose column (Type III, Collaborative Research). Complementary DNA

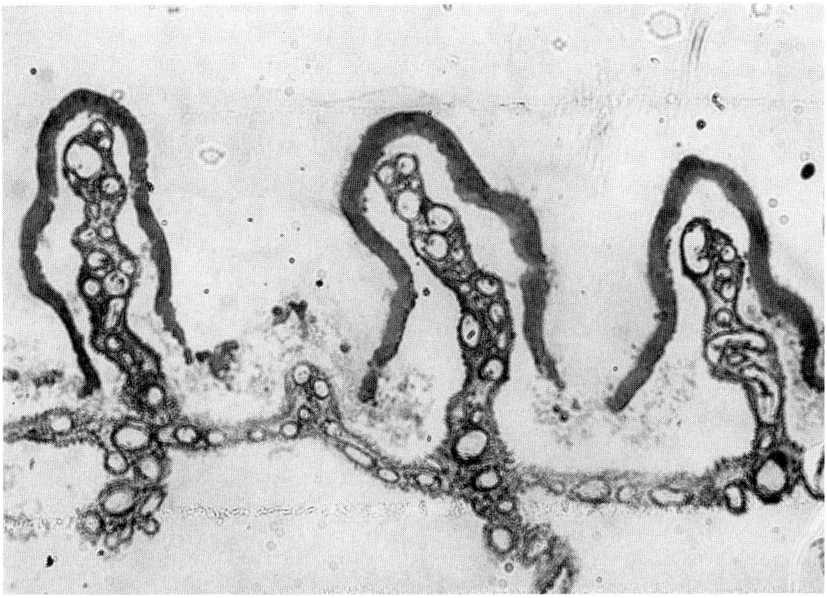

FIGURE 3 Stained section cut from fixed paraffin embedded ciliary processes, extracted during perfusion, shows progressive isolation of the bilayered epithelium from several processes with removal of iris and collagenous connective tissue and capillaries of the stroma. The latter appears to collapse on itself and retracts from the basal lamina of the pigmented epithelial layer (PE). [Reprinted from Sears *et al.* (1991) with permission generously provided by the editors and publishers of the American Ophthalmology Society and Johnson Press.]

was produced using standard techniques from CT100 and CT1300. Each of these populations of cDNA was amplified using random primers consisting of 10mers identical to the sequence used by Liang and Pardee as a 5′ primer, and oligo dT as a 3′ primer under PCR conditions that used a first single cycle with an annealing temperature of 30°C, followed by a 2-min extension, followed by 30 cycles of annealing at 50°C. As demonstrated in the accompanying schema, the tester population was amplified with a reverse nested 5′ primer, whereas the driver population was amplified with a biotinylated primer with an identical sequence as the tester primer less the flanking oligo dG 9mer. This idea allowed us to amplify selectively our subtracted product without the risk of amplifying driver amplicons.

Several factors uncovered bythe technique described and illustrated in the accompanying schema have implications for correlatinggrowth with function in the ciliary epithelium. This work is in progress. We have not yet had the opportunity to decide whether these isolates are expressed on a circadian basis nor indeed whether they are related to secretion. Two

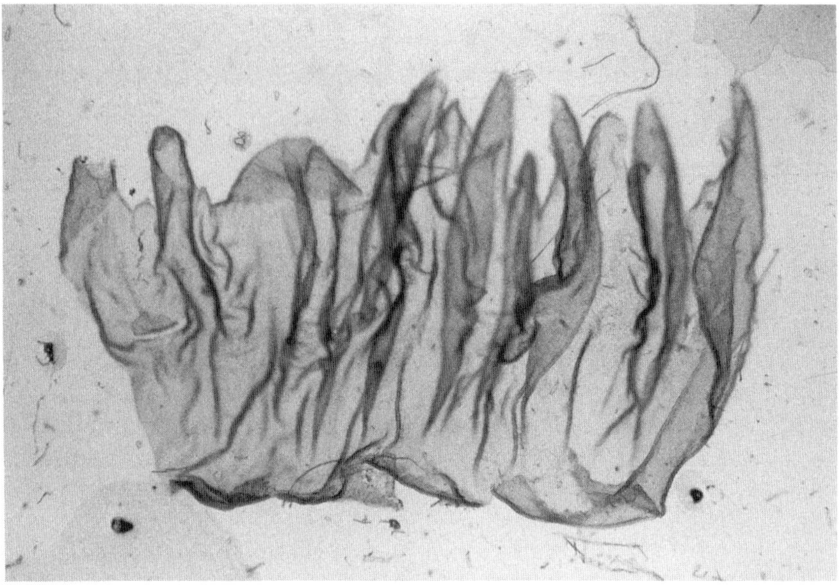

FIGURE 4 Flattening isolated ciliary processes shown immediately after perfusion and microdissection, 8× magnification. [Reprinted from Sears *et al.* (1991) with permission generously provided by the editors and publishers of the American Ophthalmology Society and Johnson Press.]

clones isolated by this method fell out during the early hours of the circadian cycle and, therefore, not surprisingly bear strong homology to genes related to the modeling of tissues during growth and development. One of these, *BMP-2A,* a member of the *TGF* supergene family, plays a role in morphogenetic growth processes (Celeste *et al.,* 1990; Francis *et al.,* 1994) along with other *BMP*s required for ocular and renal growth (Dudley *et al.,* 1995). Together with other *BMP* genes in the same family (Wozney *et al.,* 1988), it has been implicated in interdigital cell programmed death (Tabin, 1995) and has been localized to the bronchioles in the lung at branch points. The evolution of the ciliary processes into a secretory structure is, of course, of great interest here (see Fig. 1). The second clone has a nearly 100% identity on the amino acid level with a human QM gene (Dowdy *et al.,* 1991), a gene known to function as an inhibitor of transactivation of Jun (Imaki *et al.,* 1995). Light-induced phase shifts of circadian activities are associated with its expression (Wollnik *et al.,* 1995). Pursuit of QM expression patterns will be done to determine whether QM is regulated on a circadian basis, perhaps as part of the cAMP responsive nature of the ciliary epithelial bilayer.

III. HOMOLOGOUS DESENSITIZATION OF CIRCADIAN AQUEOUS FLOW

We hypothesized that homologous desensitization regulates signal transduction from the β-adrenergic receptor in the ocular ciliary epithelium to affect the circadian rhythm of aqueous humor secretion. A summary of the results to date is stated now: β-arrestin-1 was cloned from the rabbit ciliary epithelium, and the full-length cDNA used as a probe for Northern blot analysis to examine the diurnal expression of β-arrestin mRNA. Protein expression of β-arrestin at intervals during the circadian cycle of aqueous secretion showed a decrease in β-arrestin expression when maximal activation of the β-adrenergic receptor is known to increase secretion. Diurnal expression of β-arrestin suggested that homologous desensitization can regulate the circadian rhythm of aqueous flow (Wan *et al.,* 1997).

The ocular ciliary epithelium is a bilayer that secretes a fluid to supply nutrients to the avascular structures of the eye. This process of aqueous humor formation manifests a circadian rhythm that provides an ideal opportunity to uncover regulatory mechanisms for secretion. As we have seen, the molecular mechanism of aqueous secretion by the ciliary epithelium involves transmembrane signals from the β-adrenergic receptor (BAR) to the cAMP second messenger system, synchronized with the diurnal rhythm of aqueous flow. Signals from the BAR attenuate rapidly in the presence of continued stimulus (Harden, 1983; Lohse *et al.,* 1990). This phenomenon, called homologous or agonist-specific desensitization, is dependent on a β-adrenergic receptor kinase (BARK) that phosphorylates BAR (Benovic *et al.,* 1986; Freedman *et al.,* 1995), or perhaps on other kinases, GRKs, in the ciliary epithelium, enabling it to become a ligand for β-arrestin proteins (Pippig *et al.,* 1993; Söhlemann *et al.,* 1995). β-Arrestin proteins, isoforms of arrestin (S-ag) (Broekhuyse *et al.,* 1987; Wacker *et al.,* 1973), facilitate sequestration and uncoupling of G-protein-coupled receptors (Palczewski *et al.,* 1992; Ferguson *et al.,* 1996; Dawson *et al.,* 1993). To determine whether homologous desensitization contributes to the molecular mechanism underlying circadian aqueous secretion by the ciliary epithelium, we examined the expression of β-arrestin in the ciliary epithelium isolated, at different intervals, from the eyes of rabbits entrained in a light–dark cycle.

Primers specific for bovine β-arrestin successfully amplified a 5′ fragment of β-arrestin-1 from the rabbit that was virtually identical to human β-arrestin-1B isoform. (Parruti *et al.,* 1993). Sequence information from this fragment was used to construct 5′ primers in order to amplify 3′ coding sequence based on bovine arrestin sequence data (Fig. 6a). Each amplified fragment was sequenced in overlapping fashion providing terminal 3′ sequence data to enable amplification of a full-length 1.23-kb cDNA. The deduced amino acid sequence of rabbit β-arrestin-1 is shown in Fig. 6b.

a

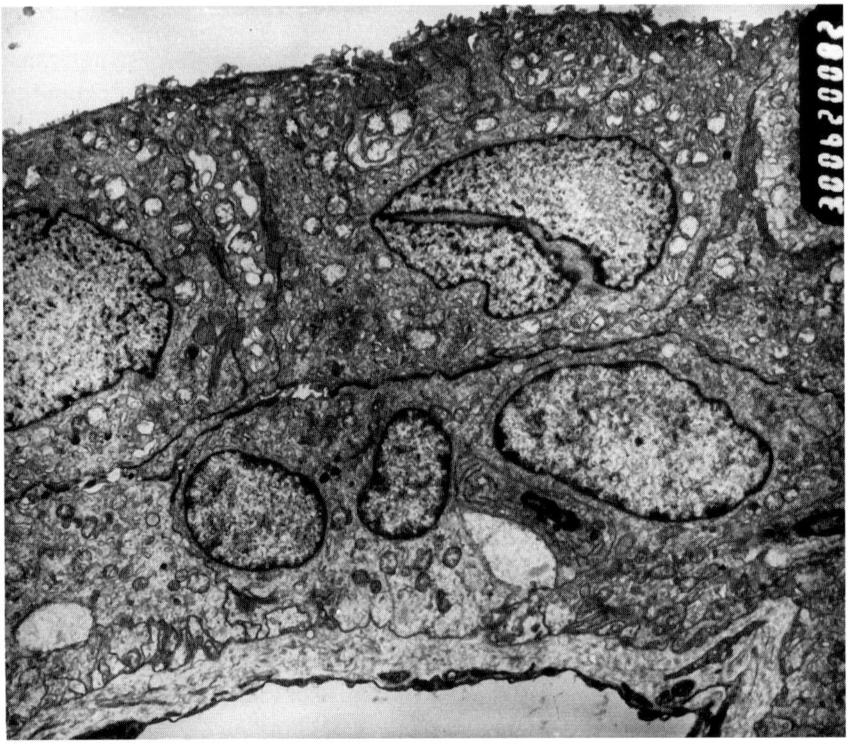

FIGURE 5 (a) Electronmicrograph of the bilayer showing preserved subcellular organelles and basal laminae and intercellular junctions between NPE and PE. 4000× magnification. (b) Higher power showing detail of a tight and several gap junctions and ciliary channels. 5000× magnification. [Reprinted from Sears *et al.* (1991) with permission generously provided by the editors and publishers of the American Ophthalmology Society and Johnson Press.]

The overall amino acid identity displayed a 97.6% homology bovine β-arrestin-1 (Lohse *et al.,* 1990) to human 97% (Parutti *et al.,* 1993) and to rat 97.4% (Attramadal *et al.,* 1992). Eight absent amino acids in rabbit β-arrestin-1 are the same ones missing in human β-arrestin-1B (LLGDLASS).

We next turned to Northern blot analysis (Fig. 7a). With a full-length antisense probe of β-arrestin-1, we identified three major mRNA species: 1.4, 2.4, and 4.4 kb. In bovine, Lohse *et al.* (1990) identified four mRNA species of approximately 1.3, 2.4, 4.1, and 7.5 kb. In rat Attramadal *et al.* (1992) identified 3 mRNA species of approximately 2.5, 4.4, and 7.5 kb with the β-arrestin-1 probe. In human three major mRNA species of 1.7, 3.0, and 7.5 kb were detected (Parutti *et al.,* 1993). The distribution of different mRNA sizes in this work and in the work of others cited suggests

b

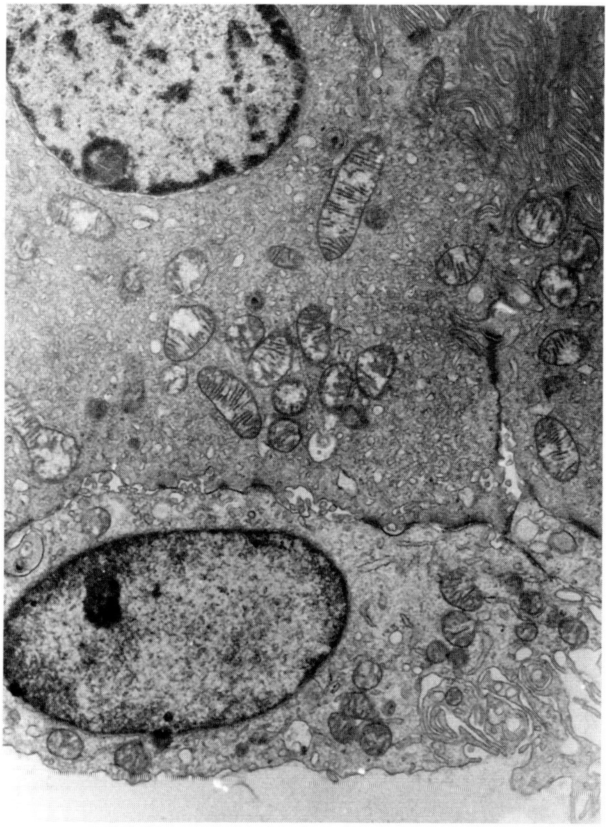

FIGURE 5—Continued

likely species and tissue differences in RNA processing. The Northern blots demonstrated a striking decrease in β-arrestin transcript at CT800 and CT1300, corresponding to late light phase and early dark phase. Signal on the Northern blot increased at CT1600, suggesting that transcription begins in the early dark phase. This signal persisted throughout the dark phase into the light phase to at least CT400. Tissue from another set of rabbits entrained to a 12-hour light/12-hour dark cycle for 2 weeks after a right unilateral cervical ganglionectomy was isolated and lyzed in high salt suspension buffer. After centrifugation, equal concentrations of soluble protein were size fractionated by 12% (SDS–PAGE) and transferred to nitrocellulose for incubation with purified β-arrestin-2 antiserum. On Western blot (Fig. 7b), a striking decrease in the expression of β-arrestin protein was demonstrated beginning at CT800 that persisted until at least CT1300, early in the dark phase. Significant levels of β-arrestin protein were detected beginning at CT1600. Prominent signals on Western blot were observed throughout the remainder of the dark phase into the light phase, to at least CT400. There were apparent differences between the right and left eyes at CT1 and CT4, but it is clear that there is significant expression of protein on both right denervated and left normally innervated sides at these intervals when compared with the CT8 and CT13 intervals. Both RNA and protein expression revealed a similar time course. The expression of β-arrestin gene products throughout the circadian cycle was equivalent by RNA and protein analysis. Two bands were detected, one at MW 47,000 and a second at 46,000, suggesting that one band may be β-arrestin-1, the other β-arrestin-2.

Indirect immunofluorescence, using purified β-arrestin-2 antiserum, showed that β-arrestin could be found within the stroma of human ciliary processes, likely within capillary vascular endothelium, as found by Parutti *et al.* (1993). Not surprisingly, β-arrestin localizes to the epithelial site of the β-adrenergic receptor, most prominently to the nonpigmented ciliary

a

Primers	Coding region (bp)		Product length (bp)
F1 5′-ATGGGCGACAAAGGGACGCGGGTG-3′	97-120	(bovine)	363 (F_1)
R1 5′-CGGAGGGATCTCAAAGGTGAA-3′	439-459	(bovine)	
F2 5′-TGCCTACCCCTTCACCTTTGAGAT-3′	333-356	(rabbit)	490 (F_2)
R2 5′-TCAGCGTGTAGACCTTGCAGA-3′	808-831	(bovine)	
F3 5′-TACGCAGACATCTGCCTCTTCAAC-3′	712-735	(rabbit)	518 (F_3)
R3 5′-TCTGTCGTTGAGCCGCGGAGA-3′	1327-1350	(bovine)	
F1			1230 (F_4)
R3			

F and R indicate forward and reverse, respectively.

FIGURE 6 (a) PCR primers used for cloning of β-arrestin from the rabbit ciliary epithelium. (b) Deduced amino acid sequence of fragment 4, a full-length 1.23-kb cDNA, compared to human, bovine, and rat β-arrestin. Residues of BBA-1, HBA-1A, HBA-1B, and RATBA-1, which are identical to the RBA-1 at each position, are designated with a period. The presence or absence of the eight amino acid sequence among RBA-1, BBA-1, and the splice variants HBA-1A, HBA-1B, and RATBA-1 are shown for comparison purposes only. The complete nucleotide sequence is deposited in GenBank (accession number U75838) or is available from the authors. [Reprinted from Wan *et al.* (1997) with permission from Academic Press.]

b **Alignment of deduced amino acid sequence of β-arrestin 1 among rabbit ciliary epithelia, bovine, human, and rat.**

```
RBA-1     MGDKGTRVFKKASPNGKLTVYLGKRGFVDHIDLVDPVDGVVLVDPEYLKERRV   53
BBA-1     .........................D........E..................   53
HBA-1A    .........................D...........................   53
HBA-1B    .........................D...........................   53
RATBA-1   .........................D...........................   50

RBA-1     YVTLTCAFRYGREDLDVLGLTFRKDLFVANVQSFPPAPEDKKPLTRLQERLIK  106
BBA-1     .....................................................  106
HBA-1A    .....................................................  106
HBA-1B    .....................................................  106
RATBA-1   .....................................................  106

RBA-1     KLGEHAYPETFEIPPKLPCSVTLQPGPEDTGKACGVDYEVKAFCAENLEEKIH  159
BBA-1     ...............N.....................................  159
HBA-1A    ...............N.......................A.............  159
HBA-1B    ...............N.......................A.............  159
RATBA-1   ...............N.....................................  159

RBA-1     SVRLVIRKVQYAPERPGPHPTAETTRLFLMSDKPLHLEASLDKEIYYHGEPII  212
BBA-1     ..................Q.......Q.........................S  212
HBA-1A    ..G...............Q.......Q.........................S  212
HBA-1B    ..G...............Q.......Q.........................S  212
RATBA-1   ..................Q.......Q.........................S  212

RBA-1     VNVHVTNNTNKTVKKIKISVRQYADICLFNTAQYKCPVAMEEADDTVAPSSTF  265
BBA-1     .....................................................  265
HBA-1A    .............E.......................................  265
HBA-1B    .............E.......................................  265
RATBA-1   .....................................................  265

RBA-1     CKVYTLTPFLANNREKRGLALDGKLKHEDTNLASSTLMREGANREILGIIVSY  318
BBA-1     .....................................L...............  318
HBA-1A    .....................................L...............  318
HBA-1B    .....................................L...............  318
RATBA-1   .....................................L...............  318

RBA-1     KVKVKLVVSRGG........DVAVELPFTLMHPKPKEEPPHREVPENETPVDT  371
BBA-1     ............LLGDLASS..........................H......  379
HBA-1A    ............LLGDLASS.................................  379
HBA-1B    .....................................................  371
RATBA-1   ............LLGDLASS..........................S......  379

RBA-1     NLIELDTNDDDIVFEDFARQRLKGMKDDKEEEDDVTGSPRLNDR   410
BBA-1     .................................E.G........   418
HBA-1A    ..........................E......E.G........   418
HBA-1B    ..........................E......E.G........   410
RATBA-1   ..............................D....G...H..N.   418
```

FIGURE 6—Continued

epithelial (NPE) cell layer (Figs. 8a and b), abutting the posterior chamber. It is possible that fluorescence in the pigment epithelial (PE) cell layer may have been partially quenched by pigment in this layer.

These results show that β-arrestin gene products are tightly regulated in time and space throughout the circadian cycle of aqueous flow. The pattern

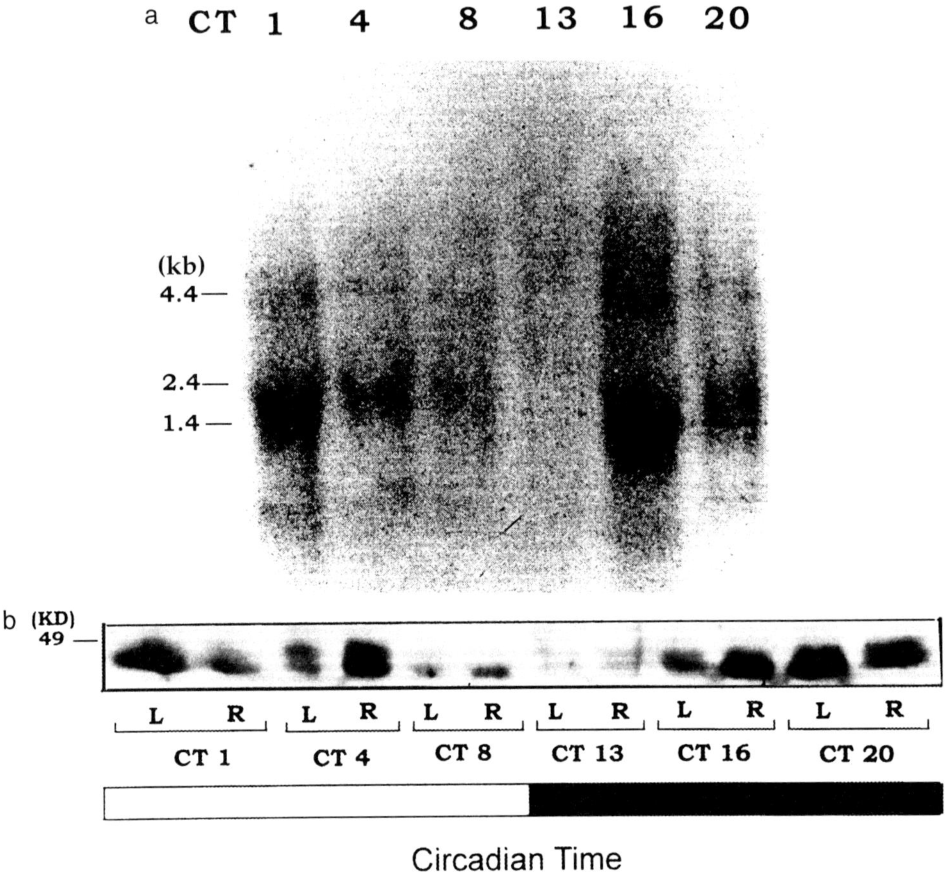

FIGURE 7 (a) Northern blot analysis of the expression of rabbit β-arrestin-1 mRNA harvested at six different intervals during the circadian cycle from isolated cilary epithelium demonstrates decreased transcription of β-arrestin-1 during CT800 and CT1300. A single-strand antisense probe was used to hybridize with total cellular RNA. Autoradiographic exposure times was 3 days with intensifying screen at −80°C. A single-strand, full-length antisense probe of β-arrestin-1 was used to hybridize with total cellular RNA prepared from isolated intact ciliary epithelium prepared at different intervals in the circadian cycle. To examine mRNA expression of rabbit β-arrestin-1 at six intervals during the circadian cycle, the cDNA, fragment 4 (1230 bp) was used as a probe for Northern blot analysis. Total RNA from rabbit ciliary bilayer was prepared by the guanidinium thiocyanate-phenol-chloroform extraction method (Chomczynski and Sacchi, 1987) and quantified by spectrophotometry at OD_{260}. Twenty micrograms of RNA from each sample was subjected to electrophoresis with ethidium bromide and visualized with UV light. A full-length cDNA corresponding to β-arrestin-1 was labeled with P^{32} by the random primer method. The total RNA was denatured in 50% formamide and 6.5% formaldehyde, size fractionated on formaldehyde–agarose gel using 20 μg/lane and transferred to a nylon membrane by capillary blotting. Molecular weight was estimated with an RNA ladder (0.24–9.5 kb range) (Gibco). For the β-arrestin-1 probe,

of expression of both mRNA and protein coincides with the interval during the circadian cycle of aqueous humor flow when the rate of secretion is either constant or decreasing, CT1600 to CT400 (Fig. 9). Then, between the hours of CT800 and CT1300, the slope of increasing aqueous humor secretion is steepest when the BAR is activated and β-arrestin expression is absent (Figs. 7a and b). Activation of the BAR during this time permits circadian secretion of aqueous humor to increase intraocular pressure; desensitization of BAR by persistent expression of β-arrestin from the early dark phase into the early light phase is therefore synchronized with the reported rhythmic decrease in cAMP, aqueous flow, and IOP (Yoshitomi *et al.*, 1991). RNA and protein extracted at identical absolute times (12 hours CT) from animals housed in rooms 12 hr out of phase in the light–dark cycle demonstrated a 12-hr lag or phase shift in peak and trough expression

---

an antisense single-strand DNA was generated from cDNA fragment 4 by Taq polymerase. Fifty nanograms of the described cDNA fragment 4 was used as a template. The reaction mixture (50 μL final volume) contained 2 m*M* each of dATP, dCTP, dTTP, and 50 μ*M* dCTP; 50 μCi of [^{32}P] dCTP (3000 Ci/*M*); 200 p*M* of the antisense primer R3; 5 μL of 10 × PCR buffer, and 2 units of Taq polymerase. The mixture was cycled 30 cycles (94, 55, and 72°C each for 1 min). The result was confirmed in two separate experiments. (b)Western blot of cytosolic proteins from isolated ciliary epithelium harvested at six different intervals during the circadian cycle 2 weeks after right unilateral ganglionectomy. R and L indicate ciliary bilayers from right and left eyes. A decrease in protein expression occurs during CT800 and CT1300 that matches a decrease in mRNA expression seen in part (a). An antibody raised against the last 75 amino acids of the carboxyl terminus of rat β-arrestin-2 was kindly provided by R. J. Lefkowitz. The antibody bears a 55% homology to β-arrestin-1 and therefore binds it and β-arrestin-2 at a titer of 1:5000. The ciliary epithelial bilayer of right ganglionectomized eyes and left normal eyes was harvested at six different circadian intervals and lyzed in buffer containing 0.5% NP40; 50 m*M* Tris, pH 7.4; 150 m*M* NaCl; 100 m*M* NaF; 0.5 m*M* phenylmethylsulfonyl fluoride; and 5 μg/mL of each aprotinin, leupeptin, and trypsin inhibitors. The tissues were homogenized by passing them through a syringe fitted with a 25-gauge needle. Unbroken tissues were pelleted by centrifugation, 800g × 5 min and discarded. The supernatant was spun at 13K for 5 min at 4°C. Protein in the cytosolic fractions was measured using Bio-Rad DC protein assay reagent. Samples containing 1 μg of protein were suspended in an equal volume 2% SDS/10% 2-mercaptoethanol Tris loading buffer. The samples were heated for 3 min at 95° and size fractionated on 12% SDS–PAGE minigels. Proteins were transferred overnight in Tris/glycine/20% methanol buffer to nitrocellulose filters with a tank transfer system by electrophoresis. An efficiency of transfer of 99% was verified by Ponceau red staining of the blots and Coomassie blue staining of gels after transfer. The blots were blocked overnight with 3% nonfat dry milk in Tris-buffered saline, 20 m*M* Tris, pH 7.6, + 0.01% Tween-20 (TBS-T) and 0.05% sodium azide, and incubated with purified β-arrestin antisera at 1:500 dilution in TBS, for 3 hr. The blots were rinsed with TBS-T several times to remove excess antibody. The immunoreactive bands were visualized with anti-rabbit antibody conjugated to horseradish peroxidase (Amersham) using enhanced chemiluminescence (ECL) for detection ECL (Amersham). The results were confirmed in duplicate experiments. [Reprinted from Wan *et al.* (1997) with permission from Academic Press.]

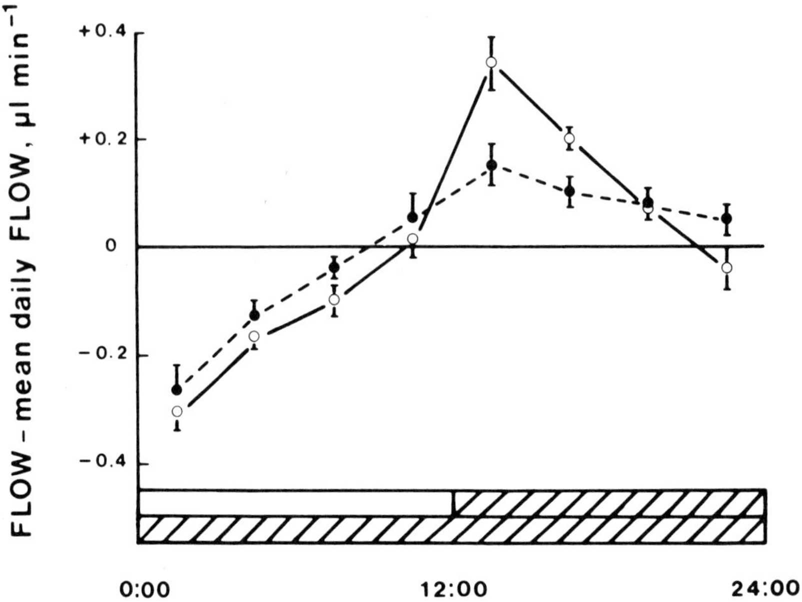

FIGURE 9 Aqueous flow in rabbits in light–dark (solid line) or constant dark (dashed line). The rhythm of flow persists in constant dark with slight blunting. [Reprinted from Wan *et al.* (1997) with permission from Academic Press.]

of β-arrestin. The expression of β-arrestin-1 regulated at the transcriptional level suggests, but does not prove, that the responsiveness of the ciliary epithelial BAR is modulated by homologous desensitization, and that this system of receptor desensitization may follow a diurnal rhythm. Candidate genes for control of desensitization are obviously either β-arrestin, BARK, or both.

A rhythmic expression of β-arrestin was also found in lyzed extracts from denervated ciliary epithelium of entrained animals (Fig. 7b). Noradrenergic innervation reaches the rabbit eye after synapse in the cervical ganglion (Sears, 1975). The stromal vessels of the ciliary processes are richly supplied (Ehinger, 1966), but the bilayered epithelium itself contains only a rare noradrenergic nerve terminal (Yamada, 1988, 1989). Examination of studies of the response of the circadian rhythm of aqueous flow to cervical sympathetic ganglionectomy (Yoshitomi and Gregory, 1991) indicates a slight effect. The signal for aqueous secretion is transduced at the BAR, but it is not clear that noradrenergic innervation is required. In fact, in sympathetically denervated human eyes, reduction of aqueous flow and IOP are induced by β-adrenergic blockade (Wentworth and Brubaker, 1981). We performed unilateral cervical ganglionectomy in entrained rabbits because we wondered whether the expression of β-arrestin in the ciliary epithelium was faithful to a central pacemaker. No change in expression of β-arrestin in the postsynaptic denervated compared to normally innervated tissue was found. Thus it is likely that sympathetic innervation solely provides the pathway for the zeitgeber (i.e., light), but the signal for the endogenous circadian rhythm of aqueous flow is very largely independent of innervation, and more directly under local and/or humoral control.

Discovering the possible regulation of β-arrestin expression in the ciliary epithelium raises new questions. Do the changes in the steady-state level of β-arrestin protein and mRNA reflect a change in the rate of synthesis, degradation, or both? There is little information about this issue. McGinnis *et al.* (1992), using information derived the analysis of change in protein and mRNA levels done by Berlin and Schimke (1965) and by Almagor and Paigen (1988), have calculated that the turnover rate of retinal arrestin (s-Ag) mRNA is about 3.5 hr, a constant rate of degradation, whether the amount of s-Ag mRNA increases or decreases. If these findings can be applied to ciliary epithelial β-arrestin, it would seem that increases in ciliary epithelial β-arrestin mRNA arise from increases in transcriptional activity of the β-arrestin gene and not from increases or decreases in degradation rates. At this time there is very little known about the turnover of β-arrestin protein. The circadian intervals used in the experiments here were too large to allow any calculations of turnover rates for messenger or protein. In this regard there is an apparent coincidence or match between the appearance of mRNA and protein expression. In reality, there is undoubtedly an undetected short time lag between the two, one considerably shorter than the intervals measured.

It is unclear at this time which modulator of BAR sensitivity is rate limiting. Either substrate (BAR) or reactant (β-arrestin) exerts control in a concentration-dependent manner. The entrainment of cell types that respect a circadian pacemaker likely involves well-timed G-protein receptor inactivation by β-arrestin in response to agonist binding. Is this cyclic change in β-arrestin caused by light and/or a circadian clock? A light-dependent regulation of transcriptional activity of the mammalian (mouse) gene for retinal arrestin has been reported (McGinnis *et al.*, 1994). The expression of β-arrestin by the ciliary epithelium reported here follows a light–dark entrained circadian clock and can be phase shifted. It is barely possible that a light-inhibited and/or dark-induced expression of ciliary epithelial β-arrestin occurs. Experiments in progress placing entrained rabbits in constant darkness will complete the answer to this question. In addition, further work is required to address the expression and role of β-arrestin-2. Thus far, the data imply that β-arrestin gene products are regulated throughout the circadian cycle of aqueous secretion. If homologous desensitization of ciliary epithelial BAR respects a circadian pacemaker, the phenomenon places the timely use of β-blockade in glaucoma therapeusis (Topper and Brubaker, 1985) on a firm molecular basis and, in general, emphasizes the utility of pharmacotherapy based on knowledge of endogenous mechanisms regulating circadian rhythms.

IV. DO GAP JUNCTIONS PARTICIPATE IN THE CIRCADIAN RHYTHM OF AQUEOUS FLOW?

The importance of intercellular communication via gap junctions for the metabolic cooperativity of adjacent cells has been well established for cells of a wide variety of tissues, many of which exhibit secretory functions. For example, both in pancreas and in pituitary a close relationship exists between control of junctional permeability and secretion (Meda, 1991). In the ciliary epithelium, the apical membranes of PE and NPE cells become apposed after invagination of the optic vesicle during development of the eye to form gap junctions (Townes-Anderson and Raviola, 1981) that constitute a continuous hydrophilic channel, permitting small molecules, such as ions, metabolites, and perhaps those involved in signal transduction, to travel from layer to layer. While gap junctions among adjacent NPE and adjacent PE are not uniform in size, Cx43 is the main structural protein of these particular junctions between NPE and PE (Coca-Prados *et al.,* 1992). The cells of the two layers make anatomical (Reale and Spitznas, 1975; Raviola and Raviola, 1978) and functional contact (Green *et al.,* 1985; Edelman *et al.,* 1994; Oh *et al.,* 1994; Shi *et al.,* 1996) by way of these gap junctions, as in other tissues (Loewenstein, 1966). Thus ciliary gap junctions facilitate communication and amplify and synchronize signals (Yamada, 1989) between and within each cell layer for the formation of aqueous.

Junctional coupling is not static, but is susceptible to significant plasticity, in its assembly and function. The phosphorylation of serine and threonine residues of junctional proteins *in vitro* and *in vivo* (Saez *et al.,* 1986; Musil *et al.,* 1990; Crow *et al.,* 1990; Laird and Revel, 1990; Musil and Goodenough, 1991; Shi *et al.,* 1996) is controlled by kinases and phosphatases, and alters the electrophoretic mobility of Cx43 (Oh *et al.,* 1991; Brissette *et al.,* 1991; Saez *et al.,* 1986; Laird and Revel, 1990). Kwak *et al.* (1995), Mikalsen *et al.* (1995), Puranam *et al.* (1993), Stagg and Fletcher (1990), and Warn-Cramer *et al.* (1996) have shown phosphorylation-induced conductance changes. For these reasons we explored the relationship between the circadian cycle and the PE:NPE gap junction coupling to decide whether these junctions participate in a physiologic process known to regulate aqueous formation. We asked two questions: (1) Are there changes in the level of expression of Cx43, the major structural protein of the PE:NPE gap junctions (Coca-Prados *et al.*, 1992), during the endogenously regulated circadian rhythm of intraocular pressure and aqueous flow. (2) Is Cx43 subject to regulatory phosphorylation?

A quantitative polymerase chain reaction (PCR) using cDNA synthesized by a standard technique from poly-A selected mRNA as the template was used to determine whether changes in the mRNA expression of Cx43 occur during the circadian cycle. Expression of Cx43 protein during the circadian cycle was determined by Western immunoblot analysis. It turned out that PCR-directed analysis of mRNA levels did not demonstrate appreciable circadian differences in Cx43 transcription. Consistent with the reverse transcriptase PCR analysis, Cx43 protein levels in isolated bilayer membranes harvested during the circadian cycle were relatively constant.

The molecular mechanism for aqueous secretion includes several transduction signals but clearly involves transmembrane signals from the β-adrenergic receptor complex. To learn whether modulators of the cAMP second messenger system influence aqueous flow by an effect on the phosphorylation state of Cx43 located between the PE and NPE, we treated isolated epithelial bilayers with forskolin, isoproterenol, and phorbol esters, phosphorylating agents with distinctly separate known targets (Nishizuka, 1984). A summary of techniques is stated in the figure legends for Figs. 10–13 and details have been given elsewhere (Sears *et al.,* 1997).

With fluorescence immunohistochemistry, it was first established that the monoclonal mouse anti-Cx43 antibody (Zymed Laboratories Inc., San Francisco, CA) to be used in Western blots recognized the Cx43 protein species comprising the junction between the PE and NPE (Fig. 10). Ciliary bilayers then were treated with: (1) a ligand of the β-adrenergic receptor, isoproterenol; (2) a direct activator of adenylate cyclase, forskolin; and/or (3) an activator of protein kinase C (PKC), 12-tetradecanoyl phorbol-13-acetate (TPA). The phosphorylation state of Cx43 was determined. Phosphorylation of Cx43 occurs after exposure of intact cells to TPA, and is demonstrated by an upward shift in electrophoretic mobility, as in other systems (Oh *et al.,* 1991; Brissette *et al.,* 1991), that is prevented by exposure to exogenous phosphatase (data not shown). The rate of phosphorylation of Cx43 in explants exposed to TPA compared faster than in controls, but control explants (no adrenergic mediators) showed "autophosphorylation" (Fig. 11). Either forskolin or isoproterenol alone induced rapid phosphorylation. A lack of synergy among TPA and forskolin or isoproterenol and forskolin is demonstrated by the appearance of two bands corresponding to phosphorylated and dephosphorylated species (Figs. 12 and 13). These data suggest that the additive effect of two adrenergic agents acting via the cAMP second messena

b

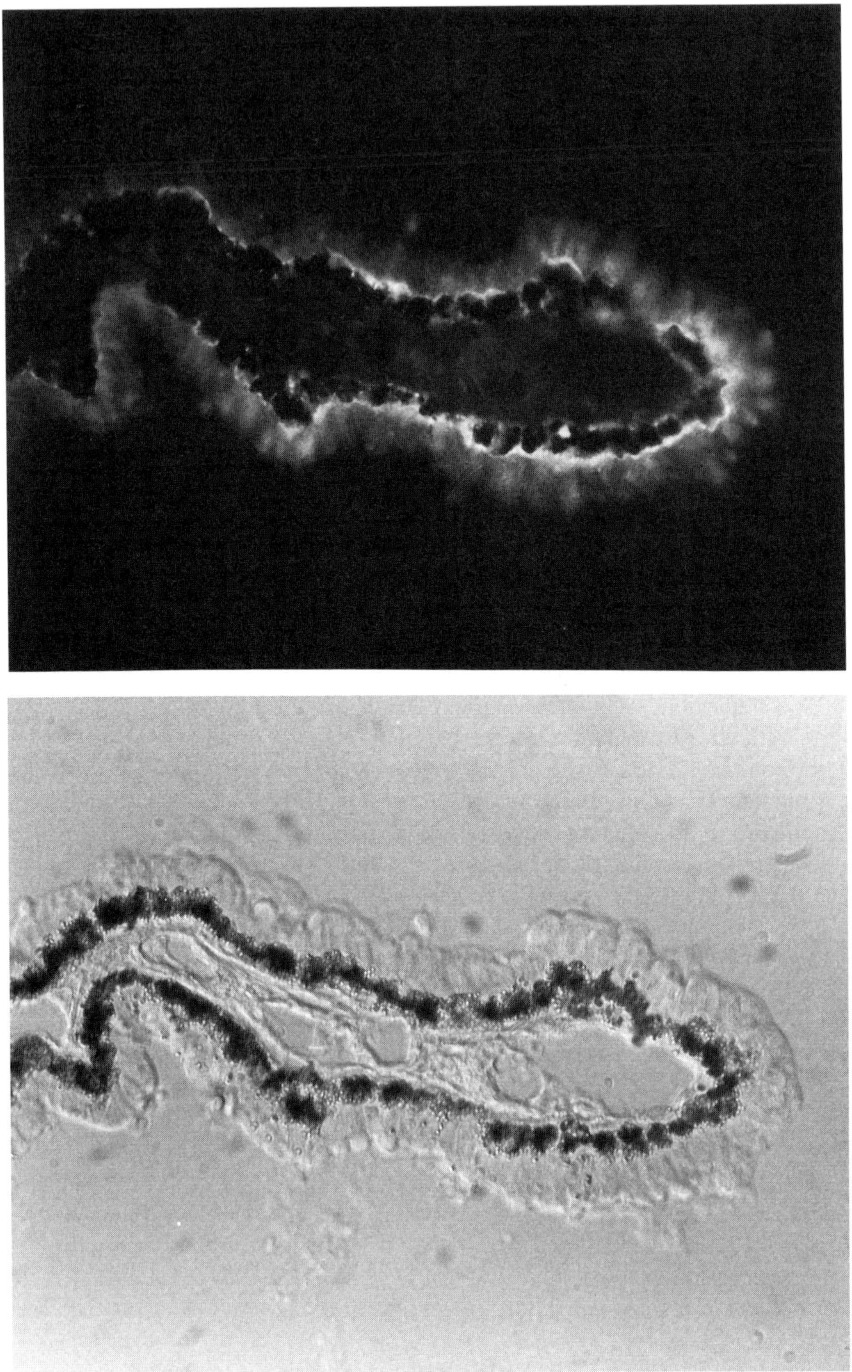

ger system may promote a cAMP-inducible phosphatase. This paradox is not too different from one noted by both bench and clinical scientists: Aqueous suppression is mediated by β-adrenergic receptor blockade, yet forskolin, an activator of adenylate cyclase, also induces aqueous suppression (Sears, 1984). The effects are undoubtedly related to the tone of a system at any given moment (Sears, 1985; Begum and Ragolia, 1996). The data demonstrate that ciliary epithelial Cx43 phosphorylation is responsive to mediators of the β-adrenergic G-protein-coupled receptor system *in vitro* but do not prove the physiologic role, if any, of cAMP-stimulated phosphorylation of Cx43. Consistent with our findings are those of Yoshimura *et al.* (1989) who used calcium and other phosphorylation activators, calmodulin, or phorbol myristate acetate. Activation of either calcium-dependent protein kinase system caused phosphorylation of multiple proteins in the ciliary body but also caused dephosphorylation.

Aqueous secretion obviously requires the integrity of the gap junctions between PE:NPE. Can the process be regulated via these junctions? Data from mRNA analysis and Western blot demonstrated that the Cx43 is constitutively expressed in the ciliary epithelia. Neither was it possible to correlate Cx43 phosphorylation with the circadian cycle of aqueous flow. The plethora of PE:NPE gap junctions may simply imply that the numbers themselves are regulatory, evenly speeding an amplified and homogeneous response to transmitters, or perhaps that these gap junctions respond in a graded rather than all-or-none mode, making it difficult to detect regulatory phosphorylation.

V. SUMMARY

Diurnal variations in intraocular pressure were first noted in 1899. The circadian nature of this rhythm was later documented by clinical investiga-

---

FIGURE 10 (a) A strong fluorescent signal can be seen between the PE and NPE of the ciliary bilayer. (b) Differential interference contrast image of a ciliary tip, treated as described herein. Eyes of pigmented rabbits fixed with 2% formaldehyde in 0.1 *M* phosphate buffer (pH 7.4) at 4°C for 1 hr, immersed in phosphate-buffered 20% glucose overnight at 4°C, were embedded in OCT compound and frozen with liquid nitrogen. Six-micrometer sections were cut with a cryostat and collected on poly-L-lysine coated glass slides. After washing with phosphate-buffered saline (PBS) containing 50 m*M* glycine for 10 min, the sections were incubated in 5% normal goat serum for 1 hr. The sections were then incubated with primary antibody using a monoclonal mouse anti-Cx43 antibody (Zymed Laboratories Inc.). After three 10-min washes in PBS, sections were counterstained with fluorescein-conjugated goat anti-mouse IgG (Vector Laboratories Inc., Burlingame, CA). The sections were mounted with an anti-photobleaching medium (Vectashieldä, Vector Laboratories Inc., Burlingame, CA) after washing in PBS, and observed under a Zeiss Axioskop fluorescence microscope.

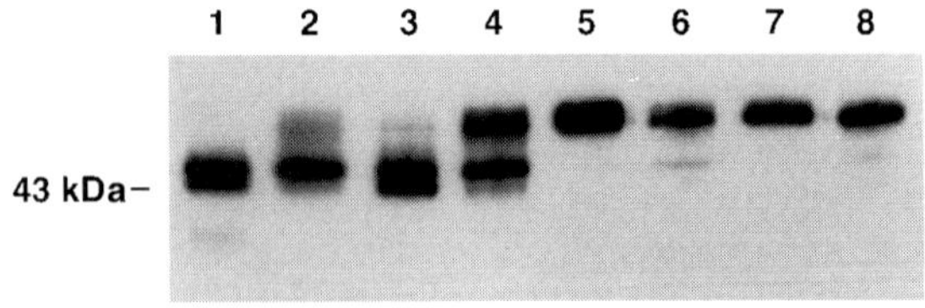

FIGURE 11 Western blot analysis of phosphorylated and dephosphorylated Cx43 was done by identifying electrophoretic mobilities using 12.5% polyacrylamide gel electrophoresis (PAGE) employing a monoclonal antibody to Cx43, described in Fig. 10, that recognized both the phosphorylated and dephosphorylated species. At intervals throughout the circadian cycle isolated ciliary epithelial bilayers were prepared (Sears *et al.*, 1991; Matsui *et al.*, 1996) free from contamination with iris, ciliary muscle, or stroma. Bilayers were incubated in M199-HEPES medium (Sigma) with isoproterenol, forskolin, or TPA in a 37°C shaker incubator at 200 rpm. Isoproterenol was added directly to the M199 medium. Forskolin was prepared as a 4 mg/mL solution in DMSO and diluted to 1/1000 in M199-HEPES medium (Sigma). TPA was prepared as a 10 m*M* solution in DMSO and diluted 1/1000 in M199-HEPES medium (Sigma). After incubation, bilayers were homogenized in 2 m*M* sodium bicarbonate, 2× phosphatase inhibitors (20 m*M* of each: sodium pyrophosphate, sodium fluoride, sodium orthovanadate, and ammonium molybdate), and100 m*M* phenylmethylsulfonyl fluoride (PMSF). The homogenate was incubated for 30–45 min on ice then spun at 12,000 rpm at 4°C and the pellets resuspended in the original buffer. Centrifugation was repeated, followed by a final resuspension in 2× phosphatase inhibitors, 1× PBS, in addition to an equal volume of 2× loading buffer (200 m*M* DTT; 4% SDS; 0.2% bromophenol blue; 20% glycerol dissolved in 100 m*M* Tris-Cl, pH 6.8). The samples were run on 12.5% PAGE to ensure that protein concentrations were similar. Gels were transferred overnight in bicarbonate buffer, stained with Ponceau S /10% TCA, marked and then blocked with 10% nonfat dry milk in TBS-T (Tris-buffered saline with Tween-20). A monoclonal antibody specific for Cx43 (Zymed Laboratories Inc.), was diluted at 1:1000 in TBS-T and used to probe the filters. After three alternating washes of TBS/TBS-T, sheep anti-mouse IgG horseradish peroxidase (Amersham) was used to recognize the anti-connexin antibody. Amersham ECL reagents were used according to standard protocol for developing the membranes on Kodak XAR film. Control explants exhibit phosphorylation, but TPA induces a phosphorylation at a faster rate. Lanes 1, 3, 5, 7 correspond to control bilayers treated for 15, 30, 60, and 120 min with M199 media without TPA; lanes 2, 4, 6, 8 correspond to 15, 30, 60, and 120 min of treatment with TPA.

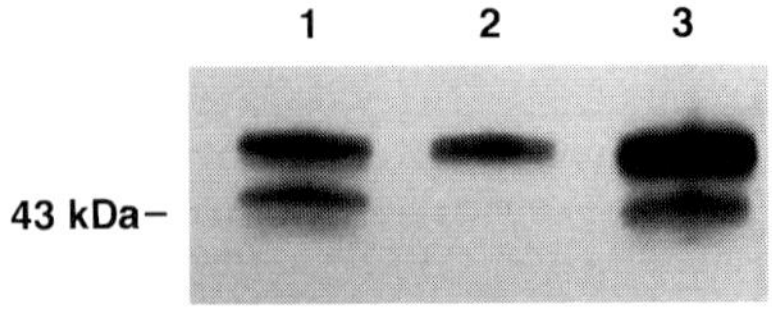

FIGURE 12 Western blot, as in Fig. 10, of ciliary epithelial explants, treated with isoproterenol demonstrates complete phosphorylation after 10-min incubation. This effect is not synergistic with forskolin. Lane 1, 10 μ *M* isoproterenol; lane 2, 50 μ*M* isoproterenol; lane 3, 1 m*M* isoproterenol; lane 4, 10 μ*M* isoproterenol and 5 μ*M* forskolin. All treatments were for maximum of 10 min.

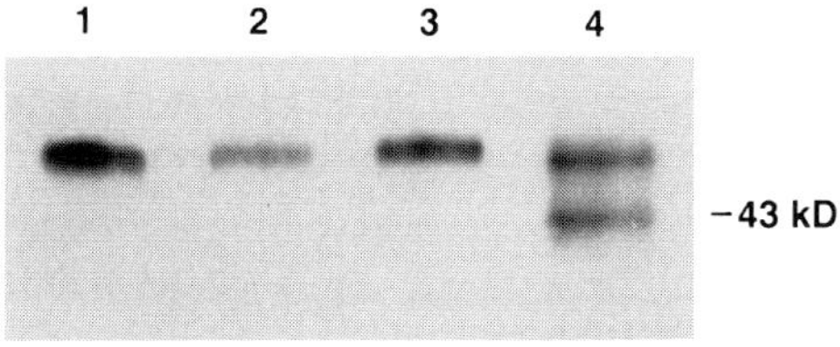

FIGURE 13 Western blot demonstrates rapid forskolin induced phosphorylation (lane 2) that is neither synergistic with TPA nor with isoproterenol. Lane 1, 1 μM isoproterenol and 5 μM forskolin; lane 2, 5 μM forskolin alone; lane 3, TPA and 5 μM forskolin.

tors. The dependence of this rhythm of intraocular pressure on changes in aqueous flow was discovered in a clinical investigation by Rosengren in 1958 and was later put on a firm foundation by a series of basic research studies. A molecular source for the circadian rhythm of aqueous flow was shown to be the β-adrenergic receptor in the ciliary epithelium, undoubtedly interfacing with other signal transduction pathways. In the current work, the circadian rhythm of aqueous flow was using to uncover molecular controls for aqueous humor formation.

No evidence was found for the participation of the Cx43 gap junctions between PE and NPE in this physiologic process. Employing either mRNA tissue extractions of isolated ciliary epithelium done at different circadian intervals or the technique of subtractive hybridization, we have isolated several factors, at least one of which, β-arrestin-1, may participate in the control of circadian rhythm of aqueous flow by the process of homologous desensitization. This phenomenon places the timely use of β-blockade in glaucoma therapeusis on a firm molecular basis.

Acknowledgments

Robert Lefkowitz of Duke University School of Medicine generously supplied antiserum to β-arrestin-2. Shan Chen, Douglas Gregory, Naoki Nagata, Tohru Nakano, Hiroki Nii, Xiaolin Wan, and Eichi Yamada are active colleagues and collaborators. Alexandra Franciscus prepared the manuscript. This work was supported by USPHS grants NIH EY-08879-07 and EY-00785-24, by the Herbert and Karen Lotman foundation, and by the E. Matilda Ziegler Foundation.

References

Almagor, H., and Paigen, K. (1988). Chemical kinetics of induced gene expression: Activation of transcription by noncooperative binding of multiple regulatory molecules. *Biochemistry* **27,** 2094–2300.

Anjou, C. I. N. (1966). Influence of light on the 24-hr variation in aqueous flare density and intraocular pressure in normal rabbit eyes. *Acta Ophthamol.* **39,** 1.

Attramadal, H., Arriza, J. L., Aoki, C., Dawson, T. M., Codina, J., Kwatra, M. M., Snyder, S. H., Caron, and M. G., and Lefkowitz, R. J. (1992). *J. Biol. Chem.* **267,** 16882–17890.

Bausher, L., and Gregory, D. (1989). Alpha$_2$ adrenergic and VIP receptors in rabbit ciliary processes interact. *Curr. Eye Res.* **8,** 47–54.

Bausher, L. P., Gregory, D. S., and Sears, M. L. (1987). Interaction between alpha$_2$ and beta$_2$ adrenergic receptors in rabbit ciliary processes. *Curr. Eye Res.* **6,** 497–505.

Begum N., and Ragolia L. J. (1996). cAMP counter-regulates insulin-mediated protein phosphatase-2A inactivation in rat skeletal muscle cells. *J. Biol. Chem.* **271,** 31166–31171.

Benovic, J. L., Strasser, R. H., Caron, M. G., and Lefkowitz, R. J. (1986). β-Adrenergic receptor kinase: Identification of a novel protein kinase that phosphorylates the agonist-occupied form of the receptor. *Proc. Natl. Acad. Sci. U.S.A.* **83,** 2797–2801.

Berlin, C. M., and Schimke, R. T. (1965). Enzyme synthesis and degradation in animal tissues. *Mol. Pharmacol.* **1,** 149–157.

Breebart, A. C. (1966). A case of Rieger's anomaly with glaucoma. Influence of sleep. *Arch. Ophthalmol.* **76,** 285.

Brissette, J. C., Kumar, N. M., Gilula, N. B., and Dotto, P. (1991). The tumor promoter 12-*o*-tetradecanoylphorbol-13-acetate and the *ras* oncogene modulate expression and phosphorylation of gap junctional proteins. *Mol. Cell Biol.* **105,** 5364–5371.

Broekhuyse, R. M., Janssen, A. P. M., and Tolhuizen, E. F. J. (1987). Effect of light adaptation on the binding of 48-kDa protein (S antigen) to photoreceptor all membranes. *Curr. Eye Res.* **6,** 607–610.

Bromberg, B. B., Gregory, D. S., and Sears, M. L. (1980). Beta-adrenergic receptors in ciliary processes of the rabbit. *Invest. Ophthalmol.* **19,** 203.

Celeste, A. J., Iannazzi, J. A., Taylor, R. C., Hewick, R. M., Rosen, V., Wang, E. A., and Wozney, J. M. (1990). Identification of transforming growth factor β family members present in bone-inductive protein purified from bovine bone. *Proc. Natl. Acad. Sci. U.S.A.* **87,** 9843–9847.

Chomczynski, P., and Sacchi, N. (1987). Single-step method of RNA isolation by acid guanidium thiocyanate-phenol-chloroform extraction. *Anal. Biochem.* **162,** 156–159.

Coca-Prados, M., Ghosh, S., Gilula, N. B., and Kumar, N. M. (1992). Expression and cellular distribution of the α1 gap junction gene product in the ocular pigmented ciliary epithelium. *Curr. Eye Res.* **11,** 113–122.

Crow, D. S., Beyer, E. C., Paul, D. L., Kobe, S. S., and Lau, A. F. (1990). Phosphorylation of connexin43 gap junction protein in uninfected and Rous sarcoma virus-transformed mammalian fibroblasts. *Mol. Cell Biol.* **10,** 1754–1763.

Davson, H. (1956). "Physiology of the Ocular and Cerebrospinal Fluids," p. 9. Little, Brown, Boston.

Dawson, T. M., Arriza, J. L., Jaworsky, D. E., Borisy, F. F., Attramadal, H., Lefkowitz, R. L., and Ronnett, G. V. (1993). β-adrenergic receptor kinase-2 and β-arrestin-2 as mediators of odorant-induced desensitization. *Science* **259,** 825–828.

DeVenecia, G., and Davis, M. D. (1963). Diurnal variation of intraocular pressure in the normal eye. *Arch. Ophthalmol.* **69,** 752.

Dowdy, S. F., Lai, K. M., Weissman, B. F., and Matsui, Y. (1991). The isolation and characterization of a novel cDNA demonstrating and altered mRNA level in nontumorigenic Wilms' microcell hybrid cells. *Nucl. Acids Res.* **19,** 5763–5769.

Drance, S. M. (1960). The significance of the diurnal variation in normal and glaucomatous eyes. *Arch. Ophthalmol.* **64,** 494.

Dudley, A. T., Lyons, K. M., and Robertson, E. J. (1995). A requirement for bone morphogenetic protein-7 during development of the mammalian kidney and eye. *Genes Dev.* **9,** 2795–2807.

Edelman, J. L., Sachs, E., and Adorante, J. S. (1994). Ion transport asymmetry and functional coupling in bovine pigmented and unpigmented ciliary epithelial cells. *Am. J. Physiol.* **266,** C1210–C1221.

Ehinger, B. (1966). Ocular and orbital negative nerves. *Acta Physiol. Scand.* **71,** Suppl., 268.

Ericson, L. A. (1958). Twenty-four hourly variations in the inflow of the aqueous humor. *Acta. Ophthalmol. (Copenhagen)* **36,** 381.

Ferguson, S. S. G., Downey III, W. E., Colapietro, A., Barak, L. S., Menard, L., and Caron, M. G. (1996). Role of β-arrestin in mediating agonist-promoted G protein-coupled receptor internalization. *Science* **271,** 363–366.

Flagg-Newton, J. L., Dahl, G., and Lowenstein, W. R. (1981). Cell junction and cyclic AMP: I. Upregulation of junctional membrane permeability and junctional membrane particles by administration of cyclic nucleotide or phosphodiesterase inhibitor. *J. Membr. Biol.* **63,** 105–121.

Francis, P. H., Richardson, M. K., Brickell, P. M., and Tickle, C. (1994). Bone morphogenetic proteins and a signaling pathway that controls patterning in the developing limb bud. *Development.* **120,** 209–218.

Freedman, N. J., Liggett, S. B., Drachman, D. E., Pei, G., Caron, M. G., and Lefkowitz, R. J. (1995). Phosphorylation and desensitization of the human β1–adrenergic receptor. *J. Biol. Chem.* **270,** 17953–17961.

Godwin, A. J., Green, L. M., Walsh, M. P., McDonald, J. R., Walsh, D. A., and Fletcher, W. H. (1993). *In situ* regulation of cell–cell communication by the cAMP-dependent protein kinase and protein kinase C. *Mol. Cell Biochem.* **127/128,** 293–307.

Goldmann, H. (1955). *In* "Glaucoma, A Symposium Organized by the Council for International Organizations of Medical Sciences, Part vi, Clinical Aspects, General Discussion and Conclusions," pp. 292–311. Charles C Thomas, Springfield, IL.

Graham, A., Francis-West, P., Brickell, P., and Lumsden, A. (1994). The signaling molecule BMP4 mediates apoptosis in the rhombencephalic neural crest. *Nature* **372,** 684–686.

Green, K., Bountra, C., Georgiou, P., and House, C. R. (1985). An electrophysiologic study of rabbit ciliary epithelial cells. *Invest. Ophthalmol. Vis. Sci.* **26,** 371–381.

Gregory, D. S., Bausher, L. P., Bromberg, B. B., and Sears, M. L. (1981). The beta adrenergic receptor and adenyl cyclase of rabbit ciliary processes. *In:* "New Directions in Eye Research" (M. L. Sears, ed.), pp. 127–146. New Haven, CT: Yale University Press.

Gregory, D. S., Aviado, D. G., and Sears, M. L. (1985). Cervical ganglionectomy alters the circadian rhythm of intraocular pressure in New Zealand rabbits. *Curr. Eye Res.* **4,** 1273–1279.

Harden, T. L. (1983). Agonist-induced desensitization of the β-adrenergic receptor-linked adenylate cyclase. *Pharmacol. Rev.* **35,** 5–26.

Henkind, P., Leitman, M., and Weitzman, E. (1973). The diurnal curve in man: New observations. *Invest. Ophthalmol.* **12,** 705.

Holmberg, A. (1959). Some characteristic components of the ciliary epithelium. *Am. J. Ophthalmol.* **48,** 426–429.

Horio, B., Sears, M. L., Mead, A., Matsui, H. and Bausher, L. P. (1996). Regulation and bioelectrical effects of cyclic adenosine monophosphate production in the ciliary epithelial bilayer. *Invest. Ophthalmol. Vis. Sci.* **37,** 607–612.

Imaki, J., Yamashita, K., Yamakawa, A., and Yoshida, K. (1995). Expression of Jun family genes in rat retinal cells: Regulation by light/dark cycle. *Brain Res. Mol. Brain Res.* **30,** 48–52.

Katavisto, M. (1964). The diurnal variation of ocular tension in glaucoma. *Acta. Ophthalmol. (Copenhagen)*, Suppl., **126.**

Kinsey, V. E., and Reddy, D. V. N. (1964). Chemistry and dynamics of aqueous humor. *In:* "The Rabbit in Eye Research" (C. Thomas, ed.). J. H. Prince, Springfield, IL.

Kitazawa, Y., and Horie, T. (1975). Diurnal variation of intraocular pressure in primary open angle glaucoma. *Am. J. Ophthalmol.* **79,** 557.

Köllner, H. (1916). Ueber die regelm, ssigen t, glichen Schwankungen des Augendruckes und ihre Ursachen. *Arch. Augenh.* **81,** 120.

Kwak, B. R., Hermans, M. M., De Jonge, H. R., Lohmann, S. M., Jongsma, H. J., and Chanson, M. (1995). Differential regulation of distinct types of gap junction channels by similar phosphorylating conditions. *Mol. Biol. Cell* **12,** 1707–1719.

Laird, D. W., and Revel, J.-P. (1990). Biochemical and immunochemical analysis of connexin43 in rat heart gap junction membranes. *J. Cell Sci.* **97,** 109–117.

Langley, D. A., and Swanljung, H. (1951). Ocular tension in glaucoma. *Br. J. Ophthalmol.* **35,** 445.

Liang, P., and Pardee, A. B. (1992). Differential display of eukaryotic messenger RNA by means of the polymerase chain reaction. *Science* **257,** 967–970.

Linnew, E., and Prijot, E. (1955). Cervical sympathetic ganglionotomy and aqueous flow. *Ann. Ophthalmol.* **54,** 831–836.

Loewenstein, W. R. (1966). Permeability of membrane junctions. *Ann. N.Y. Acad. Sci. U.S.A.* **137,** 441–472.

Lohse, M. J., Benovic, J. L., Codina, J., Caron, M. G., and Lefkowitz, R. J. (1990). β-Arrestin: A protein that regulates β-adrenergic receptor function. *Science* **248,** 1547–1550.

Lyons, K., Pelton, R. W., and Hogan, B. L. M. (1990). Organogenesis and pattern formation in the mouse: RNA distribution patterns suggest a role for bone morphogenetic protein 2A (BMP-2A). *Development,* **109,** 833–844.

Maslenikow, A. (1904). Ueber Tagesschwankungen des intraokularen Drukes bei Glaukom. *Augenh. Kunde* **11,** 564.

Matsui, H., Murakami, M., Wynns, G. C., Conroy, C. W., Mead, A., Maren, T. H., and Sears, M. L. (1996). Carbonic anhydrase activity is present in the basolateral membranes of the nonpigmented ciliary epithelium of rabbit eyes. *Exp. Eye Res.* **62,** 409–417.

McGinnis, J. F., Whelan, J. P., and Donoso, L. A. (1992). Transient, cyclic changes in mouse visual cell products during the light dark cycle. *J. Neurosci. Res.* **31,** 584-590.

McGinnis, J. F., Austin, B. J., Stepanik, P. L., and Lerious, V. (1994). Light-dependent regulation of the transcriptional activity of the mammalian gene for arrestin. *J. Neurosci. Res.* **38,** 479–482.

Meda, P., Bosco, D., Giordano, E., and Clauson, M. (1991). Junctional coupling modulation by secretagogues in two cell pancreatic systems. *In:* "Biophysics of Gap Junction Channels." (C. Peracchia, ed.), pp. 191–208. CRC Press, Boca Raton, FL.

Mikalsen, S. O., Husoy, T., and Sanner, T. (1995). Modulation of gap junctional intercellular communication by phosphorylation: Effects of growth factors, kinase activators and phosphatase inhibitors. *Prog. Clin. Biol. Res.* **391,** 425–438.

Mittag, T. W., and Tormay, A. (1985). Adrenergic receptor subtypes in rabbit iris-ciliary body membranes: Classification by radioligand studies. *Exp. Eye Res.* **40,** 239–249.

Musil, L. S., and Goodenough, D. A. (1991). Biochemical analysis of connexin43 intracellular transport, phosphorylation, and assembly into gap junctional plaques. *J. Cell Biol.* **115,** 1357–1374.

Musil, L. S., Beyer, E. C., and Goodenough, D. A. (1990). Expression of the gap junction protein connexin-43 in embryonic chick lens: Molecular cloning, ultrastructural localization, and post-translational phosphorylation. *J. Membr. Biol.* **116,** 163–175.

Nishizuka, Y. (1984). The role of protein kinase C in cell surface signal transduction and tumor promotion. *Nature* **368,** 693–698.

Oh, J., Krupin, T., Tang, L. Q., Sveen, J., and Lahlum, R. A. (1994). Dye coupling of rabbit ciliary epithelial cells *in vitro. Invest. Ophthalmol. Vis. Sci.* **35,** 2509–2514.

Oh, S. Y., Grupen, C. G., and Murray, A. W. (1991). Phorbol ester induces phosphorylation and down regulation of connexin 43 in WB cells. *Biochem. Biophys. Acta.* **1094,** 243–245.

Okisaka, S., Kuwabara, T., and Rapoport, S. I. (1974). Selective destruction of the pigmented epithelium in the ciliary body of the eye. *Science,* **184,** 1298–1299.

Palczewski, K., Rispoli, G., and Detwiler, P. B. (1992). The influence of arrestin (48K Protein) and rhodopsin on visual transduction. *Neuron* **8,** 117–126.

Parruti, G., Peracchia, F., Sallese, M., Ambrosini, G., Masini, M., Rotilio, D., and Deblasi, A. (1993). Molecular analysis of human β-arrestin-1: Cloning, tissue distribution, and regulation of expression. Identification of two isoforms generated by alternative splicing. *J. Biol. Chem.* **268,** 9753–9761.

Pippig, S., Andexinger, S., Daniel, K., Puzicha, M., Caron, M. G., and Lefkowitz, R. J. (1993). Overexpression of β-arrestin and β-adrenergic receptor kinase augment desensitization of β_2–adrenergic receptors. *J. Biol. Chem.* **268,** 3201–3208.

Prober, J. M., Trainor, G. L., Dam, R. J., Hobbs, F. W., Robertson, G. W., Zagursky, R. J., Cocuzza, A. J., Jensen, M. A., and Baumeister, K. (1987). A system for rapid DNA sequencing with fluorescent chain-terminating dideoxynucleotides. *Science* **238,** 336–341.

Puranam, K., Laird, D. W., and Revel, J.-P. (1993). Trapping an intermediate form of connexin43 in the golgi. *Exp. Cell Res.* **206,** 85–92.

Raviola, G., and Raviola, E. (1978). Intercellular junctions in the ciliary epithelium. *Invest. Ophthalmol. Vis. Sci.* **17,** 958–981.

Reale, E., and Spitznas, M. (1975). Freeze-fracture analysis of junctional complexes in human ciliary epithelia. *Albrecht von Graefes Arch. Klin. Exp. Ophthalmol.* **195,** 1–16.

Reiss, G. R., Lee, D. A., Topper, J. E., and Brubaker, R. F. (1984). Aqueous humor flow during sleep. *Invest. Ophthalmol. Vis. Sci.* **25,** 776.

Reynhout, J. K., Lampe, P. D., and Johnson, R. G. (1992). An activator of protein kinase C inhibits junction communication between cultured bovine epithelial cells. *Exp. Cell Res.* **198,** 337–342.

Rowland, J. M., Potter, D. E., and Reiter, R. J. (1981). Circadian rhythm in intraocular pressure: A rabbit model. *Curr. Eye Res.* **1,** 169–173.

Rowland, J. M., Sawyer, W. K., Tittel, J., and Ford, C. J. (1986). Studies on the circadian rhythm of IOP in rabbits: Correlation with aqueous inflow and cAMP content. *Curr. Eye Res.* **5,** 201–206.

Saez, J. C., Spray, D. C., Nairn, A. C., Hertzberg, E., Greengard, P., and Bennett, M. V. L. (1986). cAMP increases junctional conductance and stimulates phosphorylation of the 27-kDa principal gap junction polypeptide. *Proc. Natl. Acad. Sci. U.S.A.* **83,** 2473–2477.

Schmerl, E., Dietz, A. A., and Steinberg, B. (1955). Mechanism of miopiesin formation. *Am. J. Ophthalmol.* **39,** 684.

Sears, J. E., Nakano, T., and Sears, M. L. (1997) Adrenergic mediated connexin-43 phosphorylation in ocular ciliary epithelium. *Curr. Eye Res.* in press.

Sears, M. L. (1975). Catecholamines in relation to the eye. *In* "Handbook of Physiology–Endocrinology" (E. Astwood and R. Creep, eds.), pp. 553–590. American Physiological Society, Washington, DC.

Sears, M. L. (1984). Autonomic nervous system: Adrenergic agonists. *In* "Handbook of Experimental Pharmacology" (M. L. Sears, ed.), pp. 193–248. Springer Verlag, Berlin.

Sears, M. L. (1985). Regulation of aqueous flow by the adenylate cyclase receptor complex in the ciliary epithelium. *Am. J. Ophthalmol.* **100,** 194.

Sears, M. L., and Bárány, E. H. (1960). Effects of cervical sympathectomy with adrenergic inhibitors. *Arch. Ophthalmol.* **64,** 839.

Sears, M. L., Gregory, D.S., Bausher, L. P., Mishima, H., and Stjernschantz, J. (1981). A receptor for aqueous humor formation. *In* "New Directions in Ophthalmic Research" (M. L. Sears, ed.), pp. 127–145, 163–183. Yale University Press, New Haven, CT.

Sears, M. L., Yamada, E., Cummins, D., Mori, N., Mead, A., and Murakami, M. (1991). The isolated ciliary bilayer is useful for studies of aqueous humor formation. *Trans. Am. Ophthalmol. Soc.* **89,** 131–154.

Shi, X. P., Zamudio, A. C., Candia, O. A., and Wolosin, J. M. (1996). Adreno-cholinergic modulation of juctional communications between the pigmented and non-pigmented layers of the ciliary body epithelium. *Invest. Ophthalmol. Vis. Sci.* **37,** 1037–1046.

Shimizu, H., Riley, M. V., and Cole, D. F. (1967). The isolation of whole cells from the ciliary epithelium together with some observations on the metabolism of the two cell types. *Exp. Eye Res.* **6,** 141–151.

Smith, S. D., and Gregory, D. S. (1989). A circadian rhythm of aqueous flow underlies the circadian rhythm of IOP in NZW rabbits. *Invest. Ophthalmol. Vis. Sci.* **30,** 775–778.

Söhlemann, P., Hekman, M., Puzicha, M., Buchen, C., and Lohse, M. J. (1995). Binding of purified recombinant β-arrestin to guanine-nucleotide-binding-protein-coupled receptors. *Eur. J. Biochem.* **232,** 464–472.

Spray, D. C., White, R. L., Campos de Carvalho, A., Harris, A. L., and Bennett, M. V. L. (1984). Gating of gap junction channels. *Biophys. J.* **45,** 219–230.

Stagg, R. B., and Fletcher, W. H. (1990). The hormone-induced regulation of contact-dependent cell–cell communication by phosphorylation. *Endocr. Rev.* **11,** 302–325.

Stone, H. H., and Sears, M. L. (1958). Ocular pressure in experimental cross circulation—A preliminary report. *Arch. Ophthalmol.* **61,** 102.

Tabin, C. (1995). The initiation of the limb bud: Growth factors, flox genes, and retinoids. *Cell* **80**(5), 671–674.

Thiel, R. (1925). Die physiologischen und experimentell erzeugten Schwankungen des intraokularen Druckes des gesunden and glaukomat''sen. *Auges. Arch. Augenh.* **96,** 331.

Topper, J. E., and Brubaker, R. F. (1985). Effects of timolol, epinephrine, and acetazolamide on aqueous flow during sleep. *Invest. Ophthalmol. Vis. Sci.* **26,** 1315–1319.

Townes-Anderson, E., and Raviola, G. (1981). The formation and distribution of intercellular junctions in the rhesus monkey optic cup: The early development of the cilio-iridic and sensory retinas. *Dev. Biol.* **85,** 209–232.

Wacker, W. B., Donoso, L. A., Kalsow, C. M., Yankeelov, J. A., and Organisciak, D. T. (1973). Experimental allergic uveitis. Isolation, characterization, and localization of a soluble uveitopathogenic antigen from bovine retina. *J. Immunol.* **119,** 1949–1958.

Wan, X. L., Sears, J., Chen, S., and Sears, M. L. (1997). Circadian aqueous flow modified by β-arrestin induced homologous desensitization. *Exp. Eye Res.* **64,** 1005–1011.

Wang, Z., and Brown, D. D. (1991). A gene expression screen. *Proc. Natl. Acad. Sci. U.S.A.* **88,** 11505–11509.

Warn-Cramer, B. J., Lampe, P. D., Kurata, W. E., Kanemitsu, M. Y., Lool, W., Eckhart, W., and Lau, A. F. (1996). Characterization of the mitogen-activated protein kinase phosphorylation sites on the connexin-43 gap junction protein. *Biol. Chem.* **271**(7), 3779–3786.

Wentworth, W. O., and Brubaker, R. F. (1981). Aqueous humor dynamics in a series of patients with third neuron Horner's eye syndrome. *Am. J. Ophthalmol.* **92,** 407–415.

Wollnik, F., Brysch, W., Uhlmann, E., Gillardon, F., Bravo, R., Zimmermann, M., Schlingensiepen, K. H., and Herdegen, T. (1995). Block of c-Fos and JunB expression by antisense oligonucleotides inhibits light-induced phase shifts of the mammalian circadian clock. *Eur. J. Neurosci.* **7,** 388–393.

Wozney, J. M., Rosen, V., Celeste, A. J., Mitsock, L. M., Whiters, M. J., Kriz, R. W., Hewick, R. M., and Wang, E. (1988). Novel regulators of bone formation: Molecular clones and activities. *Science,* **242,** 1528–1534.

Yamada, E. (1988). Intraepithelial nerve fibers in the rabbit ocular ciliary epithelium. *Arch. Histol. Cytol.* **51,** 43–51.

Yamada, E. (1989). Further observation on the intraepithelial nerve fibers of rabbit ocular ciliary epithelium. *Arch. Histol. Cytol.* **52,** 191–195.

Yoshimura, N., Mittag, T., and Podos, S. M. (1989). Calcium-dependent phosphorylation of proteins in rabbit ciliary processes. *Invest. Ophthalmol. Vis. Sci.* **30,** 723–730.

Yoshitomi, T., and Gregory, D. S. (1991). Ocular adrenergic nerves contribute to control of the circadian rhythm of aqueous flow in rabbits. *Invest. Ophthalmol. Vis. Sci.* **32,** 523–528.

Yoshitomi, T., Horio, B., and Gregory, D. S. (1991). Change in aqueous norepinephrine and cAMP during the circadian cycle in rabbits. *Invest. Ophthalmol. Vis. Sci.* **32,** 1609–1613.

CHAPTER 9

Clinical Measurements of Aqueous Dynamics: Implications for Addressing Glaucoma

Richard F. Brubaker
Department of Ophthalmology, Mayo Clinic and Mayo Foundation, Rochester, Minnesota 55905

1063-5823/98 $25.00

I. CLINICAL COMPONENTS OF AQUEOUS DYNAMICS

Clinical interest in the circulation of aqueous humor has been stimulated to a considerable extent by the problem of glaucoma, a condition in which increased intraocular pressure can lead to damage of the optic nerve. Attempts by clinicians to understand how to control intraocular pressure as a treatment for glaucoma has spawned numerous studies of the components of the aqueous circulation. Clinicians have developed a consensus about the major physiological properties of this circulation and the therapeutic opportunities and challenges that arise from its understanding.

Four components of the aqueous system can be measured in humans and are studied clinically. These are intraocular pressure, outflow resistance, episcleral venous pressure, and the rate of aqueous humor flow. Each of these components can be measured noninvasively, but the methods and the accuracy of the measurements vary considerably from one component to another.

A. *Intraocular Pressure*

The procedure of measuring intraocular pressure noninvasively in humans and animals is called *tonometry*. Tonometers are instruments that exploit the shape and physical properties of the eye to measure the pressure within the eye. They began to appear early in the twentieth century. Virtually all tonometers are durometers. They measure the force required to deform the eye in some specific way. Calibration and standardization of tonometers has been possible because of the physical characteristics of the cornea, an ideal tissue through which the force of a tonometer can be coupled to the pressure within the anterior chamber (Fig. 1). An applanation tonometer flattens a portion of the cornea and derives the pressure from the force per unit area flattened. A low-displacement tonometer causes a negligible elevation of the pressure in the eye during tonometry. The Goldmann tonometer, both an applanation and a low-displacement tonometer, is considered the most accurate instrument and is the standard against which the performance of other devices is measured.

Intraocular pressure is not steady, but varies due to a number of factors such as the tension of the extraocular muscles, the pressure of the lids, the arterial and venous blood pressure, and the effects of gravity. Clinicians are primarily interested in measuring intraocular pressure in its steady state, when the rates of aqueous humor inflow and outflow are exactly equal and the pressure in the eye is stable. The steady-state intraocular pressure in normal human subjects ranges from approximately 10–22 mm Hg with a median of 16–17 mm Hg. The distribution is not Gaussian, but is skewed toward higher pressures (Schottenstein, 1996).

The steady-state intraocular pressure is regulated primarily by three variables, the three other components of aqueous dynamics that are of interest to clinicians. These are the outflow resistance, the episcleral venous pressure, and the rate of aqueous humor flow.

B. Outflow Resistance

As described in Chapter 7, aqueous humor flows out of the anterior chamber through a series of tortuous channels in the trabecular meshwork. The microscopic channels it encounters en route present hydrodynamic resistance that must be overcome by the hydrostatic pressure gradient between the inner compartment of the eye, the anterior chamber, and the outer receptacle for aqueous humor, the episcleral plexus of veins. The clinical terms for this hydrodynamic effect are *outflow resistance* or its reciprocal, *facility of outflow,* sometimes called the *tonographic C value.* The outflow resistance is the major determinant of intraocular pressure. Increased outflow resistance is the primary cause of elevated intraocular pressure in nearly all glaucomas.

The method of quantification of outflow resistance in the human eye was introduced into clinical medicine by Grant (1950). Grant's technique, called *tonography,* was based on the mechanical properties of the Schiötz tonometer and the rheological behavior of the globe, worked out by Friedenwald (1937). During clinical tonography, the intraocular pressure is raised above its steady state by means of a weighted tonometer. The higher pressure

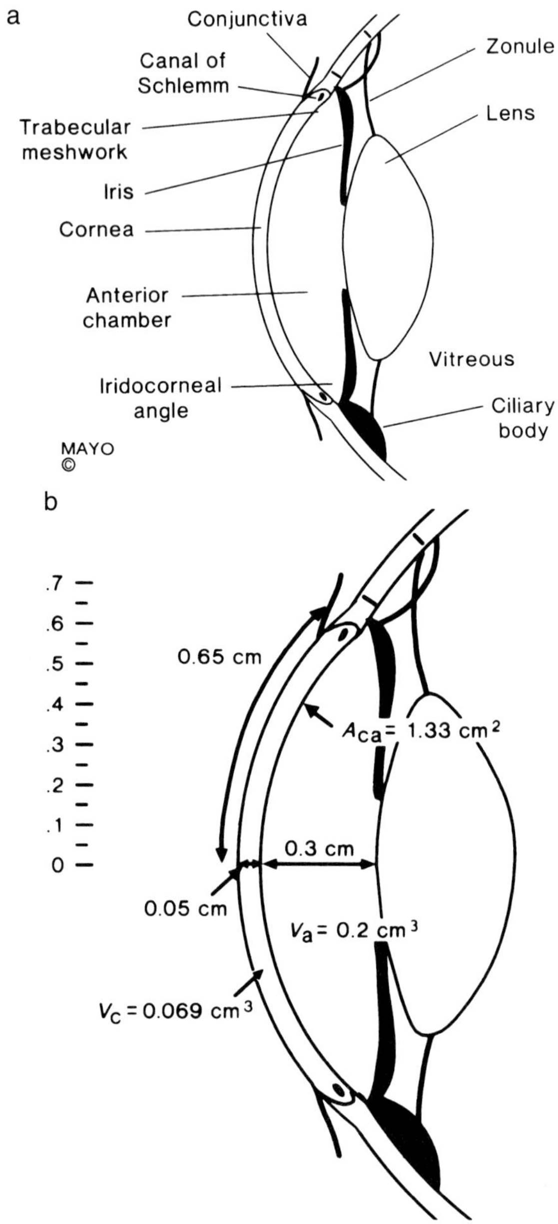

FIGURE 1 (a) Cross-sectional diagram of anterior segment of human eye. [Reprinted from Johnson and Brubaker (1986), Fig. 1, p. 60, with permission from the Mayo Foundation.] (b) Dimensions of the anterior segment of the human eye. V_c, volume of corneal stroma; V_a, volume of the anterior chamber; A_{ca}, area of the interface between the cornea and the anterior chamber. The scale is in centimeters. [Reprinted from Brubaker (1982), Fig. 4, p. 412, with permission of The American Ophthalmological Society.]

causes the rate of outflow to exceed the rate of inflow. Over a period of several minutes, the volume of the globe diminishes. Outflow resistance is the pressure elevation caused by the tonometer divided by the reduction in volume of the globe per unit time. The technique is based on the assumption that neither the outflow resistance itself nor the rate of aqueous humor formation is altered by the temporary increase of intraocular pressure above its steady state and that the volumes of other compartments within the eye, such as the intravascular space, remain unchanged. Of the four clinical measures of aqueous dynamics, tonography is the most consistent method of differentiating a normal eye from a potentially glaucomatous eye, but there is considerable overlap between the two. In the normal eye, the mean outflow resistance has been found to be 3.2 mm Hg · min/μL (Becker, 1956), but 6% of their subjects had a resistance of greater than 5 mm Hg · min/μL and 1% had a resistance greater than 6.7 mm Hg · min/μL, resistances that are commonly seen in eyes with glaucoma.

C. Episcleral Venous Pressure

In primates aqueous humor leaves the eye via the canal of Schlemm. Subsequently, it enters the venous network that encircles the sclera just beyond the limbus of the cornea. These episcleral veins drain into the plexus of vessels that drain the globe and the adnexal tissues of the eye and face. The pressure in these veins, the episcleral venous pressure, is determined by gravitational and vasoactive factors and largely independent of the flow of aqueous humor.

Microscopic observation of the surface of the living eye of humans permits visualization of aqueous veins, channels that connect the canal of Schlemm with the episcleral plexus of vessels. These aqueous veins were discovered by slit lamp examination of the living eye of humans (Ascher, 1942). The number of these aqueous veins is small in comparison to the number of recipient blood-containing vessels; the rate of flow of aqueous humor is small in comparison to the rate of flow of blood carried by the episcleral veins. As a consequence, the pressure in the episcleral plexus is largely independent of the pressure in the eye or the rate of flow of aqueous humor that leaves the eye. On the other hand, the pressure in the episcleral plexus is one of the major determinants of the steady-state pressure of the eye.

Several methods have been described for the measurement of episcleral venous pressure in humans. However, none is satisfactory for routine clinical use, and all methods are highly subjective. These methods depend on observation of the collapse of a selected vessel that results from external pressure applied by a rigid device, a flexible membrane, or a jet of air.

These methods have two problems. First, the endpoint is very subjective. Second, there is bias of selection of vessel segments for measurement. Examiners necessarily select large and superficially located vessels. However, the vessels that are easy to measure may not be typical of the vessels into which aqueous humor is flowing. As a consequence, most clinical studies of aqueous dynamics must be interpreted in the absence of quantifiable and verifiable measurements of episcleral venous pressure. For most calculations of aqueous dynamics, episcleral venous pressure is assumed to range between 5 and 10 mm Hg when subjects are sitting or standing, and several mm Hg higher when subjects are prone or supine.

The effect of elevated episcleral pressure on intraocular pressure is easy to demonstrate by means of a tilt-table. When the body is tilted 50 degrees head down, intraocular pressure doubles from its normal level of 17 mm Hg to 34 mm Hg (Carlson *et al.*, 1987). Also, if a blood pressure cuff around the neck is inflated to 20–30 mm Hg, an immediate and nearly equal rise of episcleral venous pressure and intraocular pressure occurs (Kupfer and Ross, 1971).

D. Aqueous Humor Flow

The rate of flow of aqueous humor is the third major determinant of steady-state intraocular pressure. Numerous methods of measuring aqueous humor flow have been developed that are applicable for studies in animal eyes, but few of them are useful for studies in human eyes. The few that are suitable for studies of flow in humans can be carried out under a variety of conditions, some of which are difficult or impossible to achieve in animal studies.

Two fundamentally different methods of measuring aqueous humor flow in humans have been used. Hyperpiestic methods, like tonography, depend on raising intraocular pressure and deducing aqueous flow from outflow resistance, episcleral venous pressure, and steady-state intraocular pressure. Isopiestic methods are carried out without alteration of intraocular pressure and generally depend on the observation of the rate of disappearance of a tracer. Isopiestic methods are regarded as the most accurate. In the normal eye, most aqueous humor that is formed by the ciliary body passes through the anterior chamber. Thus, the rate of flow through the anterior chamber is usually a good measure of the rate of formation of aqueous humor, but it is a good idea to remember that the former is not identical to the latter.

Holm (1968) described an isopiestic method of measuring aqueous flow in humans that does not depend on the disappearance of a tracer but rather the rate of appearance of newly formed aqueous humor as it emerges from the posterior chamber into the anterior chamber through the pupil. In his method, the pupil is constricted pharmacologically, and the aqueous humor in the anterior chamber is made visible by the topical application of fluorescein. After the aqueous humor is mixed by voluntary eye movements, the temperature of the cornea is increased with warm air in order to eliminate convection currents in the anterior chamber. Photographs are then made of the emerging pupillary "bubble" as unstained aqueous humor from the posterior chamber displaces the fluorescent fluid of the anterior chamber. The rate of increase of the volume of this bubble is determined by photogrammetry (Fig. 2).

A second isopiestic method of measuring aqueous flow was introduced by O'Rourke and Macri (1970). In this method, a gamma-emitting nuclide is introduced by injection into the anterior chamber, and the rate of disappearance of radioactivity from the eye is measured with an external gamma counter. This method has been used in animal and human eyes, but has limited use in humans because of the need to puncture the cornea. Neverthe-

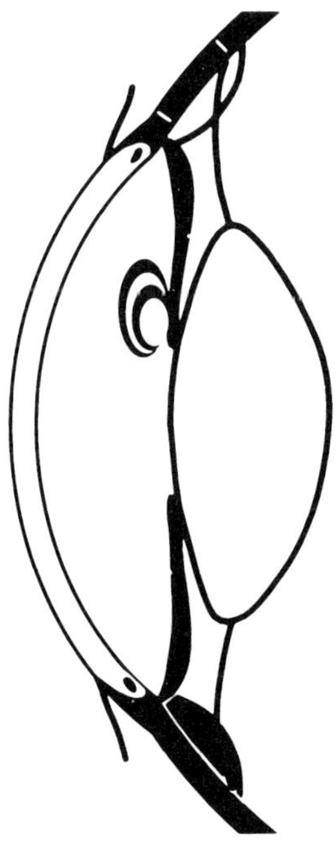

FIGURE 2 Photogrammetric procedure of Holm for measuring the rate of flow of aqueous humor into the anterior chamber. The rate of growth of the pupillary "bubble" is measured with slit lamp photography. [Reprinted from Brubaker (1984), Fig. 3–4, p. 43, with permission from Grune & Stratton.]

less, the method is reliable and can serve as a standard against which to judge the accuracy of other methods (Fig. 3). There is good agreement between the method of Holm (1968) and the method of O'Rourke and Macri (1970).

A third method, employing the fluorescent tracer fluorescein, has gained wide acceptance for clinical studies. This method was first introduced by Goldmann who administered the tracer intravenously (Goldmann, 1950, 1951a). When administered systemically, fluorescein enters the anterior chamber via the vessels of the iris and the limbus as well as from the posterior chamber. The pharmacokinetics of systemically administered fluorescein is influenced by plasma binding as well as the conjugation of fluorescein to fluorescein glucuronide in the liver. These factors complicate the interpretation of Goldmann's method, and it has not been employed by other investigators because of the technical and interpretive difficulties. Later, Jones and Maurice (1966) described a simpler method in which fluorescein is introduced into the cornea by iontophoresis or drop instillation, and the rate of loss of fluorescein (fluorescein washout) is observed (Fig. 4).

The topical fluorescein method of Jones and Maurice (1966) permits the measurement of aqueous flow over several hours and under almost any conditions. Most of what is known about the rate of aqueous flow in humans has been learned by this method. A thorough understanding of the basis and limitations of the method is needed before attempting to use it or to interpret its results.

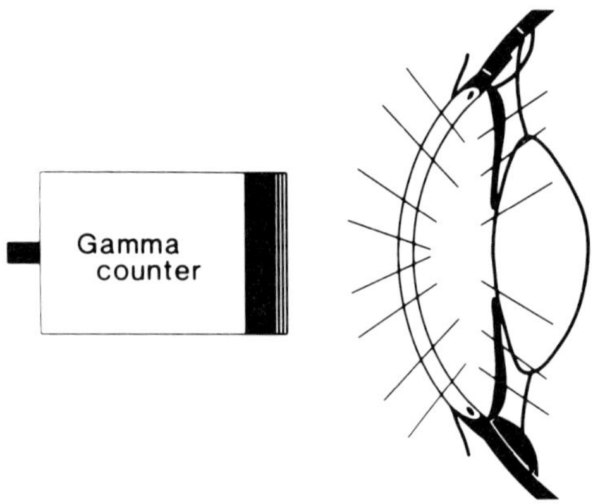

FIGURE 3 Radioactive tracer technique as applied by O'Rourke for measuring the rate of flow of aqueous humor. [Reprinted from Brubaker (1984), Fig. 3–5, p. 43, with permission from Grune & Stratton.]

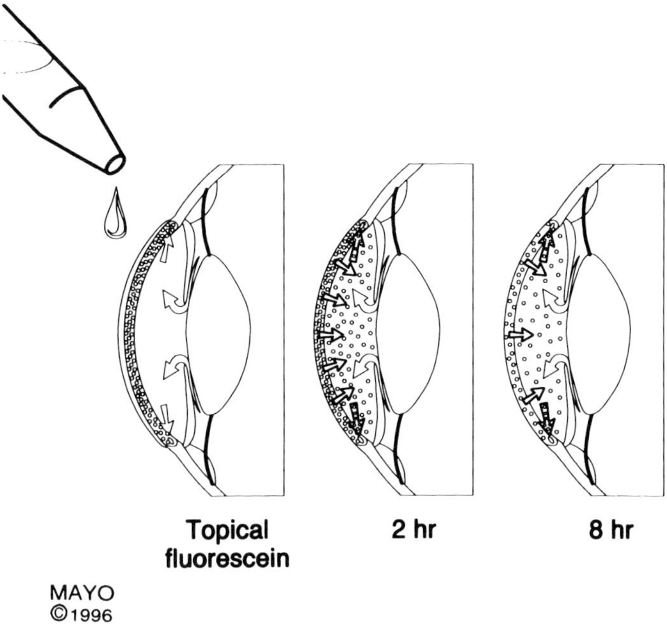

FIGURE 4 Topical fluorescein method of Jones and Maurice (1996) (fluorescein washout method) for measuring aqueous humor flow. (With permission from the Mayo Foundation.)

II. FLUORESCEIN WASHOUT METHOD OF MEASURING AQUEOUS FLOW

A. Principles and Assumptions

When a high concentration of fluorescein is instilled into the conjunctival cul-de-sac, a small fraction of the applied dose enters the corneal stroma through the epithelium. The remainder is quickly washed away by the flow of tears across the surface of the cornea and conjunctiva. After entering the stroma, fluorescein's preferential route of escape is into the anterior chamber through the corneal endothelium, a layer of cells that is much more permeable to small water-soluble organic molecules than is the corneal epithelium. The corneal stroma acts as a depot from which fluorescein is gradually released into the anterior chamber. A small and negligible fraction of the stromal depot of fluorescein will diffuse to the periphery of the cornea where it can enter the vessels of the limbus.

From the anterior chamber, the dye can leave the eye by two routes. In the normal eye, a minor fraction diffuses into the vessels of the iris stroma. The major portion leaves the eye along with the aqueous humor in which it is dissolved. Posterior egress of the tracer is blocked in the normal eye by the combined surfaces of the iris and the crystalline lens that together form a barrier to posterior loss. Experiments show that at least 95% of the loss of topically applied fluorescein from eyes with an intact lens–iris barrier occurs as a result of bulk outflow of aqueous humor as it passes from the pupil to the iridocorneal angle (Goldmann, 1951a; Nagataki, 1975; Araie *et al.,* 1980; McLaren and Brubaker, 1986; McLaren *et al.,* 1993). Most of the remaining loss is due to diffusion into vessels of the iris. Thus, the rate of clearance from the anterior chamber of topically applied fluorescein is an excellent measure of the rate of aqueous humor flow.

The pharmacokinetics of topically applied fluorescein provides the basis for a clinically accurate method to measure aqueous humor flow as long as fluorescence is an accurate measure of molarity of fluorescein and an instrument is available to make accurate measurements of fluorescein's fluorescence in the living eye. Such instruments are available, and experiments show that fluorescence, except in certain pathologic conditions of the eye, is a good measure of molarity over a wide range of concentrations of fluorescein in the cornea and anterior chamber (Maurice, 1967; Brubaker, 1982).

B. Instruments

An *ocular fluorometer,* or *fluorophotometer,* is a special instrument that permits measurements of fluorescence in the living eye rather than in a stationary cuvette. Such an instrument requires optical systems that focus an excitation beam on the target and collect the emitted fluorescent signal from the same target. In addition, the instrument must include some provision for aligning the focal diamond of the instrument (the intersection of the excitation beam and the light pathway of the detector) with the eye. Some instruments employ a slit lamp biomicroscope for excitation, emission, and observation of the site of measurement. The excitation beam is created by inserting a blue interference filter in the slit beam. The fluorescent light is collected by one of the two microscope pathways and delivered to a photomultiplier tube. The operator observes the eye, and the placement of the focal diamond is carried out manually. Other instruments are comprised of mechanical devices that scan the focal diamond across a contiguous series of voxels (volume elements) within the eye. These scanning fluorophotometers measure fluorescence along an axis or plane of the eye and are designed to minimize the problem of eye movement (Figs. 5a and b). Both types of instruments have been used to measure aqueous humor flow (Brubaker *et al.,* 1990).

Good fluorophotometers are able to measure the concentration of fluorescein from the lower limit permitted by the autofluorescence of the cornea, approximately 10^{-9} g/mL, to the upper limit permitted by the inner filter effects of concentrated fluorescein in the cornea and anterior chamber, approximately 10^{-5} g/mL (Maurice, 1967). In addition, the focal diamond must be small enough to fit within the 0.5-mm thickness of the corneal stroma. Measurements must be made rapidly since motion of the living eye can seriously degrade the spatial resolution of the instrument. Given an ideal instrument, the accuracy of measurement is determined by the optical properties of the cornea. The cornea scatters light and reflects light from its surfaces. These effects degrade the accuracy of measurements in the anterior chamber when the stroma is brightly fluorescent. Some investigators measure the effect of stromal scattering in individual eyes and make corrections for it. In eyes with normal corneas, a good fluorophotometer is able to make repeated measurements of stromal fluorescence within 4% and of cameral fluorescence within 6%, limited mainly by the lack of uniformity of the distribution of fluorescence (Brubaker, 1996).

C. Procedures

The procedure of measuring aqueous humor flow by the fluorescein washout method begins with the instillation of fluorescein in the conjunctival cul-de-sac. Generally, 2% sodium fluorescein is instilled at least 5 hr before the measurement is to begin. The delay permits the tracer to distribute itself evenly in the stroma and allows the concentration in the anterior chamber to rise to measurable levels. A convenient procedure is to ask the subject to awaken in the middle of the night and to instill fluorescein several times at 5-min intervals, then to return to sleep. The next morning, the eye is ready for measurements of flow. Since the half-life of fluorescein in the cornea is approximately 4 hr, the initial concentration in the stroma must be doubled for each 4 hr of delay between instillation and measurement. Because of the large range over which measurements are linear, an initial stromal concentration of 10^{-6} g/mL permits measurements of flow for nearly 24 hr.

Before commencing with measurements of flow, the lids are cleansed in order to remove extraneous sources of fluorescence near the eye that might interfere. The stromal concentration is measured to ensure that it is high enough to permit several hours of measurement yet not so high that the fluorescein absorbs a significant proportion of the excitation beam or the emitted fluorescence. The stromal fluorescence is also checked for uniformity of distribution. For most accurate measurements, the subject will have

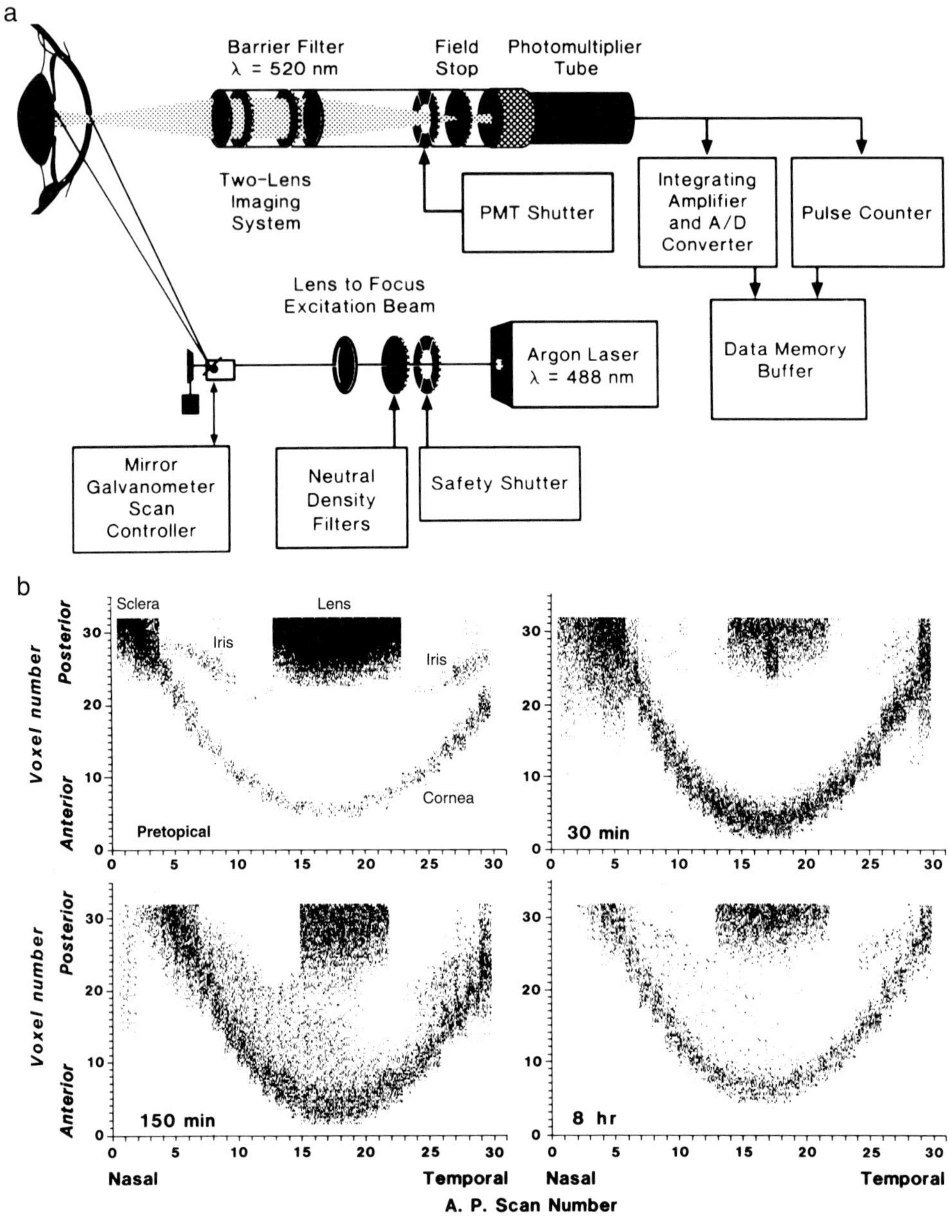

FIGURE 5 (a) Diagram of a scanning ocular fluorophotometer. [Reprinted from McLaren and Brubaker (1985), Fig. 2, p. 146, with permission from Association for Research in Vision and Ophthalmology.] (b) Fluorescence of the anterior chamber and cornea before and after topical fluorescein as measured with the scanning ocular fluorophotometer of McLaren and

had a measurement of autofluorescence of the cornea at an earlier time, before fluorescein instillation, and this value is subtracted from the corneal signal.

At some point in the procedure, the volume of the anterior chamber of each eye is measured by one of several methods. One method is to measure the diameter and depth of the anterior chamber and to calculate its volume on the assumption that its geometry is approximated by a spherical segment. Another is to make a Polaroid photograph of an optical section of the anterior chamber and to calculate the volume as a series of cylindrical sleeves (Johnson *et al.*, 1978). In the human eye, the volume of the anterior chamber is 200 ± 50 μL, and it varies according to the refractive state of the eye and the age of the subject (Fontana and Brubaker, 1980). The variance in cameral volume among different individuals is great enough that individual measurements are needed for accurate measurements of flow.

After the initial measurement of fluorescence in the cornea and anterior chamber, the measurements are repeated at regular intervals, usually every hour or two. The precision of the measurement of fluorescence in the eye as stated earlier is ±4–6%. The rate of loss of fluorescein from the eye is approximately 17% per hour. Thus, measurements of flow at intervals of less than 30 min are of uncertain utility.

During the interval between measurements, the subject is unencumbered. For this reason, the subject can be assigned any task, from sedentary to vigorous, depending on the conditions under which the flow is being investigated. It is prudent for the investigator to control the activity and conditions of the subject during this interval, since many things can perturb the flow or its measurement. In particular, the subject must be cautioned against any activity, including eating, drinking, or exercise, that might alter the osmotic pressure of the plasma. Also, subjects must be carefully instructed not to use any pharmacologic agent that might affect flow during the interval. In some experiments, all subjects are asked to participate in an activity or take a pharmacologic agent in order to measure its effect on aqueous flow in comparison to a control period. By this means, the responses of the aqueous system have been studied under a variety of circumstances.

Brubaker. Autofluorescence in the anterior segment is illustrated in the "pretopical" scan. Fluorescence is shown at 30 min, 150 min, and 8 hr after one drop of 0.25% fluorescein (Fluress) was applied. The intensity scale has a range of 1–10 ng/mL (<1 ng/mL, white; ≥10 ng/mL, black) in the pretopical scan and 1–100 ng/mL (<1 ng/mL, white; ≥100 ng/mL, black) in the other scans. The density of the gray scale is logarithmically related to fluorescein concentration. Note the visibility of the pupillary "bubble" in the scans at 150 min and 8 hr. [Reprinted from McLaren and Brubaker (1985), Fig. 6, p. 149, with permission from Association for Research in Vision and Ophthalmology.]

After an interval of 1 or 2 hr, the fluorescences in the cornea and the anterior chamber are measured. The *rate of clearance* is the rate of fluorescein loss from the combined cornea and anterior chamber relative to the average concentration of fluorescein in the anterior chamber during the interval. The fluorophotometer measures fluorescence, which is used in turn to calculate concentration of fluorescein in the cornea and the anterior chamber. The concentration in each tissue, times its volume, is the amount of fluorescein in each tissue. The total amount of fluorescein in the eye at any time is the sum of the amounts in the cornea and the anterior chamber. The loss during the interval is the difference in the total amount present at the beginning and at the end of the interval. The rate of loss is the loss divided by the time. For intervals of 1 to 2 hr or less, the average concentration of fluorescein in the anterior chamber is very nearly equal to the mean of its concentration at the beginning and the end of the interval. The formula that is used to calculate clearance is then:

$$\text{Clearance} = (\Delta F/\Delta t)/C,$$

where ΔF is the amount of fluorescein lost during the interval, Δt is the length of the interval, and C is the average concentration of fluorescein in the anterior chamber during the interval. The flow of aqueous humor is the clearance minus the loss due to diffusion of fluorescein into the iridal and limbal vessels, usually assumed to be 0–0.25 μL/min.

D. Normal Values

The rate of aqueous flow through the anterior chamber has been measured in human subjects with a variety of techniques. Table I gives a summary of the results of many of the published studies.

The rate of aqueous flow in active persons between 8 a.m. and noon is 2.97 ± 0.77 μL/min (mean $\pm$ SD, $n = 519$, one eye per subject). The distribution appears to be Gaussian (Fig. 6). In the afternoon, the rate of flow decreases 10% to 2.68 ± 0.64 μL/min ($n = 490$). The coefficient of variation in flow from one eye of one person to one eye of another person is approximately 25%. Thus, 95% confidence limits of normal would include flows that were $\pm$50% of the mean of the normal population. In the case of morning flows, this normal range is approximately from 1.5 to 4.5 μL/min.

E. Repeatability and Accuracy

The repeatability of the fluorescein washout method has been measured in two ways. One way has been to compare the right to the left eye when

TABLE I

Aqueous Humor Flow in Human Eyes Determined by Various Isopiestic Methods

Investigator	Year	Method	Number of subjects	Aqueous flow (μL/min mean ± SD)
Goldmann	1951a	Intravenous fluorescein	36	2.2 ± 0.4
Jones and Maurice	1966	Fluorescein iontophoresis	10	2.4 ± 0.5
Holm	1968	Photogrammetry	17	3.1 ± 1.6
O'Rourke	1974	Radioactive albumin washout	10	3.0 ± 0.5[a]
Bloom *et al.*	1976	Fluorescein iontophoresis	19	2.8 ± 0.6
Yablonski *et al.*	1978	Topical fluorescein	15	2.3 ± 0.7[a]
Brubaker	Unpublished observation	Topical fluorescein	519	3.0 ± 0.8[b]
			490	2.7 ± 0.6[c]

[a] Flow calculated from turnover assuming volume of anterior chamber of 200 μL.
[b] Measured 8 A.M. to noon.
[c] Measured noon to 4 P.M.

flows are measured simultaneously. In this comparison, the assumption is made that the flows in the two eyes of normal subjects are very nearly the same. The coefficient of variation of flow from one eye to the other of normal subjects is 16% when both measurements are made simultaneously from 8 a.m. to 4 p.m. Some of this variation is due to differences between the two eyes and some is due to the error of measurement.

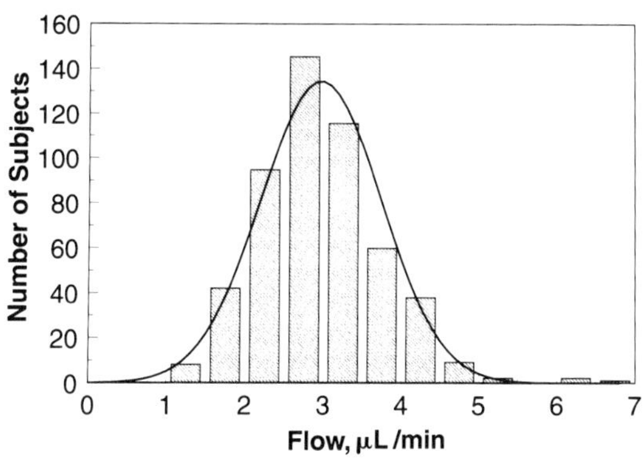

FIGURE 6 Distribution of aqueous humor flow in normal human subjects (n = 519) measured by fluorescein washout between 8 A.M. and 12 noon. (From Brubaker, unpublished.)

The second method of studying repeatability is to measure flow of an eye on two separate occasions at the same time of day and under the same conditions. Again, assuming that the aqueous humor flow does not vary from day to day, this comparison gives an idea of the repeatability of the procedure. The coefficient of variation of the measurements of the same eye on two different days from 8 a.m. to 4 p.m. is 20%. These two methods of measuring repeatability suggest that the flow of aqueous humor varies from day to day but that the flow in one eye tends to mimic the flow in the fellow eye. For this reason topical drug effects can be detected more easily by comparing one eye simultaneously to its fellow eye.

The accuracy of the method is more difficult to measure, for there is no method that is considered to be considerably more accurate than all others. However, the methods that are based on the fewest assumptions are the pupillary bubble method of Holm (1968) and the radioactive method of O'Rourke and Macri (1970). Table I gives the comparison of these two methods to the fluorescein washout method. All three methods give similar results. However, there is no study in which the same group of subjects has been measured by all three methods.

F. *Implications for Single-Cell Secretory Activity*

The pars plicata of the ciliary body has a robust circulation. Of the plasma that enters the ciliary body, 4% is filtered into its tissue spaces and is available to the ciliary epithelium for secretion of aqueous humor. As discussed in previous chapters, a large portion of this fluid, nearly devoid of its soluble protein, is transported into the posterior chamber as aqueous humor. The final step of the secretory process is carried out by the nonpigmented ciliary epithelial cells, which in the human eye number approximately 4 million and have an estimated total volume of 8 μL. Since the rate of aqueous formation during the day is approximately 3 μL/min, it is necessary for each cell on the average to secrete a volume equal to one-third of its own volume every minute (Fig. 7). The rate of metabolic work performed by the ciliary epithelium in forming aqueous humor may determine the upper limit of the rate of secretion and explain why hypersecretion has never been found to be a cause of glaucoma. This notion, that the normal rate of aqueous humor formation is the maximum rate, is logical when one considers that the metabolic demands of the tissues that depend on aqueous humor probably vary within narrow limits. Other tissues, such as skeletal muscle, have a much broader range of metabolic needs and exhibit a much larger range of circulatory flow rates.

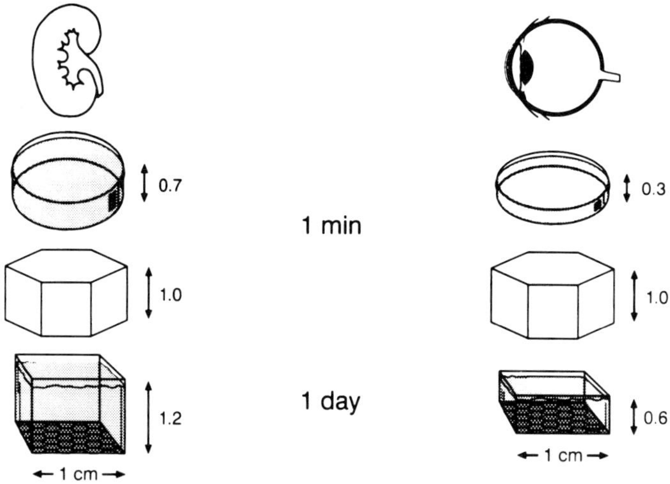

FIGURE 7 Comparison between the rate fluid transport by the renal tubular system on the left and the pars plicata of the ciliary body on the right. In 1 min, the average renal tubular cell transports 70% of its cell volume; the average ciliary epithelial cell transports 30% of its volume. In 1 day, renal tubular cells transport 1.2 cm^3/cm^2 area; ciliary epithelial cells transport 0.6 cm^3/cm^2 area. [Reprinted from Brubaker (1991), Fig. 3, p. 3149, with permission from Association for Research in Vision and Ophthalmology.]

III. OBSERVATIONS BASED ON CLINICAL MEASUREMENTS IN VOLUNTEERS

A number of aspects of aqueous formation have been studied in human subjects. Some of these include the relation of aqueous humor flow to age, sex, time of day, intraocular pressure, hormones, and nervous activity.

A. Age Dependence of Flow

Some of the anatomic and physiologic properties of the eye change with advancing age. For example, the crystalline lens grows during life, and its growth causes the anterior chamber to become shallower. Figure 8 illustrates the change in volume of the anterior chamber with age. Becker, using an indirect method of calculating flow, showed that flow diminished appreciably after the age of 60 (Becker, 1958). Other workers using fluorophotometry, showed that flow decreased gradually from age 20 to 80 at approximately 2.4% per decade (Brubaker *et al.,* 1981). The results of a more recent study of 300 subjects were similiar, showing a loss of 3.2% per

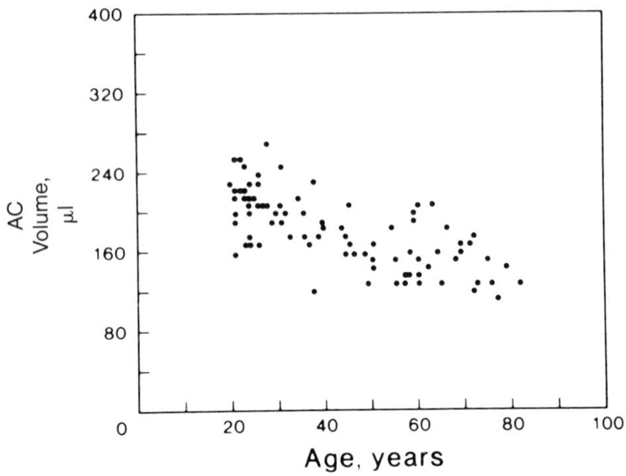

FIGURE 8 The volume of the anterior chamber (in microliters) and age. The anterior chamber volume decreases with age significantly ($p < 0.001$). [Reprinted from Brubaker *et al.*, (1981), Fig. 1, p. 285, with permission from the American Academy of Ophthalmology.]

decade (Brubaker, 1991). A more recent analysis of 519 subjects (Brubaker, unpublished) shows a slowing of 3.3%–3.5% per decade of age when all age groups are included (Fig. 9). However, it appears that persons under the age of 10 may have lower flows, and if these persons are excluded from the analysis, the rate of decline after age 20 would be slightly greater, 4%/ decade (Fig. 9). At any rate, it appears that between the ages of 20 and

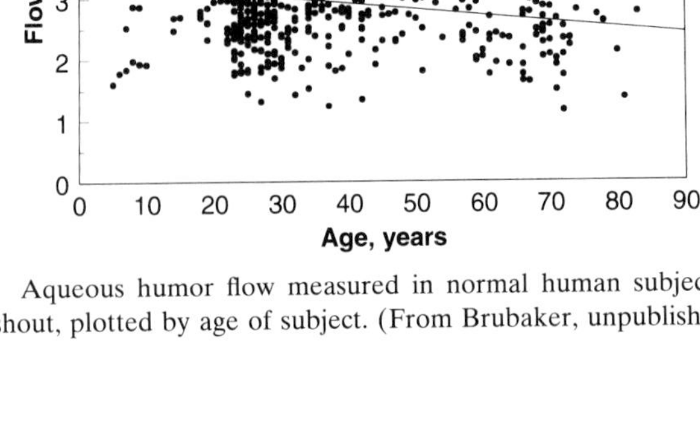

FIGURE 9 Aqueous humor flow measured in normal human subjects (n = 519) by fluorescein washout, plotted by age of subject. (From Brubaker, unpublished.)

80 the aqueous flow of a normal person would decline 25% and the volume of their anterior chamber would decline 40%. Thus, the rate of fluid turnover in the anterior chamber actually increases with age despite the reduction in the rate of flow of aqueous humor.

The age-dependent changes in turnover rate and flow are not known to have any adverse effect on the health of the eye. Because the resistance to outflow of aqueous humor rises slowly with age (Becker, 1958), the reduction in flow may be advantageous. The reduction in flow minimizes the rise of intraocular pressure that results from the rise in outflow resistance.

B. Sex Independence of Flow

Fontana and coworkers found in a study of 196 men and 276 women that the volume of the anterior chamber, when adjustments were made for age, was 5% larger in the men than in the women (Fontana and Brubaker, 1980). However, no comparable difference has been found in the rate of aqueous humor flow in the two sexes. In one study of 51 men and 62 women, the flow in women was 4% greater, but the difference was not statistically significant (Brubaker, 1982). In a much more extensive comparison, of 262 women versus 202 men of comparable age, the age-adjusted flow for women from 8 a.m. to 4 p.m. was 2.81 ± 0.67 and for men was 2.89 ± 0.68, a 3% difference ($p = 0.09$). Though no statistically significant difference has been found, it is likely that the rate of aqueous flow may be related to the size of the eye where a small difference has been shown. It would require comparison of 700 men and 700 women to be 95% confident of finding a 5% difference between them!

C. Pressure Independence of Flow

It is well known that the level of intraocular pressure is dependent in part on the rate of flow of aqueous humor into the anterior chamber. The theory of this relationship was first expressed in mathematical terms by Goldmann (1950). Whether the rate of aqueous humor formation is dependent in any way on the intraocular pressure has been the subject of speculation for many years. Investigators have entertained two mutually exclusive hypotheses about the relation between intraocular pressure and aqueous flow. One hypothesis is that aqueous formation is not sensitive to moderate changes of intraocular pressure. The other is that the eye regulates intraocular pressure at a steady level by making compensatory changes in the rate of aqueous formation.

A number of experiments have been devised to study this relationship in human eyes. Experiments employing a combination of tonometry, tonography, and episcleral venous manometry made indirect calculations of aqueous flow at normal and elevated episcleral pressure (Kupfer *et al.,* 1971; Kupfer and Ross, 1971; Gaasterland *et al.,* 1973, 1975). These experiments found that the rate of aqueous flow slows slightly as intraocular pressure rises. Subsequent experiments, however, have failed to confirm a clinically significant relationship. Carlson and associates measured aqueous flow with fluorophotometry in subjects whose intraocular pressure was altered by gravitational influences on a tilt-table (Carlson *et al.,* 1987). The intraocular pressure was alternated between its normal level (17 mm Hg) in the head-up position to a higher level (34 mm Hg) in the head-down position. Despite the large change in intraocular pressure, there was no measurable change in aqueous flow.

A number of studies have been carried out in persons who have spontaneous elevation of intraocular pressure because of an increased resistance to outflow. Of special note was a study carried out in persons with pigment dispersion syndrome. In this condition there is mechanical disruption of the delicate posterior pigment layer of the iris. This disruption causes pigment particles to scatter into the aqueous humor. The particles then flow with the aqueous humor and become trapped in the trabecular meshwork as aqueous flow leaves the eye. The trapped particles cause an increase in outflow resistance and consequently an increase of intraocular pressure. In the study, eyes with pigment dispersion and an average intraocular pressure of 26 mm Hg had an average flow of 2.86 μL/min. Eyes with pigment dispersion and an average intraocular pressure of 17 mm Hg had the same rate of aqueous flow (Brown and Brubaker, 1989).

In three other studies, the aqueous flow was measured before and after the intraocular pressure of glaucomatous eyes was lowered by a procedure known as *laser trabeculoplasty.* This procedure consists of treating the inner surface of the trabecular meshwork with a series of laser burns. The procedure reduces the resistance to outflow and lowers the intraocular pressure. In all three studies, the rate of aqueous flow was the same before and after the procedure (Brubaker and Liesegang, 1983; Araie *et al.,* 1984; Yablonski *et al.,* 1985).

Myotonic dystrophy is a condition that is associated with very low intraocular pressures. These low pressures are presumed to occur because the aqueous humor is able to leak into the supraciliary space through the atrophic ciliary muscle. A study by Walker compared 26 patients with this condition whose average pressure was 7 mm Hg to 37 normal controls whose average pressure was 15 mm Hg. The rate of aqueous flow in the two groups was almost identical (Walker *et al.,* 1982). This finding was confirmed in a subsequent study (Khan and Brubaker, 1993).

The cited studies do not support the notion that the eye regulates its level of intraocular pressure by compensatory adjustments in the rate of aqueous humor formation. Rather, the data support the idea, expressed by Bárány a half-century ago, that the rate of aqueous flow, in the hierarchy of regulated variables, must be controlled within narrow limits to maintain the health of the crystalline lens and other avascular structures within the eye (Bárány, 1947). Thus occupied, the rate of aqueous formation is not available to have any regulatory effect on the intraocular pressure. Convincing evidence that intraocular pressure is regulated at all is lacking. If regulation does occur, it is mediated by other factors, such as outflow resistance or episcleral venous pressure.

D. Circadian Rhythm

The first evidence that aqueous formation might undergo regular changes with the circadian cycle was presented by Ericson (1958). He calculated aqueous flow by means of a suction cup and found that flow during sleep was much lower than flow during waking hours. His work has been confirmed by means of the fluorescein washout method. Several studies have shown that the rate of aqueous flow during sleep is approximately half the rate during the morning hours after awakening (Reiss *et al.,* 1984; Koskela and Brubaker, 1991a) (Fig. 10). The rate of aqueous flow measured from midnight to 6 a.m. in sleeping subjects is 1.28 ± 0.43 μL/min ($n = 180$, one eye per subject) (Fig. 11). Here the range of normal includes 0.42–2.1 μL/min. The rate at night during sleep is only 43% of the rate in the morning, and the night rate also shows a regression with age.

Attempts have been made to understand the factors that drive the circadian rhythm of aqueous flow. In one study, flow was measured when one eye was closed and the other open during daylight hours. The flows were the same in the open and closed eye (Topper and Brubaker, 1985). Flow was also measured during a period when the subjects were recumbent but awake during the day and was compared to measurements when the subjects were participating in normal activities on another day. The flow on the two days was nearly the same, only 12% lower on the day of recumbency (Topper and Brubaker, 1985). In other studies, subjects were studied during an evening of sleep deprivation at night (Reiss *et al.,* 1984; Maus *et al.,* 1996b). When subjects were not permitted to sleep, their aqueous flow was depressed during their normal hours of sleep, but only to half the extent as observed on a control night during sleep. In another study, subjects were

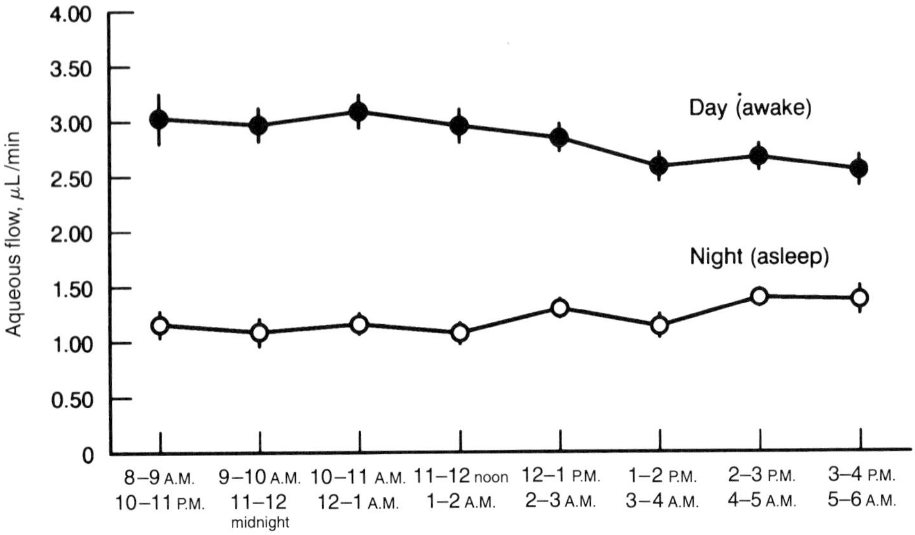

FIGURE 10 Rate of aqueous flow measured hourly during the day and at night during sleep to show the normal circadian rhythm in humans. [Reprinted from Koskela and Brubaker (1991a), Fig. 1, p. 2505, with permission from Association for Research in Vision and Ophthalmology.]

allowed to sleep in the dark during one sleep cycle and under bright lights during another sleep cycle. The rate of flow was the same in the dark and in the light (Koskela and Brubaker, 1991a). Thus, the phenomenon of

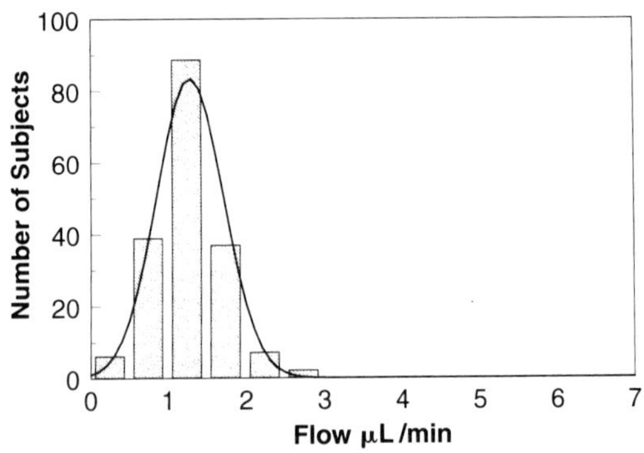

FIGURE 11 Distribution of aqueous humor flow in normal human subjects ($n = 180$) measured by fluorescein washout between midnight and 6 A.M. during sleep. (From Brubaker, unpublished.)

aqueous suppression that occurs rhythmically seems to be a true circadian rhythm and that is not overcome entirely by light, ambulation, or the level of activity of the subject.

Studies in rabbits have shown that this mammal also has a circadian rhythm of aqueous flow that can be entrained and persists through several cycles of complete darkness (Rowland *et al.*, 1981, 1986; Smith and Gregory, 1989). Similarly, rabbits have been shown to have a circadian rhythm of intraocular pressure, monitored continuously by means of telemetry, that follows the same pattern of their cycle of aqueous flow (McLaren *et al.*, 1996a). A corresponding experiment in humans has not been possible because of the technical inability to monitor intraocular pressure in the undisturbed eye during sleep.

It has been concluded that there is a very consistent circadian rhythm of aqueous humor flow in humans. The depth of the suppression of aqueous flow during sleep is as great as the inhibitory effect of any of the pharmaceutical agents employed for the treatment of glaucoma. The mechanisms that control this rhythm are only partly understood and still occupy the interest of current investigators (e.g., see Chapter 8).

E. Hormonal and Nervous Influences on Flow

The fact that aqueous humor flow has a circadian rhythm has stimulated investigation into the regulatory mechanisms that bring about this rhythm. The source of this regulation has been elusive. With the single exception of the catecholamines, no hormone has been found to have a consistent and physiologically relevant effect on aqueous humor flow in human eyes. The following paragraphs outline some of the studies that have been carried out.

1. Melatonin

Viggiano and coworkers (1994) gave a series of 19 human volunteers large oral doses of melatonin during the daylight hours. The level of melatonin in the plasma rose to more than twice the normal level that occurs naturally during sleep. Despite the high melatonin level on the treated day, the mean flow was not significantly different from the control period, 2.71 ± 0.64 μL/min (mean ± SD) during melatonin treatment and 2.80 ± 0.66 μL/min during placebo treatment. Some subjects were given oral melatonin at night in order to see if excess metabolites of melatonin would interfere with the normal nocturnal suppression. These subjects were observed to have suppression to 39% of the daytime rate, at least as great and perhaps even greater than without the exogenous melatonin. Melatonin might be partly responsible for the nocturnal nadir of aqueous flow, but this hormone cannot be solely responsible for its circadian rhythm.

2. Antidiuretic Hormone

In another study, Viggiano and coworkers (1993) measured the rate of aqueous humor flow in patients suffering from neurogenic diabetes insipidus. These subjects lacked the hormone arginine vasopressin (antidiuretic hormone) and were on replacement therapy with nasal desmopressin to prevent the severe diuresis associated with the clinical condition. When no desmopressin was used, aqueous humor flow in the patients was 2.34 ± 0.69 μL/min. When desmopressin was used, the flow was 8% higher, 2.53 ± 0.79 μL/min. The small difference could have been due entirely to the difference in the osmotic pressure of the plasma on the two days. On the day of desmopressin deprivation, the accompanying diuresis caused the plasma to become 8 mOsm hypertonic in comparison to the day when desmopressin replacement therapy was used. Had this difference in osmotic pressure not occurred, the flows on the 2 days would have been nearly identical. The authors concluded that this hormone has no clinically significant effect on the aqueous humor flow and does not influence its circadian rhythm.

3. Hormones of Pregnancy

Ziai and coworkers (1994) have studied a series of women during the trimesters of pregnancy and afterward to determine if the hormonal changes that occur during pregnancy have any effects on aqueous humor flow. The plasma concentration of progesterone increased from near zero in the nonpregnant state to more than 400 ng/mL in the third trimester, and β_1-chorionic gonadotrophin, which was not measurable in the nonpregnant state, rose to 122,000 IU/mL during the first trimester. Despite these large changes in hormone concentrations, the rate of aqueous humor flow remained stable through pregnancy and afterward. A similar study of the relation between serum progesterone and aqueous flow was carried out in 20 women during their normal menstrual cycle (Gharagozloo and Brubaker, 1991a). Again, no correlation was found between the aqueous flow and the concentration of the hormone in plasma.

4. Epinephrine and Norepinephrine

Epinephrine has been found in a number of studies to have a stimulatory effect on aqueous humor flow. In one study, topically applied epinephrine was found to stimulate aqueous flow 19% during the day (Townsend and Brubaker, 1980). In another study, a greater effect, 47% stimulation, was observed when the hormone was given at night, and the flow was measured during sleep (Topper and Brubaker, 1985). The closely related drug isoproterenol was found to have no effect when given topically during the day in two studies, but was found to stimulate aqueous humor flow 34% during sleep when given with the phosphodiesterase inhibitor, theophylline (Larson and Brubaker, 1988). The β_2-selective adrenergic agonist terbutaline also had no effect when applied topically during waking hours but stimulated aqueous flow 15% in sleeping subjects (Gharagozloo *et al.,* 1988a). Kacere *et al.* (1992) showed that intravenously infused epinephrine stimulated aqueous humor flow by 27% in sleeping subjects. It has been shown that the epinephrine effect can be blocked by a β-adrenergic antagonist (Rettig *et al.,* 1994). These studies suggest that aqueous humor flow is stimulated by β-adrenergic agonists.

The discovery of the effect of epinephrine on aqueous prompted a study of patients who completely lack circulating epinephrine because of previous surgical adrenalectomy. Interestingly, these patients are observed to have the normal rhythm of aqueous humor flow with high rates during the day and low rates during sleep (Maus *et al.,* 1994). Thus, it can be concluded that epinephrine, like melatonin, cannot be solely responsible for the circadian rhythm of flow.

Recently, a study was conducted of intravenously administered norepinephrine. This catecholamine was given during sleep while aqueous flow was measured. The results of the study suggest that norepinephrine also is a stimulator of aqueous humor flow, but much less effective than epinephrine. These studies are still under way.

5. Corticosteroids

Sheridan and coworkers (1994) studied the effect of dexamethasone on the rate of aqueous flow. This exogenous corticosteroid was given both during the day and at night in order to produce a steady effect for 24 hr. During the daytime, the addition of the steroid increased aqueous flow 9% as compared to placebo. No effect was observed at any other time of the circadian cycle. In another study, Jacob studied the combined effect of hydrocortisone and intravenous epinephrine on the rate of aqueous humor flow during sleep (Jacob *et al.,* 1996). When hydrocortisone was given with epinephrine, aqueous flow was stimulated to a greater extent, 42% than when placebo was given with epinephrine, 27%. It was concluded that the corticosteroid had no direct effect but augmented the effect of the catecholamine. The same reasoning explains why dexamethasone was observed to have an effect in the daytime, when catecholamine levels are high, and no effect at night, when they are low (Sheridan *et al.,* 1994). Topically applied dexamethasone did not have any effect on aqueous flow (Anselmi *et al.,* 1968; Rice *et al.,* 1983).

6. Horner's Syndrome

Horner's syndrome is a condition in which the sympathetic innervation of the eye and face is reduced or absent on one side. When the terminal neuron of the sympathetic pathway to the eye is absent, the clinical signs include miosis of the pupil, ptosis of the lids, and anhydrosis around the eye. Wentworth and Brubaker (1981) and later Larson and Brubaker (1988) studied patients affected with Horner's syndrome in one eye. In both studies, the rate of flow was faster in the denervated eye, although not statistically significantly greater. In Larson's study, the flow was measured during sleep, and the affected eye also was observed to have a circadian rhythm. In both studies, the denervated eye was found to have an abnormal response to topical catecholamines. These studies are interpreted as showing that aqueous formation is not dependent on adrenergic innervation or can adapt to loss of this innervation.

At present, the catecholamines epinephrine and norepinephrine are the only hormones or neurotransmitters that have been found to have consistently measurable effects on the rate of aqueous humor formation in human eyes. It is possible that each of these catecholamines, whether reaching ocular receptors via the general circulation or via sympathetic nerve terminals, stimulates aqueous flow during the daytime. Absence of these stimuli is a potential cause for the lower rate during sleep. Corticosteroids augment this effect and may play an additional role in the circadian rhythm when catecholamines are present. Much more experimental work will be required before a clear understanding of the control of aqueous formation in humans is understood.

IV. OBSERVATIONS IN CLINICAL SYNDROMES

It is well known that increased resistance to outflow is a prominent finding in most glaucomas. Grant showed that a group of patients with glaucoma had distinctively different tonographic tracings than a group of normal persons (Grant, 1951). Subsequently, tonography has been used by investigators to study the resistance to outflow in glaucoma and to determine the mechanism of action of ocular hypotensive drugs. Relatively less attention has been given to measurement of aqueous humor flow, except as derived indirectly from tonography. Some tonographic estimates of aqueous humor flow have led to the conclusion that the elevation of intraocular pressure in rare cases is due to hypersecretion rather than to increased resistance to outflow. More recent studies of aqueous flow employing fluorescein washout have not confirmed this notion. Rather, these studies have been consistent in showing that the rate of aqueous humor formation in glaucoma is normal, in contrast to outflow resistance that is often abnormal.

A. Chronic Simple Glaucoma

Larsson (Larsson *et al.,* 1995a) studied 20 patients with chronic simple glaucoma and 20 age-matched controls. The average age of both groups was 64 years. All patients discontinued the use of aqueous suppressing drugs 6 weeks before being studied. The mean intraocular pressure of the glaucomatous group was 26 mm Hg, nearly twice the mean pressure of the control group, 14 mm Hg. The outflow resistance of the glaucomatous group was 7.1 mm Hg · min/μL, approximately twice that of the normal group, 4.4 mm Hg · min/μL. The daytime flow of aqueous humor of the glaucomatous group was 2.70 μL/min and 2.39 μL/min for the normal group (p = 0.11). The flow of aqueous humor during sleep in the glaucomatous group was 1.29 μL/min, and 1.02 μL/min for the normal group (p = 0.02) (Fig. 12). This finding suggests that aqueous humor formation may be somewhat higher during sleep in patients with chronic simple glaucoma, but insufficient to explain the higher intraocular pressure. The difference in pressure is explained adequately by the higher outflow resistance, as shown in many earlier studies.

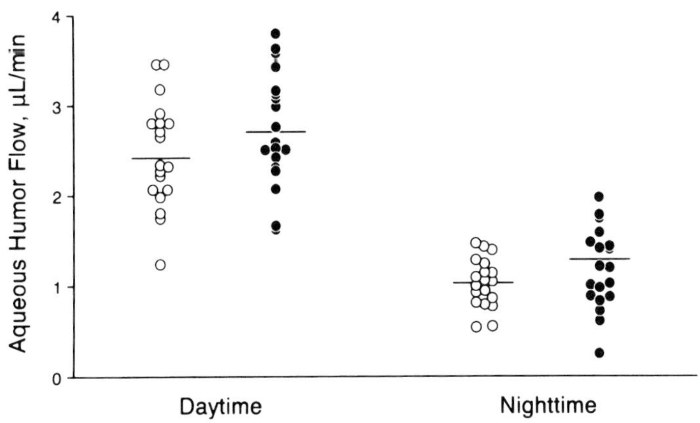

FIGURE 12 Flow of aqueous humor at daytime and nighttime in patients with open-angle glaucoma and normal controls. The difference between the groups was not statistically significant at daytime (p = .11) but was statistically significant at nighttime (p = .02). Open circles indicate normal controls; solid circles, patients with open-angle glaucoma. [Reprinted from Larsson *et al.* (1995a), Fig. 1, p. 285, *Arch. Ophthalmol.,* 113:283–286. Copyright 1995, American Medical Association.]

B. Normal Tension Glaucoma

Normal tension glaucoma, also called *low-tension glaucoma,* is a condition in which cupping and field loss characteristic of glaucoma is found in persons who lack evidence of increased intraocular pressure. Larsson *et al.,* (1993) studied 10 persons with normal tension glaucoma and 10 age-matched controls. The average age of both groups was 70–71 years. The intraocular pressure was nearly the same in both groups, 15 mm Hg for the affected patients and 14 for the controls. Likewise, the outflow resistance was similar, 5.6 mm Hg · min/μL for the patients and 4.6 mm Hg · min/μL for the controls. Aqueous flow during the day was nearly identical in the two groups, 2.48 uL/min for the patients and 2.45 μL/min for the controls. At night the flow seemed to be higher in the glaucoma patients, 1.24 μL/min compared to 0.96 μL/min for the controls, but the difference was not statistically significant ($p = 0.09$). The authors concluded that the dynamics of aqueous humor inflow and outflow in these patients was normal.

C. Ocular Hypertension

Ocular hypertension is the term used to describe patients who have elevated intraocular pressures without evidence of cupping and field loss characteristic of glaucoma. Ziai *et al.* (1993) studied 20 persons with ocular hypertension and 20 normal controls. The mean intraocular pressure of the ocular hypertensives was 20 mm Hg and of the normal controls was 13 mm Hg. The outflow resistance for the ocular hypertensives was 5.0 mm Hg · min/μL, nearly twice the value of the normals, 3.0 mm Hg · min/μL. Aqueous flow during waking hours of the day was nearly the same in the two groups, 2.92 μL/min in the ocular hypertensives and 3.15 μL/min in the normals. These subjects were not tested at night during sleep. Whereas the intraocular pressure and the resistance to outflow were both high in the ocular hypertensives, the flow was normal.

D. Pigment Dispersion Syndrome

The pigment dispersion syndrome is a condition that occurs most frequently in young myopic males. In this condition, the posterior surface of the iris makes contact with the anterior zonules of the crystalline lens. As mentioned earlier in this chapter, friction between the two surfaces causes the disruption of pigment cells on the posterior surface of the iris. These cells disperse their content of pigment granules into the aqueous humor.

The pigment then travels with the flowing aqueous humor and is trapped in the trabecular cells of the trabecular meshwork as aqueous flow leaves the eye. In some eyes, the pigment particles obstruct the outflow of aqueous humor, resulting in an increase of intraocular pressure that may lead to glaucomatous damage of the disk. The pathophysiology of the increased pressure in such cases is well understood, and their study is an ideal way to learn about the relative influence of obtructed outflow and raised intraocular pressure on the rate of aqueous humor flow.

Brown and Brubaker (1989) studied 40 persons with pigment dispersion syndrome, both with and without pressure elevation, and 40 age-matched normal controls. The average age of the two groups was 42 years. The average intraocular pressure of the normal controls was 16 mm Hg, and the average aqueous flow was 2.60 μL/min during the day. The average intraocular pressure among normotensive persons with pigment dispersion was 17 mm Hg and their average aqueous flow was 2.85 μL/min. The average intraocular pressure among hypertensive persons with pigment dispersion was 26 mm Hg, and their average flow was 2.86 μL/min. The authors concluded that the small difference in aqueous flow between persons with pigment dispersion and normals was due to the larger dimensions of eyes with pigment dispersion, an anatomic feature that predisposes to the condition. They also concluded that the increase in pressure is not due to an increase of aqueous formation nor does the higher intraocular pressure suppress the rate of aqueous humor formation.

E. Exfoliation Syndrome with Glaucoma

Exfoliation syndrome is a condition that can be recognized by a characteristic pattern of white deposits on the anterior capsule of the lens, the anterior portion of the zonule, and other structures within the eye such as the ciliary body, the iris, the cornea, and the trabecular meshwork. The source and nature of the material are unknown. The condition can occur with or without glaucoma. There is evidence from electron microscopy that the ciliary epithelium is adversely affected in this condition. Ultrastructural changes occur that indicate increased protein synthesis and decreased active transport in these cells (Lutjen-Drecoll *et al.,* 1988).

Johnson and Brubaker (1982) studied a group of 10 patients who were affected with this condition in one eye but not the other. The mean intraocular pressure in the affected eye was 32 mm Hg and in the unaffected eye was 18 mm Hg. The flow in the affected eye was 2.27 μL/min and in the unaffected eye was 2.85 μL/min, significantly lower in the affected eye. The authors concluded that the lower flow was due to the disease process that adversely affected the ciliary epithelium.

It was later discovered that the result of this study may have been due not to the disease process but rather to the use of the aqueous suppressing drug timolol in the affected eye (Gharagozloo *et al.,* 1992). The authors realized that an insufficient time had been allowed after the discontinuation of the drug for its effect to disappear. Gharagozloo *et al.* (1992) subsequently studied another series of 18 untreated patients with unilateral exfoliation syndrome. The intraocular pressure in the affected eyes was 14 mm Hg and in the unaffected eyes 12 mm Hg. The authors found the same rate of aqueous formation in the affected and in the unaffected eye, 2.40 μl/min. This flow was not significantly different from the flow in an age-matched group of normal controls, 2.60 μL/min. The difference in the results of the two studies could have also been due in part to the degree of severity of the cases, being more severe in the study of Johnson and Brubaker (1982). Nevertheless, it is clear that persons with well-established exfoliation syndrome can have normal aqueous humor flow. Whether the finding of normal flow is due to normal secretion by morphologically abnormal cells or whether it is due to hypersecretion by normal unaffected cells to make up a deficiency created by hyposecretion of affected cells is unknown.

F. Fuchs's Uveitis Syndrome

Fuchs' uveitis syndrome is characterized by heterochromia of the irides, uveitis, cataract, and glaucoma. Atrophy of the stroma of the iris can occur late in the condition, and abnormal vessels in the iris are commonly seen. Unlike inflammatory uveitis, this syndrome is not associated with synechiae in the iridocorneal angle. The condition is usually unilateral.

Johnson *et al.* (1983) studied 10 patients who were affected in one eye. The intraocular pressure in the affected eye was 17 mm Hg and in the unaffected eye 13 mm Hg. These authors did not report aqueous flow, but rather the clearance of fluorescein from the anterior chamber. This choice was made because the affected eye was found to have a greater inward leakage rate of systemically administered fluorescein, a fivefold greater rate than the unaffected eye. The clearance of fluorescein was 3.74 μL/min in the affected eye and 3.49 μL/min in the unaffected eye. The 7% greater clearance in the affected eye could be explained by an increased diffusional clearance of fluorescein. The authors concluded that aqueous flow in the affected eye could be somewhat slower than in the unaffected eye but that it could not be significantly greater.

G. Myotonic Dystrophy

Myotonic dystrophy is a progressive hereditary disease that is characterized by myotonia, muscular wasting, frontal baldness, testicular atrophy, and ocular abnormalities including ocular hypotony. In some cases, extremely low intraocular pressures are observed, sometimes lower than what is presumed to be the normal episcleral venous pressure. The simplest clinical explanation for this hypotony would be that the rate of aqueous formation is extremely low.

Walker *et al.* (1982) and later Khan and Brubaker (1993) studied series of 26 and 17 patients affected with myotonic dystrophy. In neither study did it appear that the rate of aqueous formation was low enough to explain the hypotony. In the study by Walker, the rate of aqueous flow was estimated to be 10% lower than in normal controls during waking hours. In the study by Khan, the flow was measured during waking hours in the morning, in the afternoon, and at night during sleep. The most reasonable supposition about diffusional losses of fluorescein at the three time periods in this study suggests that aqueous formation in myotonic dystrophy is approximately 10% lower than normal, in agreement with the earlier study. The authors presume that the hypotony—the mean value of intraocular pressure in the two studies was 7–8 mm Hg—is due primarily to an abnormal rate of uveoscleral outflow through the atrophic ciliary muscle.

H. Diabetes Mellitus

A well-known complication of long-standing diabetes mellitus is diabetic retinopathy. This condition begins with the loss of capillary pericytes and can progress to proliferative retinopathy and blindness. There are many other ocular effects of diabetes mellitus, including effects on aqueous dynamics including intraocular pressure, aqueous humor formation, aqueous humor flare, and permeability of the blood–ocular barrier. All of these effects are dependent on the age of onset of the disease, the duration of diabetes, the age of the patient, and the severity of the retinopathy. Though agreement from one study to another is not consistent, most investigators have found that the intraocular pressure tends to be slightly lower in diabetes mellitus than in age-matched controls. This finding is consistent with the reports in other studies that the rate of aqueous formation is lower in patients with diabetic retinopathy. The notion that aqueous formation would be impaired in diabetes mellitus is also appealing in light of the fact that this condition is associated with problems of the general circulation including increased blood viscosity, decreased tissue oxygenation, vascular leakiness, capillary shunting, and capillary nonperfusion.

Larsson and co-workers (1995b) made a systematic study of the aqueous dynamics in 61 patients with type 1 diabetes mellitus (early age onset) and 60 patients with type 2 diabetes mellitus (late age onset). Intraocular pressure, outflow resistance, and aqueous humor flow were measured and correlated with severity of diabetic retinopathy, age of onset, duration of diabetes, and age of the patient at the time of measurement. These patients were also compared to 32 age-matched control subjects. In this study, diabetics were found to have surprisingly normal values of all parameters, especially when allowance was made for the age of the patient. There was a statistically significant inverse relation between the degree of retinopathy and the rate of aqueous humor flow ($p = 0.04$), which was 22% lower in patients with grade 4 retinopathy compared to patients with grade 1 retinopathy or normal age-matched controls (Fig. 13). This study did not include patients with end-stage proliferative retinopathy, and it is likely that such patients may have shown more severe changes in their parameters of aqueous dynamics (Hayashi *et al.*, 1989). Nevertheless, it appears from this study that the vascular changes that occur in diabetes mellitus have much less of a clinically significant effect on the parameters of aqueous dynamics than on the health of the retina.

I. Cystic Fibrosis

Cystic fibrosis is the most common serious inherited disorder of Caucasian populations. The disease is characterized by hyperviscosity of secretions

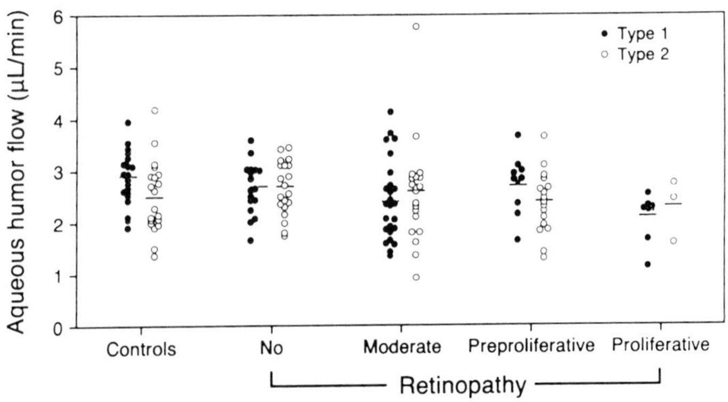

FIGURE 13 Aqueous humor flow in patients with type 1 and type 2 diabetes mellitus who have different grades of retinopathy (mean is indicated by the bar). [Reprinted from Larsson *et al.* (1995b), Fig. 2, p. 366, with permission of Ophthalmic Publishing Company.]

and associated with elevated concentrations of sodium and chloride in sweat. The mutation leads to a defect in epithelial transport that is thought to affect a wide variety of organs, including the salivary glands, the nasal, cervical, and tracheal epithelium, and the epithelia of the pancreas. The defect in the genome has been shown to cause an abnormality in a regulator protein of a chloride channel that prevents proper binding of adenosine triphosphate or prevents the conformation change necessary for proper gating of the channel. Such channels are insensitive to activation by protein kinase A or C (Fig. 14).

McCannel *et al.*, (1992b) studied the dynamics of aqueous humor in 12 adult survivors of this disease in order to determine if this genetic defect affects aqueous humor formation. Aqueous humor flow in these patients during the hours from 8 a.m. to 12 noon was 2.84 ± 0.16 μl/min (mean ± SD); between the hours of 12 noon to 4 p.m., was 2.40 ± 0.14 μL/min; and during sleep from midnight to 6 a.m., 1.16 ± 0.25 μL/min. The flows at the same times and under the same conditions for a group of normal controls were 2.86 ± 0.86, 2.63 ± 0.57, and 1.18 ± 0.41 μL/min. The affected patients were also tested to determine their responses to the β-adrenergic antagonist timolol. Intraocular pressure and aqueous humor flow were found to diminish to the same extent as had been observed in other studies of normal subjects. The authors concluded that the specific channel that is affected in cystic fibrosis is not essential for the formation of aqueous humor or

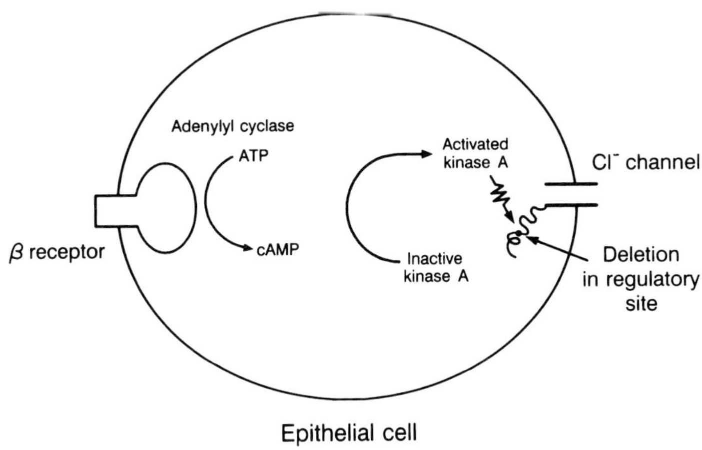

FIGURE 14 Hypothesized defect in cystic fibrosis. Activation of β receptor of eptihelial cell results in synthesis of cAMP and activation of kinase A, but chloride-selective channel fails to open in response to the stimulus. [Reprinted from Brubaker (1991), Fig. 4, p. 3157, with permission of Association for Research in Vision and Ophthalmology.]

that the mutation does not affect the normal gating of that channel in the ciliary epithelia.

V. EFFECTS OF PHARMACEUTICAL AGENTS

Most pharmaceutical agents that are used clinically and administered either systemically or topically to the eye that have been tested for effects on aqueous humor formation have been found to have no measurable effect. Three very significant exceptions to this generalization are carbonic anhydrase inhibitors, β-adrenergic antagonists, and α_2 selective adrenergic agonists.

A. Carbonic Anhydrase Inhibitors

Acetazolamide, a sulfonamide derivative, was the first clinically useful drug that was found to lower intraocular pressure due to its effect on the rate of aqueous humor formation. The synthesis of this compound was described in 1950, and within two years, reports of its clinical effects had appeared in print. Within four years, three groups had described its ocular hypotensive effects (Becker, 1954; Breinin and Gortz, 1954; Grant and Trotter, 1954). Systemic administration of acetazolamide was associated with a reduction of intraocular pressure without a corresponding change in outflow resistance. It was concluded that acetazolamide's effect was to suppress aqueous humor formation. This conclusion was confirmed in many subsequent experiments both in animals and in humans employing a variety of methods of measurement. The same effect was observed for other carbonic anhydrase inhibitors, including methazolamide, ethoxzolamide, and dichlorphenamide.

Dailey and coworkers (1982) used fluorophotometry to measure the effect of acetazolamide on the rate of aqueous humor flow. They tested 21 normal subjects and found that acetazolamide, compared to placebo, reduced the daytime rate of aqueous humor flow by 27%.

McCannel and coworkers (1992) also used fluorophotometry to measure the effect of systemically administered acetazolamide on the rate of aqueous humor flow in normal human subjects both during normal daytime activities and during sleep at night. The drug was found to reduce the daytime rate of aqueous flow by 21%. The rate of flow in these subjects without any drug was 59% lower at night during sleep than during waking hours. The addition of acetazolamide at night reduced the night rate an additional

24% to 0.82 ± 0.32 μL/min, one of the lowest rates observed with this technique of measurement in normal human subjects.

Since the time of the discovery of the ocular hypotensive effect of acetazolamide, there was considerable interest in discovering a related compound that would produce the same effects when administered topically to the eye. Despite this effort, several decades passed without success. The problem was to find a drug that was nontoxic, that could penetrate the cornea, and that had very high affinity to ocular carbonic anhydrase. Maren and coworkers (1977) pointed out that any useful drug would be required to inhibit 99% of the activity of the enzyme in order to cause clinically useful suppression of aqueous formation.

The problems of penetration and affinity were solved with the discovery of a new class of water-soluble sulfonamides by Ponticello and coworkers (1987). One of these sulfonamides, dorzolamide hydrochloride has been successfully developed into a topical formulation that is now used for the treatment of glaucoma. Maus and coworkers (1996b) have measured the ability of topical dorzolamide to suppress aqueous humor flow in human subjects and have compared its efficacy to that of systemically administered acetazolamide. In this study, acetazolamide was able to suppress the daytime flow of aqueous humor by 30%, whereas topical dorzolamide suppressed flow by 17%. The addition of systemic acetazolamide to eyes treated with topical dorzolamide increased the suppression from 17 to 30%, whereas the addition of topical dorzolamide to eyes of patients treated with systemic acetazolamide caused no additional suppression of aqueous flow. Thus, it appears that the best clinically available topical carbonic anhydrase inhibitor has half the efficacy of systemic acetazolamide. The reason for this difference in efficacy is not known, but could be due to lack of sufficient penetration of the topical agent to all relevant targets within the eye or due to extraocular effects of systemically administered carbonic anhydrase inhibitors. Whether the flow-suppressing effect of systemic administration can ever be attained by any topically applied drug is unknown.

B. β-Adrenergic Antagonists

In 1967 it was discovered that systemically administered propranonol, a β-adrenergic antagonist, lowered intraocular pressure (Phillips *et al.,* 1967) and the following year, it was reported that topically applied propranolol also lowered intraocular pressure of humans (Bucci *et al.,* 1968). Nearly a decade later, it was shown that the β-adrenergic antagonist timolol lowered intraocular pressure when applied topically to humans (Katz *et al.,* 1976). Timolol subsequently was approved for the treatment of glaucoma and quickly became the most frequently prescribed drug for this disorder. Timolol was found to achieve its ocular hypotensive effect by suppression of the rate of aqueous humor formation. This effect is very consistent, and ranges from 47% suppression in one study (Yablonski *et al.,* 1978) to 35% in another study (Coakes and Brubaker, 1978), and between 30 and 39% in other studies (Gharagozloo *et al.,* 1992; Larsson *et al.,* 1993, 1995a). Subsequent studies of other drugs of the same class such as betaxolol and levo-bunolol have shown 32% suppression of aqueous flow, comparable to that of timolol (Topper and Brubaker, 1985; McCannel *et al.,* 1992a).

Studies of glaucoma patients who have been treated chronically with timolol show that the initial effect gradually wanes so that approximately half of the initial aqueous suppressing effect is lost after a year's treatment (Brubaker *et al.,* 1982). When timolol treatment is discontinued after chronic use there is very slow but complete recovery of the rate of aqueous humor flow. The terminal half-life of action of the drug after chronic use is approximately 9 days (Schlecht and Brubaker, 1988). Thus, timolol treatment has no permanent effect on aqueous humor flow, but recovery is slow. Similar results have been found with levo-bunolol from which recovery is slow but complete (Gaul *et al.,* 1989). Recovery from betaxolol is also complete, but more rapid (Gaul *et al.,* 1989).

The studies cited were all carried out in human subjects during daytime hours. When measured at night during sleep, timolol, unlike acetazolamide, has been found to have no measurable effect in otherwise untreated normal human subjects (Topper and Brubaker, 1985; Brubaker *et al.,* 1987; McCannel *et al.,* 1991). As discussed previously, intravenously infused epinephrine is able to stimulate the rate of aqueous humor flow in sleeping human subjects, and timolol is able to block this effect (Maren, 1976). These data suggest that the action of timolol, and by analogy other β-adrenergic antagonists, is to antagonize the effect of hormonal or nervous stimulators of aqueous humor flow, endogenous effects that are presumably present during normal daytime activities but absent during sleep. Adrenally derived epinephrine is a good candidate for one of these endogenous stimulators, but other endogenous regulators of aqueous humor flow may be of equal or greater importance.

C. *Selective* α_2-Adrenergic Agonists

In 1966, four years after clonidine was first synthesized, this selective α-adrenergic agonist was reported to cause ocular hypotension after systemic administration (Makabe, 1966). Subsequent studies showed that topically applied clonidine also lowered intraocular pressure, but its mechanism of action was not clear. A fluorophotometric study suggested that it suppressed aqueous humor flow by 21%, sufficient to explain its hypotensive effect (Lee *et al.,* 1984). Subsequently, topical administration of a derivative of clonidine, para-aminoclonidine (apraclonidine, aplonidine) was found to lower intraocular pressure and decrease aqueous flow in humans by 30–34% (Gharagozloo *et al.,* 1988b; Koskela and Brubaker, 1991b). Aplonidine in rabbit eyes causes an increase in the concentration of prostaglandins in aqueous humor (Wang *et al.,* 1989). In monkeys the ocular hypotensive effect of aplonidine is blocked by the prostaglandin synthetase inhibitor flurbiprofen (Wang *et al.,* 1989). However, in humans, the flow-suppressing effect of aplonidine is not blocked by pretreatment with flurbiprofen (McCannel *et al.,* 1991).

It has been shown that the suppression of aqueous flow produced by acute doses of the β-adrenergic antagonist timolol and the α_2 selective adrenergic agonist aplonidine were the same in normal volunteers who had never been treated before with either drug (Gharagozloo *et al.,* 1988b). It was also shown that neither drug was additive to the other (Koskela and Brubaker, 1991b), suggesting that the effect of both drugs is mediated by a shared pathway. Aplonidine, unlike timolol was also found to suppress the rate of aqueous humor flow in sleeping subjects, and the authors reasoned that this finding showed the two pathways could not be identical in all respects (Koskela and Brubaker, 1991b).

In another study, aplonidine was added to the treatment regimen of glaucoma patients who were on long-term treatment with timolol. The addition of aplonidine to the chronic timolol regimen was followed by a 14–18% further reduction in the rate of aqueous flow (Gharagozloo and Brubaker, 1991b). The authors concluded that aplonidine was able to overcome the portion of aqueous suppression that had been lost during adaptation of the eye to chronic treatment with timolol. It appears from these studies that the actions of α_2 selective adrenergic agonists and β-adrenergic antagonists at the clinical level are similar but not identical.

D. Other Adrenergic Agonists

Epinephrine has been used for half a century for the treatment of glaucoma. Goldmann (1951b) was the first to study the effect of epinephrine on aqueous dynamics. His study suggested that epinephrine was a suppressor of aqueous humor flow. This conclusion remained unchallenged in the scientific literature for many years. Later, it was shown that epinephine was a weak *stimulator* of aqueous humor flow and that its ocular hypotensive effect must be due to a decrease in outflow resistance (Ballintine, 1961;

Ballintine and Garner, 1961; Becker *et al.,* 1961; Kronfeld, 1963; Becker and Morton, 1966; Townsend and Brubaker, 1980). The aqueous flow-stimulating effects of epinephrine are not blocked by the α-adrenergic antagonist thymoxamine (Lee *et al.,* 1983) and are not imitated by the alpha selective adrenergic agonist phenylephrine (Lee and Brubaker, 1982). However, several β selective adrenergic agonists are also stimulators of aqueous humor flow, including salbutimol (Coakes and Siah, 1982), isoproterenol (Larson and Brubaker, 1988), and terbutaline. (Gharagozloo *et al.,* 1988). The effect of epinephrine on flow is greater at night during sleep (Topper and Brubaker, 1985) as is the effect of isoproterenol (Larson and Brubaker, 1988) and terbutaline (Gharagozloo *et al.,* 1988). The acute effect of epinephrine is blocked by the β-adrenergic antagonist timolol (Higgins and Brubaker, 1980).

These studies suggest that the relevant receptor for these effects is a β-adrenergic receptor. However, the effect of β-adrenergic agonists on flow is weak and there is rapid adaptation to the effect. For example, when epinephrine is given topically twice a day, it causes no measurable effect on flow after 24 hr and 2 weeks of administration (Schneider and Brubaker, 1991). Also, as mentioned earlier, persons who have undergone bilateral extirpation of the adrenal glands are still observed to have the normal circadian rhythm of aqueous humor flow (Maus *et al.,* 1994).

E. Cholinergics

Pilocarpine has been used for the treatment of glaucoma for more than a century. Pilocarpine lowers intraocular pressure by causing a decrease in outflow resistance of the glaucomatous eye. This effect is due to increased tension in the ciliary muscle, which is anchored to the scleral spur (Van Buskirk and Grant, 1973). If the ciliary muscle is surgically disinserted from the scleral spur (Kaufman and Barany, 1976), pilocarpine no longer lowers pressure. An offsetting effect of pilocarpine is to reduce the rate of uveoscleral flow by blocking the entry of substances from the anterior chamber into the supraciliary space (Bill and Wålinder, 1966).

Fluorophotometric studies of the effect of pilocarpine or other cholinergic agonists or of the effect of cholinergic drugs on aqueous humor flow are few in number. In one study of pilocarpine's effect on flow, the results were too scattered to draw conclusions (Anselmi *et al.,* 1968). In another study a small stimulation of flow of 14% was found (Nagataki and Brubaker, 1982). In yet another, no effect was found (Araie and Takase, 1981). Scopolamine, an anticholinergic, when administered by means of a dermal patch, has a small mydriatic effect on the pupil but has no measurable effect on aqueous flow or intraocular pressure (Brimijoin *et al.*, 1993). Most investigators believe that the cholinergic system of the eye has very little role in regulating aqueous humor formation.

F. Prostaglandins

It is known that some prostaglandins have potent ocular hypotensive action in some species (Bito, 1984a,b; Bito *et al.*, 1989; Alm and Villumsen, 1989). The mechanism of this effect is not entirely clear. Most recent studies suggest that the ocular hypotensive effect of prostaglandins is due to an enhancement of the rate of uveoscleral drainage of aqueous humor (Hayashi *et al.*, 1987; Kerstetter *et al.*, 1988; Brubaker, 1989; Gabelt and Kaufman, 1989; Kaufman and Crawford, 1989; Nilsson *et al.*, 1989; Villumsen and Alm, 1989).

Some prostaglandins have been studied for their effects on aqueous humor flow in human subjects. Prostaglandin $F_{2\alpha}$ isopropyl ester was found to lower intraocular pressure, but had no measurable effects in human subjects on aqueous flow or concentration of protein in the aqueous humor (Hayashi *et al.*, 1987; Kerstetter *et al.*, 1988; Nilsson *et al.*, 1989a,b; Gabelt and Kaufman, 1989; Villumsen and Alm, 1989). Another synthetic prostaglandin, latanoprost likewise was found to have no effects on aqueous humor flow or blood–aqueous barrier permeability to fluorescein in normal human subjects and in ocular hypertensive subjects (Toris *et al.*, 1993; Ziai *et al.*, 1993). However, in both of these studies, ocular hypotension was observed. Latanoprost has recently been approved in the United States for treatment of chronic glaucoma.

The exact mechanism of the prostaglandins that accounts for the ocular hypotensive effect is the subject of continuing study, but the consensus at present is that it enhances uveoscleral outflow of aqueous humor (Toris *et al.*, 1997).

VI. NONINVASIVE MEASUREMENTS OF OTHER PARAMETERS OF AQUEOUS HUMOR

Many studies of the composition of aqueous humor have been carried out in animal eyes, but the number of studies of the composition of human aqueous humor are limited in number and limited in experimental design. Studies of composition necessarily require the removal of a sample of aqueous from the anterior or posterior chamber, an invasive procedure that is usually carried out as part of a surgical procedure. One of the challenges of ocular physiologists is to devise methods of making measurements of the composition of aqueous humor without removing a specimen from the eye. Various methods have been attempted to measure the concentration of proteins, glucose, ascorbate, oxygen, and the pH of aqueous humor. These attempts have included measurement of such things as light scattering, optical rotation, polarization of fluorescence, phosphorescence lifetime, spectral shift of fluorescence, and spectra of magnetic resonance.

A. Measurement of Albumin and Other Proteins

Instruments have been developed that measure Rayleigh scattering from the anterior chamber and correlate the intensity of backscattering with the concentration of colloid molecules, mainly proteins, in the anterior chamber (Anjou and Krakau, 1960; Dyster-Aas and Krakau, 1963; Bengtsson *et al.*, 1978) (Fig. 15). The best and only commercially available instrument for this purpose is the Kowa Flare-Cell Meter (Sawa *et al.*, 1988). This instrument measures the backscattered signal from a helium–neon laser beam that is passed through the anterior chamber. The beam is modulated rapidly into and out of the acceptance window of a photon-counting photomultiplier, a technique that reduces the interference of stray light. The instrument can detect colloids in normal aqueous humor. It has been used to measure the normal variations in colloid concentration that occur because of the circadian rhythm of aqueous humor formation and the changes in concentration

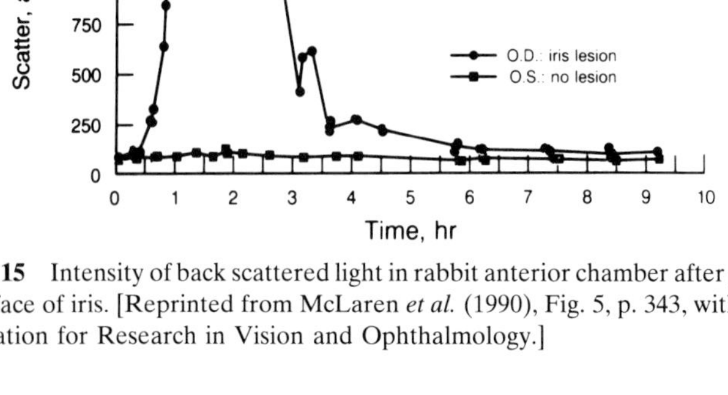

FIGURE 15 Intensity of back scattered light in rabbit anterior chamber after photocoagulation of surface of iris. [Reprinted from McLaren *et al.* (1990), Fig. 5, p. 343, with permission from Association for Research in Vision and Ophthalmology.]

that occur with aqueous-suppressing drugs such as acetazolamide (Oshika *et al.*, 1989). Quasi-elastic light scattering has been used to measure lipoproteins in rabbit plasma (Bursell *et al.*, 1986). This technique is more commonly employed in studies of the cornea and lens (Bursell *et al.*, 1990) than of aqueous humor.

A second method of measuring proteins in the anterior chamber is the measurement of the polarization of fluorescence of fluorescein (Penniston, 1982). Normally, when excited with linearly polarized light, fluorescein in aqueous solution at physiological pH and normal body temperature will emit unpolarized light. However, when bound to macromolecules such as albumin, the emitted light is linearly polarized. This effect is the basis for measuring the concentration of albumin in the anterior chamber of the living human eye (Herman *et al.*, 1988) (Fig. 16). The method suffers due to its lack of specificity for albumin and due to its relative insensitivity to low concentrations of albumin and other proteins that bind fluorescein.

B. Measurement of pH

Fluorophores that are weak acids like fluorescein, pyranine, and bis-carboxyethyl-carboxyfluorescein (BCECF) have excitation and emission

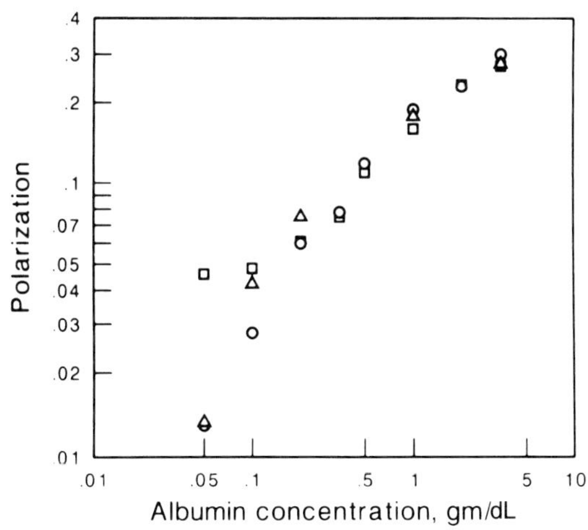

FIGURE 16 Polarization of fluorescence of fluorescein in the presence of varying concentrations of rabbit serum albumin in three kinds of chambers: ○, in a cuvette; Δ, in a chamber formed by two contact lenses; □, in the perfused anterior chamber of a living rabbit. [Reprinted from Herman *et al.* (1988), Fig. 3, p. 135, with permission from Association for Research in Vision and Ophthalmology.]

spectra that are dependent on pH. Investigators have devised methods of measuring the pH of the tears, corneal stroma, and the aqueous humor (of animals) by measuring the spectra of these fluorophores after topical administration to the eye. For example, Bonanno and Polse have used fluorescein to measure the pH of the stroma of the living human eye. Thomas and coworkers (1990) have used pyranine for measurements of pH in the cornea and the anterior chamber of rabbit eyes, but have not yet extended this work to the human eye. At present fluorescein is the only pH-sensitive fluorophore that is approved for human use. The high cost of the approval process is a barrier that is likely to preclude the commercial availability of new pH-sensitive fluorophores for human studies.

C. Measurement of Oxygen

Oxygen is capable of quenching fluorescence or phosphorescence. This property can serve as the basis of a method to measure the oxygen concentration of the aqueous humor of living animals. Roberts and coworkers (Roberts *et al.,* 1992) reported the use of a phosphorescent probe of palladium uroporphyrin for measuring the concentration of oxygen in the crystalline lens. This phosphor has been employed for measuring oxygen tension in the aqueous humor (McLaren *et al.,* 1996b). The phosphor was injected into the vitreous 3 days before measurements. The oxygen tension was highest in the anterior portion of the anterior chamber, approximately 60 mm Hg and was lower, 40 mm Hg, near the anterior surface of the lens. Placement of an oxygen-impermeable contact lens over the cornea caused a rapid fall to below 20 mm Hg in the central aqueous humor.

As with pH probes, suitable probes for measurement of oxygen tension in human eyes have not been developed.

An indirect method of measuring oxygen concentration has been described by Masters (1990) and is called *redox fluorophotometry.* This technique is based on the measurement of the intensity ratios of fluorescence of the pyridine nucleotides and the flavins that are present in living cells of the cornea. Novack and Stefánson (1990) have used a similar technique to study the oxidative metabolism of the optic nerve and retina. The advantage of these techniques is that they employ the properties of naturally occurring fluorophores; the disadvantage for the study of the aqueous is that these fluorophores are not normally found in aqueous humor.

D. Measurement of Ascorbate

Ascorbic acid is found in high concentration, relative to plasma concentrations, in diurnal mammals, but not in nocturnal mammals (Reiss *et al.,*

1986; Koskela *et al.,* 1989) (Fig. 17). The differences in ascorbate concentration found in animals has been of interest to understand the role of ascorbate in the physiology of the eye. Also, the study of ascorbate in disease states in humans could give some clue to its role in the aqueous or its role in the pathophysiology of certain disease states of the eye, including glaucoma. A noninvasive method of measuring ascorbate in the aqueous humor of living animals would be very useful to scientists who wish to study the role of this vitamin in the eye.

Chou and coworkers (1986) have attempted to measure ascorbate in the living eye using the rate of conversion of one fluorescent material to another. Resazurin, a fluorescent N-oxide, can be reduced by ascorbate to resorufin, another fluorophore. The rate of the reaction is dependent on the concentration of ascorbate in the system (Fig. 18). Both resazurin and resorufin can be excited and emit at wavelengths that pass through the cornea. Each fluorophore can be detected in the presence of the other . Chou has worked

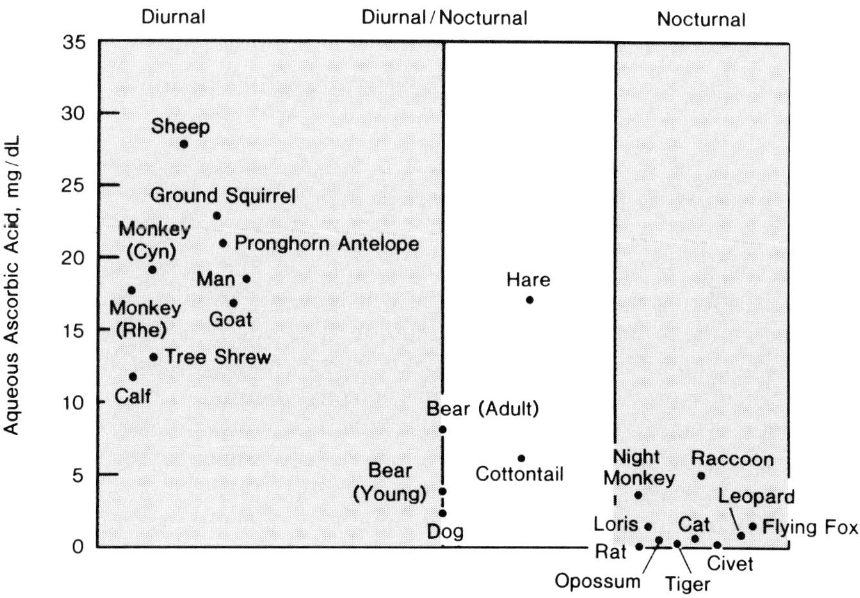

FIGURE 17 Correlation of ascorbic acid concentration in aqueous with animal's behavior in regard to light. Classification into behavior groups is based on advice from Edwin Gould, DVM (National Zoo, Washington, DC), and Robert Snyder, DVM (Philadelphia Zoo). Rhe, rhesus; Cyn, cynomolgus. [Reprinted from Reiss *et al.* (1986), p. 754, *Arch. Ophthalmol.,* 104:753–755. Copyright 1986, American Medical Association.]

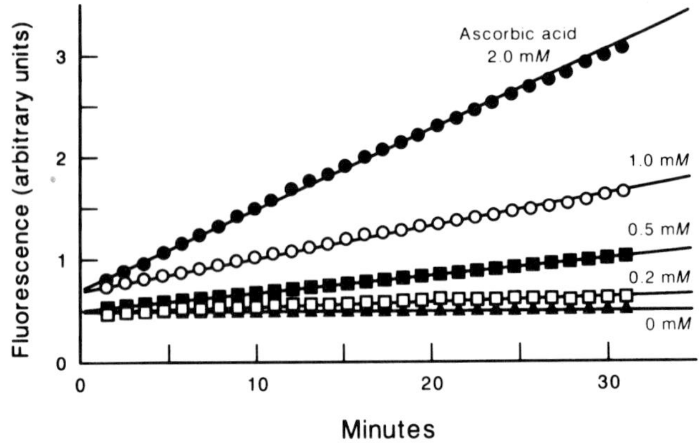

FIGURE 18 Spontaneous reaction between ascorbic acid and resazurin at neutral pH to form resorufin. [Reprinted from Chou *et al.* (1986), Fig. 3, p. 271, with permission from Mayo Foundation.]

out the kinetics in simple systems, but has not been able to get a sufficient concentration of oxidized resazurin into the anterior chamber without injecting it. In theory the technique seems to have promise, but in practice it has not worked out.

VII. SUMMARY AND FUTURE CHALLENGES

The ciliary epithelia of the eye secrete aqueous humor at a very rapid rate, considering the small number of cells devoted to this task and the oncotic gradient that opposes its formation. The rate of secretion during the morning hours appears to be the maximum rate that can occur. Elevation of intraocular pressure that occurs in glaucoma appears to be due to increased resistance to outflow of aqueous humor or to elevation of episcleral venous pressure, not to the hypersecretion of aqueous humor. During sleep, there is modest conservation of secretory activity. At present, it is not clear what hormonal or neural mechanisms control this circadian rhythm. At its lowest point of secretion during sleep, the rate of aqueous flow in the human eye is approximately half the rate during waking hours. If the rate falls much below half, there is the possibility that the crystalline lens will become cataractous due to the lack of substrate, oxygen, or due to the buildup of toxic metabolites.

Most of the studies of aqueous humor dynamics in human subjects are of intraocular pressure and aqueous humor flow. Few studies have been conducted of the composition of aqueous humor under a variety of circumstances. Such studies, were methods available, might shed light on a number of pathological conditions that are observed in clinical practice. To develop such methods is one of the current challenges of research in the physiology of the aqueous circulatory system.

References

Alm, A., and Villumsen, J. (1989). Effects of topically applied $PGF_{2\alpha}$ and its isopropylester in normal and glaucomatous human eye. *In* "The Ocular Effects of Prostaglandins and Other Eicosanoids" (L. Z. Bito and J. Stjernschantz, eds.), pp. 447–458. Alan R. Liss, New York.

Anjou, C. I. N., and Krakau, C. E. T. (1960). A photographic method for measuring the aqueous flare of the eye in normal and pathological conditions. *Acta Ophthalmol.* **38,** 178–224.

Anselmi, P., Bron, A. J., and Maurice, D. M. (1968). Action of drugs on the aqueous flow in man measured by fluorophotometry. *Exp. Eye Res.* **7,** 487–496.

Araie, M. and Takase, M. (1981). Effects of various drugs on aqueous humor dynamics in man. *Jpn. J. Ophthalmol.* **25,** 91–111.

Araie, M., Sawa. M., Nagataki, S., and Mishima, S. (1980). Aqueous humor dynamics in man as studied by oral fluorescein. *Jpn. J. Ophthalmol.* **24,** 346–362.

Araie, M., Yamamoto, T., Shirato, S., and Kitazawa, Y. (1984). Effects of laser trabeculoplasty on the human aqueous humor dynamics: A fluorophotometric study. *Ann. Ophthalmol.* **16,** 540–544.

Ascher, K. W. (1942). Aqueous veins: Preliminary note. *Am. J. Ophthalmol.* **25,** 31–38.

Ballintine, E. J. (1961). Glaucoma. *In* "Transactions of the Fifth Conference" (F. W. Newell, ed.), pp. 247–253. Josiah Macy, Jr., Foundation, New York.

Ballintine, E. J., and Garner, L. L. (1961). Improvement of the coefficient of outflow in glaucomatous eyes. Prolonged local treatment with epinephrine. *Arch. Ophthalmol.* **66,** 314–317.

Bárány, E. H. (1947). The influence of intra-ocular pressure on the rate of drainage of aqueous humour: Stabilization of intra-ocular pressure or of aqueous flow. *Br. J. Ophthalmol.* **31,** 160–176.

Becker, B. (1954). Decrease in intraocular pressure in man by a carbonic anhydrase inhibitor, Diamox: A preliminary report. *Am. J. Ophthalmol.* **37,** 13–15.

Becker, B. (1958). The decline in aqueous secretion and outflow facility with age. *Am. J. Ophthalmol.* **46,** 731–736.

Becker, B., and Christensen, R. E. (1956). Water-drinking and tonography in the diagnosis of glaucoma. *Arch. Ophthalmol.* **56,** 321–326.

Becker, B., and Morton, W. R. (1966). Topical epinephrine in glaucoma suspects. *Am. J. Ophthalmol.* **62,** 272–277.

Becker, B., Pettit, T. H., and Gay, A. J. (1961). Topical therapy of open-angle glaucoma. *Arch. Ophthalmol.* **66,** 219–225.

Bengtsson, E., Krakau, C. E. T., and Ohman, R. (1978). The inhibiting effect of indomethacin on the disruption of the blood–aqueous barrier in the rabbit eye. With a technical note: Measurement of the aqueous flare. *Invest. Ophthalmol. Vis. Sci.* **14,** 306–312.

Bill, A., and Wålinder, P.-E. (1966). The effects of pilocarpine on the dynamics of aqueous humor in a primate (Macaca irus). *Invest. Ophthalmol. Vis. Sci.* **5,** 170–175.

Bito, L. Z. (1984a). Prostaglandins, other eicosanoids, and their derivatives as potential antiglaucoma agents. *In* "Applied Pharmacology in the Medical Treatment of Glaucoma" (S. M. Drance and A. H. Neufeld, eds.), pp. 477–505. Grune & Stratton, Orlando.

Bito, L. Z. (1984b). Comparison of the ocular hypotensive efficacy of eicosanoids and related compounds. *Exp. Eye Res.* **38,** 181–194.

Bito, L. Z., Camras, C., Gum, G., and Resul, B. (1989). The ocular hypotensive effects of topically applied prostaglandins on eyes of experimental animals. *In* "The Ocular Effects of Prostaglandins and Other Eicosanoids" (L. Z. Bito and J. Stjernschantz, eds.), pp. 349–386. Alan R. Liss, New York.

Bloom, J. N., Levene, R. Z., Thomas, G., and Kimura, R. (1976). Fluorophotometry and the rate of aqueous flow in man. I. Instrumentation and normal values. *Arch. Ophthalmol.* **94,** 435–443.

Bonanno, J. A., and Polse, K. A. (1987). Measurement of *in vivo* human corneal stromal pH: Open and closed eyes. *Invest. Ophthalmol. Vis. Sci.* **28,** 522–530.

Breinen, G. M., and Gortz, H. (1954). Carbonic anhydrase inhibitor acetazolamide (Diamox): A new approach to the therapy of gluacoma. *Arch. Ophthalmol.* **52,** 333–348.

Brimijoin, M., Larsson, L.-I., Arnce, R. D., and Brubaker, R. F. (1993). The scopolamine dermal patch and its effect on intraocular pressure and aqueous humor flow. *J. Glaucoma* **2,** 203–205.

Brown, J. D., and Brubaker, R. F. (1989). A study of the relation between intraocular pressure and aqueous humor flow in the pigment dispersion syndrome. *Ophthalmology* **96,** 1468–1470.

Brubaker, R. F. (1982). The flow of aqueous humor in the human eye. *Trans. Am. Ophthalmol. Soc.* **80,** 391–474.

Brubaker, R. F. (1984). The physiology of aqueous humor formation. *In* "Applied Pharmacology in the Medical Treatment of Glaucomas" (S. M. Drance, ed.), pp. 35–70, Grune & Stratton, Orlando.

Brubaker, R. F. (1989). Fluorophotometric studies of prostaglandin effects on the human eye: The lack of association of reduced intraocular pressure with altered flow or barrier function. *In* "The Ocular Effects of Prostaglandins and Other Eicosanoids" (L. Z. Bito and J. Stjernschantz, eds.), pp. 477–481. Alan R. Liss, New York.

Brubaker, R. F. (1991). Flow of aqueous humor in humans. *Invest. Ophthalmol. Vis. Sci.* **32,** 3145–3166.

Brubaker, R. F. (1996). Measurement of aqueous flow by fluorophotometry. *In* "The Glaucomas" (R. Ritch, M. B. Shields, and T. Krupin, eds.), 2nd ed., Vol. I, Chap. 22, p. 451. Mosby, St. Louis.

Brubaker, R. F., and Liesegang, T. J. (1983). Effect of trabecular photocoagulation on the aqueous humor dynamics of the human eye. *Am. J. Ophthalmol.* **96,** 139–147.

Brubaker, R. F., Nagataki, S., Townsend, D. J., Burns, R. R., Higgins, R. G., and Wentworth, W. (1981). The effect of age on aqueous humor formation in man. *Ophthalmology* **88,** 283–287.

Brubaker, R. F., Nagataki, S., and Bourne, W. M. (1982). Effect of chronically administered timolol on aqueous humor flow in patients with glaucoma. *Ophthalmology* **89,** 280–283.

Brubaker, R. F., Carlson, K. H., Kullerstrand, L. J., and McLaren, J. W. (1987). Topical forskolin (Colforsin) and aqueous flow in humans. *Arch. Ophthalmol.* **105,** 637–641.

Brubaker, R. F., Maurice, D. M., and McLaren, J. W. (1990). Fluorometry of the anterior segment. *In* "Noninvasive Diagnostic Techniques in Ophthalmology" (B. R. Masters, ed.), Chap. 15, pp. 248–280. Springer Verlag, New York.

Bucci, M. G., Missiroli, A., Pecori Giraldi, J. P., and Virno, M. (1968). La somministrazione locale del propanololo nella terapia del glaucoma. *Bolletino D'Oculistica* **47,** 51–60.

Bursell, S.-E., Serur, J. R., Haughton, J. F., Nipper, H., Sinclair, I. N., Spears, J. R., and Paulin, S. (1986). Cholesterol levels assessed with photon correlation spectroscopy. *Proc. SPIE* **712,** 175–181.

Bursell, S.-E., Magnante, P. C., and Chylack, L. T. (1990). *In vivo* uses of quasi-elastic light scattering spectroscopy as a molecular probe in the anterior segment of the eye. *In* "Noninvasive Diagnostic Techniques in Ophthalmology" (B. R. Masters, ed.), Chap. 18, pp. 342–365. Springer Verlag, New York.

Carlson, K. H., McLaren, J. W., Topper, J. E., and Brubaker, R. F. (1987). Effect of body position on intraocular pressure and aqueous flow. *Invest. Ophthalmol. Vis. Sci.* **28,** 1346–1352.

Chou, C.-K., Penniston, J. T., and Brubaker, R. F. (1986). Ascorbic acid in the anterior chamber: Can it be measured noninvasively? *Trans. Am Ophthalmol. Soc.* **80,** 269–279.

Coakes, R. L. and Brubaker, R. F. (1978). The mechanism of timolol in lowering intraocular pressure. *Arch. Ophthalmol.* **96,** 2045–2048.

Coakes, R. L., and Siah, P. B. (1982). The effects of topical salbutamol on aqueous humor dynamics in the normal human eye. *Invest. Ophthalmol. Vis. Sci.* **22,** Suppl., 40.

Dailey, R. A., Brubaker, R. F., and Bourne, W. M. (1982). The effects of timolol maleate and acetazolamide on the rate of aqueous formation in normal human subjects. *Am. J. Ophthalmol.* **93,** 232–237.

Dyster-Aas, H. K., and Krakau, C. E. T. (1963). A photo-electric instrument for measuring the aqueous flare in the intact eye. *Ophthalmologica* **146,** 48–56.

Ericson, L. A. (1958). Twenty-four hourly variations on the aqueous flow: Examination with perilimbal suction cup. *Acta Ophthalmol.* (*Copenhagen*) **50** Suppl., 1.

Fontana, S. T., and Brubaker, R. F. (1980). Volume and depth of the anterior chamber in the normal aging human eye. *Arch. Ophthalmol.* **98,** 1803–1808.

Friedenwald, J. S. (1937). Contribution to the theory and practice of tonometry. *Am. J. Ophthalmol.* **20,** 985–1024.

Gaasterland, D., Kupfer, C., Ross, K., *et al.* (1973). Studies of aqueous humor dynamics in man. III. Measurements in young normal subjects using norepinephrine and isoproterenol. *Invest. Ophthalmol.* **12,** 267–279.

Gaasterland, D., Kupfer, C., and Ross, K. (1975). Studies of aqueous humor dynamics in man. IV. Effects of pilocarpine upon measurements in young normal volunteers. *Invest. Ophthalmol.* **14,** 848–853.

Gabelt, B. T., and Kaufman, P. L. (1989). Prostaglandin $F_{2\alpha}$ increases uveoscleral outflow in the cynomolgus monkey. *Exp. Eye Res.* **49,** 389–402.

Gaul, G. R., Will, N. J., and Brubaker, R. F. (1989). Comparison of a noncardioselective beta-adrenoceptor blocker and a cardioselective blocker in reducing aqueous flow in humans. *Arch. Ophthalmol.* **107,** 1308–1311.

Gharagozloo, N. Z., and Brubaker, R. F. (1991a). The correlation between serum progesterone and aqueous dynamics during the menstrual cycle. *Acta Ophthalmologica* **69,** 791–795.

Gharagozloo, N. Z., and Brubaker, R. F. (1991b). Effect of apraclonicine in long-term timolol users. *Ophthalmology* **98,** 1543–1546.

Gharagozloo, N. Z., Larson, R. S., Kullerstrand, L. J., and Brubaker, R. F. (1988a). Terbutaline stimulates aqueous humor flow in humans during sleep. *Arch. Ophthalmol.* **106,** 1218–1220.

Gharagozloo, N. Z., Relf, S. J., and Brubaker, R. F. (1988b). Aqueous flow is reduced by the alpha-adrenergic agonist, apraclonidine hydrochloride (AL0 2145). *Ophthalmology* **95,** 1217–1220.

Gharagozloo, N. Z., Baker, R. H., and Brubaker, R. F. (1992). Aqueous dynamics in exfoliation syndrome. *Am. J. Ophthalmol.* **114,** 473–478.

Goldmann, H. (1950). Über Fluorescein in der menschlichen Vorderkammer. Das Kammerwasser-Minutenvolumen des Menschen. *Ophthalmologica* **119,** 65–95.

Goldmann, H. (1951a). Abflussdruck, Minutenvolumen und Widerstand der Kammerwasserstromung des Menschen. *Doc. Ophthalmol.* **5,** 278–356.

Goldmann, H. (1951b). L'origine de l'hypertension oculaire dans le glaucome primitif. *Ann. Oculist.* **184,** 1086–1105.

Grant, W. M. (1950). Tonographic method for measuring the facility and rate of aqueous flow in human eyes. *Arch. Ophthalmol.* **44,** 204–214.

Grant, W. M. (1951). Clinical measurements of aqueous outflow. *Arch. Ophthalmol.* **46,** 113–131.

Grant, W. M., and Trotter, R. R. (1954). Diamox (acetazoleamide) in treatment of glaucoma. *Arch. Ophthalmol.* **51,** 735–739.

Hayashi, M., Yablonski, M. E., and Bito, L. Z. (1987). Eicosanoids as a new class of ocular hypotensive agents. II: Comparison of the apparent mechanism of the ocular hypotensive effects of A and F type prostaglandins. *Invest. Ophthalmol. Vis. Sci.* **28,** 1639–1643.

Hayashi, M., Yablonski, M. E., Boxrub, C., Fong, N., Berger, C., and Jovanovic, H. (1989). Decreased formation of aqueous humour in insulin-dependent diabetic patients. *Br. J. Ophthalmol.* **73,** 621–623.

Herman, D. C., McLaren, J. W., and Brubaker, R. F. (1988). A method of determining concentration of albumin in the living eye. *Invest. Ophthalmol. Vis. Sci.* **29,** 133–137.

Higgins, R. G., and Brubaker, R. F. (1980). Acute effect of epinephrine on aqueous humor formation in the timolol-treated normal eye as measured by fluorophotometry. *Invest. Ophthalmol. Vis. Sci.* **19,** 420–423.

Holm, O. (1968). A photogrammetric method for estimation of the pupillary aqueous flow in the living eye. I. *Acta Ophthalmol.* **46,** 254–277.

Jacob, E., FitzSimon, J. S., and Brubaker, R. F. (1996). Combined corticosteroid and catecholamine stimulation of aqueous humor flow. *Ophthalmology* **103,** 1303–1308.

Johnson, D. H., and Brubaker, R. F. (1982). Dynamics of aqueous humor in the syndrome of exfoliation with glaucoma. *Am. J. Ophthalmol.* **93,** 629–634.

Johnson, D. H., and Brubaker, R. F. (1986). Glaucoma: An overview. *Mayo Clin. Proc.* **61,** 59–67.

Johnson, D., Liesegang, T. J., and Brubaker, R. F. (1983). Aqueous humor dynamics in Fuchs' uveitis syndrome. *Am. J. Ophthalmol.* **95,** 783–787.

Johnson, S., Coakes, R. L., and Brubaker, R. F. (1978). A simple photogrammetric method of measuring anterior chamber volume. *Am. J. Ophthalmol.* **85,** 469–474.

Jones, R. F., and Maurice, D. M. (1966). New methods of measuring the rate of aqueous flow in man with fluorescein. *Exp. Eye Res.* **5,** 208–220.

Kacere, R. D., Dolan, J. W., and Brubaker, R. F. (1992). Intravenous epinephrine stimulates aqueous formation in the human eye. *Invest. Ophthalmol. Vis. Sci.* **33,** 2861–2865.

Katz, I. M., Hubbard, W. A., Getson, A. J., and Gould, A. L. (1976). Intraocular pressure decrease in normal volunteers following timolol ophthalmic solution. *Invest. Ophthalmol.* **15,** 489–492.

Kaufman, P. L., and Barany, E. H. (1976). Loss of acute pilocarpine effect on outflow facility following surgical disinsertion and retrodisplacement of the ciliary muscle from the scleral spur in the cynomolgus monkey. *Invest. Ophthalmol. Vis. Sci.* **15,** 793–807.

Kaufman, P. L., and Crawford, K. (1989). Aqueous humor dynamics: How $PGF_{2\alpha}$ lowers intraocular pressure. *Prog. Clin. Biol. Res.* **312,** 387–416.

Kerstetter, J. R., Brubaker, R. F., Wilson, S. E., and Kullerstrand, L. J. (1988). Prostaglandin $F_{2\alpha}$-1-isopropylester lowers intraocular pressure without decreasing aqueous humor flow. *Am. J. Ophthlamol.* **105,** 30–34.

Khan, A. R., and Brubaker, R. F. (1993). Aqueous humor flow and flare in patients with myotonic dystrophy. *Invest. Ophthalmol. Vis. Sci.* **34,** 3131–3139.

Koskela, T., and Brubaker, R. F. (1991a). The nocturnal suppression of aqueous humor flow in humans is not blocked by bright light. *Invest. Ophthalmol. Vis. Sci.* **32,** 2504–2506.

Koskela, T., and Brubaker, R. F. (1991b). Apraclonidine and timolol. Combined effects in previously untreated normal subjects. *Arch. Ophthalmol.* **109,** 804–806.

Koskela, T. K., Reiss, G. R., Brubaker, R. F., and Ellefson, R. D. (1989). Is the high concentration of ascorbic acid in the eye an adaptation to intense solar irradiation? *Invest. Ophthalmol. Vis. Sci.* **30,** 2265–2267.

Kronfeld, P. C., as cited by Lyle, D. J. (1963). Society Proceedings, Glaucoma Research Concerence, Del Monte, Calif., September 1962. *Am. J. Ophthalmol.* **55,** 821–832.

Kupfer, C., and Ross, K. (1971). Studies of aqueous humor dynamics in man. I. Measurements in young normal subjects. *Invest. Ophthalmol.* **10,** 518–522.

Kupfer, C., Gaasterland, D., and Ross, K. (1971). Studies of aqueous humor dynamics in man. II. Measurements in yong normal subjects using acetazolamide and l-epinephrine. *Invest. Ophthalmol.* **10,** 523–533.

Larson, R. S., and Brubaker, R. F. (1988). Isoproterenol stimulates aqueous flow in humans with Horner's syndrome. *Invest. Ophthalmol. Vis. Sci.* **29,** 621–625.

Larsson, L.-I., Rettig, E. S., Sheridan, P. T., and Brubaker, R. F. (1993). Aqueous humor dynamics in low-tension glaucoma. *Am. J. Ophthalmol.* **116,** 590–593.

Larsson, L.-I., Rettig, E. S., and Brubaker, R. F. (1995a). Aqueous flow in open-angle glaucoma. *Arch. Ophthalmol.* **113,** 283–286.

Larsson, L.-I., Pach, J. M., and Brubaker, R. F. (1995b). Aqueous humor dynamics in patients with diabetes mellitus. *Am. J. Ophthalmol.* **120,** 362–367.

Lee, D. A., and Brubaker, R. F. (1982). Effect of phenylephrine on aqueous humor flow. *Curr. Eye Res.* **2,** 89–92.

Lee, D. A., Brubaker, R. F., and Nagataki, S. (1983). Acute effect of thymoxamine on aqueous humor formation in the epinephrine-treated normal eye as measured by fluorophotometry. *Invest. Ophthalmol. Vis. Sci.* **24,** 165–168.

Lee, D. A., Topper, J. E., and Brubaker, R. F. (1984). Effect of clonidine on aqueous humor flow in normal human eyes. *Exp. Eye Res.* **38,** 239–246.

Lutjen-Drecoll, E., Futa, R., and Tamm, E. (1988). New ultrastructural findings in exfoliation glaucoma. *Invest. Ophthalmol. Vis. Sci.* **29,** ARVO Suppl., 274.

Makabe, R. (1966). Ophthalmologische Untersuchungen mit Dichlorphenylaminoimidazolin. *Dtsch. Med. Wochenschr.* **38,** 1686–1688.

Maren, T. H. (1976). The rates of movement of Na^+, Cl^-, and HCO_3^- from plasma to posterior chamber: Effect of acetazolamide and relation to the treatment of glaucoma. *Invest. Ophthalmol.* **15,** 356–364

Maren, T. H., Haywood, J. R., Chapman, S. K., and Zimmerman, T. Z. (1977). The pharmacology of methazolamide in relation to the treatment of glaucoma. *Invest. Ophthalmol.* **16,** 730–742.

Masters, B. R. (1990). *In vivo* corneal redox fluorometry. *In* "Noninvasive Diagnostic Techniques in Ophthalmology" (B. R. Masters, ed.), Chap. 14, pp. 223–247. Springer Verlag, New York.

Maurice, D. M. (1967). The use of fluorescein in ophthalmological research. *Invest. Ophthalmol.* **6,** 464–477.

Maus, T. L., Young, Jr., W. F., and Brubaker, R. F. (1994). Aqueous flow in humans after adrenalectomy. *Invest. Ophthalmol. Vis. Sci.* **35,** 3325–3331.

Maus, T. L., Larsson, L.-I., McLaren, J. W., and Brubaker, R. F. (1996a). Comparison of dorzolamide and acetazolamide as suppressors of aqueous humor flow in humans. *Ophthalmology* in press.

Maus, T. L., McLaren, J. W., Shepard, Jr., J. W., and Brubaker, R. F. (1996b). The effects of sleep on circulating catecholamines and aqueous flow in human subjects. *Exp. Eye Res.* **62,** 351–358.

McCannel, C., Koskela, T., and Brubaker, R. F. (1991). Topical flurbiprofen pretreatment does not block apraclonidine's effect on aqueous flow in humans. *Arch. Ophthalmol.* **109,** 810–811.

McCannel, C. A., Heinrich, S. R., and Brubaker, R. F. (1992a). Acetazolamide but not timolol lowers aqueous humor flow in sleeping humans. *Graefe's Arch. Clin. Exp. Ophthalmol.* **230,** 518–520.

McCannel, C. A., Scanlon, P. D., Thibodeau, S., and Brubaker, R. F. (1992b). A study of aqueous humor formation in patients with cystic fibrosis. *Invest. Ophthalmol. Vis. Sci.* **33,** 160–164.

McLaren, J. W., and Brubaker, R. F. (1985). A two-dimensional scanning ocular fluorophotometer. *Invest. Ophthalmol. Vis. Sci.* **26,** 144–152.

McLaren, J. W., and Brubaker, R. F. (1986). Measurement of fluorescein and fluorescein monoglucuronide in the living human eye. *Invest. Ophthalmol. Vis. Sci.* **27,** 966–974.

McLaren, J. W., Trocme, S. D., Relf, S., and Brubaker, R. F. (1990). Rate of flow of aqueous humor determined from measurements of aqueous flare. *Invest. Ophthalmol. Vis. Sci.* **31,** 339–346.

McLaren, J. W., Ziai, N., and Brubaker, R. F. (1993). A simple three-compartment model of anterior segment kinetics. *Exp. Eye Res.* **56,** 355–366.

McLaren, J. W., Brubaker, R. F., and FitzSimon, J. S. (1996a). Continuous measurement of intraocular pressure in rabbits by telemetry. *Invest. Ophthalmol. Vis. Sci.* **37,** 966–975.

McLaren, J. W., Dinslage, S., and Brubaker, R. F. (1996b). Measuring oxygen tension in the anterior chamber of rabbits. In preparation.

Nagataki, S. (1975). Aqueous humor dynamics of human eyes as studied using fluorescein. *Jpn. J. Ophthalmol.* **19,** 235–249.

Nagataki, S., and Brubaker, R. F. (1982). Effect of pilocarpine on aqueous humor formation in human beings. *Arch. Ophthalmol.* **100,** 818–821.

Nilsson, S. F. E., Samuelsson, M., Bill, A., and Stjernschantz, J. (1989a). Increased uveoscleral outflow as a possible mechanism of ocular hypotension caused by prostaglandin $F_{2\alpha}$-1-isopropylester in the cynomolgus monkey. *Exp. Eye Res.* **48,** 707–716.

Nilsson, S. F. E., Sperber, G. O., and Bill, A. (1989b). The effect of prostaglandin $F_{2\alpha}$-1-isopropylester ($PGF_{2\alpha}$-IE) on uveoscleral outflow. *In* "The Ocular Effects of Prostaglandins and Other Eicosanoids" (L. Z. Bito and J. Stjernschantz, eds.), pp. 429–436. Alan R. Liss, New York.

Novack, R. L., and Stefánson, E. (1990). Measurement of retina and optic nerve oxidative metabolism *in vivo* via dual wavelength reflection spectrophotometry of cytochrome a, a_3. *In* "Noninvasive Diagnostic Techniques in Ophthalmology" (B. R. Masters, ed.), Chap. 15, pp. 248–280. Springer Verlag, New York.

O'Rourke, J. (1974). Measurement of capillary function in eye diseases. *Trans. Am. Ophthalmol. Soc.* **72,** 606–649.

O'Rourke, J., and Macri, F. J. (1970). Studies in uveal physiology. II. Clinical studies of the anterior chamber clearance of isotopic tracers. *Arch. Ophthalmol.* **84,** 415–420.

Oshika, T., Araie, M., and Equichi, S. (1989). Time change of aqueous protein concentration after oral acetazolamide administration in normal human subjects. *Invest. Ophthalmol. Vis. Sci.* **30,** (suppl), 375.

Penniston, J. T. (1982). Fluorescence polarization measurement of binding of fluorescein to albumin. *Exp. Eye Res.* **34,** 435–443.

Phillips, C. I., Howitt, G., and Rowlands, D. J. (1967). Propranolol as ocular hypotensive agent. *Br. J. Ophthalmol.* **51,** 222–226.

Ponticello, G. S., Freedman, M. B., Habecker, C. N., Lyle, P.A., Schwam, H. Varga, S. L., Christy, M. E., Randall, W. C., and Baldwin, J. J. (1987). Thienothiopyran-2-sulfonamides: A novel class of water-soluble carbonic anhydrase inhibitors. *J. Med. Chem.* **30,** 591–597.

Reiss, G. R., Lee, D. A., Topper, J. E., and Brubaker, R. F. (1984). Aqueous humor flow during sleep. *Invest. Ophthalmol. Vis. Sci.* **25,** 776–778.

Reiss, G. R., Werness, P. G., Zollman, P. E., and Brubaker, R. F. (1986). Ascorbic acid levels in the aqueous humor of nocturnal and diurnal mammals. *Arch. Ophthalmol.* **104,** 753–755.

Rettig, E. S., Larsson, L.-I., and Brubaker, R. F. (1994). The effect of topical timolol on epinephrine-stimulated aqueous humor flow in sleeping humans. *Invest. Ophthalmol. Vis. Sci.* **35,** 554–559.

Rice, S. W., Bourne, W. M., and Brubaker, R. F. (1983). Absence of an effect of topical dexamethasone on endothelial permeability and flow of aqueous humor. *Invest. Ophthalmol. Vis. Sci.* **24,** 1307–1311.

Roberts, J. E., Harriman, A., Atherton, S. J., and Dillon, J. (1992). A non-invasive method to detect oxygen tensions in the lens. *Invest. Ophthalmol. Vis. Sci.* **33,** Suppl., 799.

Rowland, J. M., Potter, D. E., and Reiter, R. J. (1981). Circadian rhythm in intraocular pressure: A rabbit model. *Curr. Eye Res.* **1,** 169–173.

Rowland, J. M., Sawyer, W. K., Tittel, J., and Ford, C. J. (1986). Studies on the circadian rhythm of IOP in rabbits: Correlation with aqueous inflow and cAMP content. *Curr. Eye Res.* **5,** 201–206.

Sawa, M., Tsurimaki, Y., Tsuru, T., and Shimizu, H. (1988). New quantitative method to determine protein concentration and cell number in aqueous *in vivo. Jpn. J. Ophthalmol.* **32,** 132–142.

Schlecht, L. P., and Brubaker, R. F. (1988). The effects of withdrawal of timolol in chronically treated glaucoma patients. *Ophthalmology* **95,** 1212–1216.

Schneider, T. L., and Brubaker, R. F. (1991). Effect of chronic epinephrine on aqueous humor flow during the day and during sleep in normal healthy subjects. *Invest. Ophthalmol. Vis. Sci.* **32,** 2507–2510.

Schottenstein, E. M. (1996). Intraocular pressure and tonometry. *In* "The Glaucomas" (R. Ritch, M. B. Shields, and T. Krupin, eds.), 2nd ed., Vol. I, Chap. 20, p. 407ff. Mosby, St. Louis.

Sheridan, P. T., Brubaker, R. F., Larsson, L.-I., Rettig, E. S., and Young, Jr., W. F. (1994). The effect of oral dexamethasone on the circadian rhythm of aqueous humor flow in humans. *Invest. Ophthalmol. Vis. Sci.* **35,** 1150–1156.

Smith, S. D., and Gregory, D. S. (1989). A circadian rhythm of aqueous flow underlies the circadian rhythm of intraocular pressure in NZW rabbits. *Invest. Ophthalmol. Vis. Sci.* **30,** 775–778.

Thomas, J. V., Brimijoin, M. R., Neault, T. R., and Brubaker, R. F. (1990). The fluorescent indicator pyranine is suitable for measuring stromal and cameral pH *in vivo. Exp. Eye Res.* **50,** 241–249.

Topper, J. E., and Brubaker, R. F. (1985). Effects of timolol, epinephrine, and acetazolamide on aqueous flow during sleep. *Invest. Ophthalmol. Vis. Sci.* **26,** 1315–1319.

Toris, C. B., Camras, C. B., and Yablonski, M. E. (1993). Effects of PhXA41, a new prostaglandin $F_{2\alpha}$ analog, on aqueous humor dynamics in human eyes. *Ophthalmology* **100,** 1297–1304.

Toris, C. B., Camras, C. B., Yablonski, M. E., and Brubaker, R. F. (1997). Effects of exogenous prostaglandins on aqueous humor dynamics and blood-aqueous barrier function. *Surv. Ophthalmol.* **41,** Suppl. 2, S69–S75.

Townsend, D. J., and Brubaker, R. F. (1980). Immediate effect of epinephrine on aqueous formation in the normal human eye as measured by fluorophotometry. *Invest. Ophthalmol. Vis. Sci.* **19,** 256–266.

Van Buskirk, E. M., and Grant, W. M. (1973). Lens depression and aqueous outflow in enucleated primate eyes. *Am. J. Ophthalmol.* **76,** 632–640.

Viggiano, S. R., Abboud, C. F., and Brubaker, R. F. (1993). Effect of desmopressin on aqueous humor flow in humans. *Am. J. Ophthalmol.* **115,** 511–516.

Viggiano, S. R., Koskela, T. K., Klee, G. G., Samples, J. R., Arnce, R., and Brubaker, R. F. (1994). The effect of melatonin on aqueous humor flow in humans during the day. *Ophthalmology* **101,** 326–331.

Villumsen, J., and Alm, A. (1989). Prostaglandin $F_{2\alpha}$-isopropylester eye drops: Effects in normal human eyes. *Br. J. Ophthalmol.* **73,** 419–426.

Walker, S. D., Brubaker, R. F., and Nagataki, S. (1982). Hypotony and aqueous humor dynamics in myotonic dystrophy. *Invest. Ophthalmol. Vis. Sci.* **22,** 744–751.

Wang, R. F., Camras, C. B., Podos, S. M., Lee, K. W., Bhuyan, K. C., and Bhuyan, D. K. (1989). The role of prostaglandins in the para-aminoclonidine-induced reduction of intraocular pressure. *Trans. Am. Ophthalmol. Soc.* **87,** 94–104.

Wentworth, W. O., and Brubaker, R. F. (1981). Aqueous humor dynamics in a series of patients with third neuron Horner's syndrome. *Am. J. Ophthalmol.* **92,** 407–415.

Yablonski, M. E., Zimmerman, T. J., Waltman, S. R., and Becker, B. (1978). A fluorophotometric study of the effect of topical timolol on aqueous humor dynamics. *Exp. Eye Res.* **27,** 135–142.

Yablonski, M. E., Cook, D. J., and Gray, J. (1985). A fluorophotometric study of the effect of argon laser trabeculoplasty on aqueous humor dynamics. *Am. J. Ophthalmol.* **99,** 579–582.

Ziai, N., Dolan, J. W., Kacere, R. D., and Brubaker, R. F. (1993). The effects on aqueous dynamics of PhXA41, a new prostaglandin $F_{2\alpha}$ analogue, after topical application in normal and ocular hypertensive human eyes. *Arch. Ophthalmol.* **111,** 1351–1358.

Ziai, N., Ory, S. J., Khan, A. R., and Brubaker, R. F. (1994). β-human chorionic gonadotropin, progesterone, and aqueous dynamics during pregnancy. *Arch. Ophthalmol.* **112,** 801–806.

Index

V

W

ISBN 0-12-153345-X

9 780121 533458 90018